Natural Bioactive Products in Sustainable Agriculture

Joginder Singh • Ajar Nath Yadav
Editors

Natural Bioactive Products in Sustainable Agriculture

Springer

Editors
Joginder Singh
Department of Microbiology
Lovely Professional University
Phagwara, Punjab, India

Ajar Nath Yadav
Department of Biotechnology
Eternal University
Sirmour, Himachal Pradesh, India

ISBN 978-981-15-3023-4 ISBN 978-981-15-3024-1 (eBook)
https://doi.org/10.1007/978-981-15-3024-1

This Springer imprint is published by the registered company Springer Nature Singapore Pte Ltd.
The registered company address is: 152 Beach Road, #21-01/04 Gateway East, Singapore 189721, Singapore

Preface

Growing consumer demand, insect/pest resistance pressure and an ever-changing regulatory environment necessitate the discovery of new crop protection agents for growers of today and tomorrow. Many of the developments in agrochemical research in the last few decades have their origin in a wide range of natural products from a variety of sources. In light of the continuing need for new tools to address an ever-changing array of microbes, modern agricultural practices and evolving regulatory requirements, the needs for new agrochemical tools remain as critical as ever. In that respect, nature continues to be an important source for novel chemical structures and biological mechanisms to be applied for the development of certain control agents. Following this trend, the proposed book aims to cover important aspects of the use of microorganisms and microbial metabolites as an alternative to the chemicals in the agriculture field. Microbes, as a whole, can be exploited as eco-friendly, cost-effective and sustainable agents. The ultimate goal is to support the scientific community, professionals and enterprises that aspire latest developments and advances about the exploitation of microbial products including their wide application, traditional uses, modern practices and designing of strategies to harness their potential.

The endeavour of the book entitled *Natural Bioactive Products in Sustainable Agriculture* is to present details of cutting-edge research in the field of agriculture and development from bioactive natural products and help its readers to understand how natural product research continues to make significant contributions in the sustainable development of agriculture and environment. The book comprises 14 chapters contributed by an elite group of researchers, working in the forefront of agricultural, natural products development practices. The book covers the advanced perspective, wide applications, traditional uses and modern methods of harnessing the potential of natural biological products in agriculture for sustainable development. The latest developments and advances about the utilization of biological natural products in the field of agriculture are also covered. It also includes interesting chapters that deals with the exploitation of natural products as agrochemicals. It, therefore, serves as a useful book for students, researchers, specialists, agriculturalists, chemical engineers, professionals and strategy developers working in the agrochemical and allied sectors.

Phagwara, Punjab, India — Joginder Singh
Sirmour, Himachal Pradesh, India — Ajar Nath Yadav

Contents

Editors and Contributors

About the Editors

Joginder Singh is Professor in the Department of Microbiology, Lovely Professional University, Punjab, India. Previously, he worked as a Young Scientist in the Microbial Biotechnology and Biofertilizer Laboratory, Department of Botany, Jai Narain Vyas University, funded by the DST, Government of India. His research interests include exploring efficient strategies toward bioremediation and phytoremediation of pollutants and pesticides from water and soil. He is currently working on the designing and development of cleanup technology (biofilters) for in situ bioremediation of textile industrial effluents. He has published more than 60 research and review articles in the peer-reviewed journals, edited 4 books, and authored 30 chapters in edited books. He has chaired technical sessions and presented papers at several international and national seminars, symposia, and conferences. He serves as a reviewer for various prestigious journals and is an active member of numerous scientific societies and organizations, including the Association of Microbiologists of India, K.K. Nanda Foundation for Advancement of Plant Sciences, Indian Society for Radiation Biology, International Phytotechnology Society, and European Federation of Biotechnology.

Ajar Nath Yadav is Assistant Professor (SS) in the Department of Biotechnology, Dr. Khem Singh Gill Akal College of Agriculture, Eternal University, Baru Sahib, Himachal Pradesh, India. He has 4 years of teaching and 10 years of research experience in the fields of industrial biotechnology, microbial biotechnology, and plant-microbe interactions. He completed his doctorate degree in Microbial Biotechnology jointly from the Indian Agricultural Research Institute, New Delhi, and BIT, Mesra, Ranchi, his MSc in Biotechnology from Bundelkhand University, and his BSc in CBZ from the University of Allahabad, India. His 111 publications include 67 research/review papers, 11 books, and 33 book chapters. He has also presented papers at various international and national conferences. He has received 12 Best Paper Presentation Awards, 1 Young Scientist Award (NASI-Swarna Jyanti Purskar), and 3 Certificates of Excellence in Reviewing Awards. He also received the "Outstanding Teacher Award" at the 6th Annual Convocation at Eternal University in 2018. He serves as an editor/editorial board member and reviewer for more than 35 national and international peer-reviewed journals. He has lifetime membership of the Association of Microbiologists in India, Indian Science Congress Council, and the National Academy of Sciences, India.

Contributors

Jayanthi Abraham School of Bio Sciences and Technology, Vellore Institute of Technology VIT, Vellore, Tamil Nadu, India

Nadeem Akthar Department of Animal Biosciences, University of Guelph, Guelph, ON, Canada

Raghavendra Aminedi Laboratory of Plant-Microbe Interactions, Regional Centre for Biotechnology, NCR Biotech Science Cluster, Faridabad, Haryana, India

Debdulal Banerjee Microbiology and Microbial Biotechnology Laboratory, Department of Botany and Forestry, Vidyasagar University, Midnapore, West Bengal, India

Kanchan Bhardwaj Department of Botany, Shoolini University of Biotechnology and Management Sciences, Solan, Himachal Pradesh, India

Mohd. Danish Department of Botany, Aligarh Muslim University, Aligarh, Uttar Pradesh, India

Shivika Datta Department of Zoology, Doaba College, Jalandhar, Punjab, India

Daljeet Singh Dhanjal Department of Biotechnology, Lovely Professional University, Jalandhar, Punjab, India

Anupam Dikshit Biological Product Laboratory, Department of Botany, University of Allahabad, Prayagraj, Uttar Pradesh, India

Dhananjay Suresh Gaikwad School of Agriculture, Lovely Professional University (LPU), Phagwara, Punjab, India

Raja Husain Division of Biochemistry, ICAR-Indian Agricultural Research Institute, New Delhi, India

Kashif Hussain Gyani Inder Singh Institute of Professional Studies, Dehradun, Uttarakhand, India

Touseef Hussain Department of Botany, Aligarh Muslim University, Aligarh, Uttar Pradesh, India

Saket Jha Biological Product Laboratory, Department of Botany, University of Allahabad, Prayagraj, India

Simran Kaur Department of Biotechnology, Lovely Professional University, Phagwara, Punjab, India

Vijay Kumar Department of Chemistry, Regional Ayurveda Research Institute for Drug Development, Gwalior, Madhya Pradesh, India

Anis Mirza Department of Horticulture, School of Agriculture, Lovely Professional University, Phagwara, Punjab, India

Preeti Mishra Biological Product Laboratory, Department of Botany, University of Allahabad, Prayagraj, Uttar Pradesh, India

Yasar Nishat Department of Botany, Aligarh Muslim University, Aligarh, Uttar Pradesh, India

Anand Pandey Department of Plant Sciences, Avvaiyar Government College for Women, University of Puducherry, Karaikal, Puducherry, India

Seema Patel Bioinformatics and Medical Informatics Research Center, San Diego State University, San Diego, CA, USA

Rashid Pervez Division of Nematology, ICAR-Indian Agricultural Research Institute, New Delhi, India

Radhakrishnan Preetha Department of Food Process Engineering, School of Bioengineering, SRM Institute of Science and Technology, Kattankulathur, Tamilnadu, India

Hiran Kanti Santra Microbiology and Microbial Biotechnology Laboratory, Department of Botany and Forestry, Vidyasagar University, Midnapore, West Bengal, India

Pankaj Sharma Himachal Pradesh State Biodiversity Board, Shimla, Himachal Pradesh, India

Ruchi Sharma School of Bioengineering & Food Technology, Shoolini University of Biotechnology and Management Sciences, Solan, Himachal Pradesh, India

Ramesh Shruthy Department of Food Process Engineering, School of Bioengineering, SRM Institute of Science and Technology, Kattankulathur, Tamil Nadu, India

Jatinder Singh Department of Horticulture, School of Agriculture, Lovely Professional University, Phagwara, Punjab, India

Joginder Singh Department of Biotechnology, Lovely Professional University, Jalandhar, Punjab, India

Ravikant Singh Biological Product Laboratory, Department of Botany, University of Allahabad, Prayagraj, India

Simranjeet Singh Department of Biotechnology, Lovely Professional University, Jalandhar, Punjab, India

Punjab Biotechnology Incubator, Mohali, Punjab, India

Regional Advanced Water Testing Laboratory, Mohali, Punjab, India

Arpita Tripathi Biological Product Laboratory, Department of Botany, University of Allahabad, Prayagraj, Uttar Pradesh, India

Sharad Kumar Tripathi Biological Product Laboratory, Department of Botany, University of Allahabad, Prayagraj, India

Natural Metabolites: An Eco-friendly Approach to Manage Plant Diseases and for Better Agriculture Farming

1

Touseef Hussain, Simranjeet Singh, Mohd. Danish, Rashid Pervez, Kashif Hussain, and Raja Husain

Abstract

Natural metabolites and biocontrol agents are becoming more popular and are getting consideration to be viable replacement methods for controlling various plant diseases nowadays because the environment is safer and, in some cases, the only option is available for the protection of plants against the pathogens. In the present scenario, beyond good horticultural and agricultural practices, producers often depend mostly on chemically synthetic pesticides and fertilizers that are not only harmful but also very costly. Development of pathogen-resistant breeds becomes a worldwide problem which imposes and threatens some chemical companies to produce new pesticides with their registration process and profitability. There has been a considerable change in the perspective of farmers toward the use of pesticides for crop protection and crop production. There are several types of biological control agents and natural metabolite products are available, but for effective acquisition and future development, it will need a great understanding of complex interactions between humans, plants, and the environment. In this chapter, we will discuss wide varieties of plants and

T. Hussain (✉) · M. Danish
Department of Botany, Aligarh Muslim University, Aligarh, Uttar Pradesh, India

S. Singh
Department of Biotechnology, Lovely Professional University, Phagwara, Punjab, India

Punjab Biotechnology Incubator, Mohali, Punjab, India

Regional Advanced Water Testing Laboratory, Mohali, Punjab, India

R. Pervez
Division of Nematology, ICAR-Indian Agricultural Research Institute, New Delhi, India

K. Hussain
Gyani Inder Singh Institute of Professional Studies, Dehradun, Uttarakhand, India

R. Husain
Division of Biochemistry, ICAR-Indian Agricultural Research Institute, New Delhi, India

J. Singh, A. N. Yadav (eds.), *Natural Bioactive Products in Sustainable Agriculture*, https://doi.org/10.1007/978-981-15-3024-1_1

pathogens and their interaction and management through natural metabolites produced by microbes and plants. These interactions can affect the health of plants in several ways. There are several microbes that reside in plant roots and interact with plants that are beneficial while some can be harmful because they are involved in the development of plant diseases, which occur at various levels of interaction scale that leads to natural control.

Keywords
Biocontrol · Natural · Metabolites · Plant health · Plant pathogens

1.1 Introduction

Agriculture is one of the most growing fields, which include many crops, that might be affected by various abiotic and biotic stresses. Biotic stresses include bacteria, fungi, viruses, insects, nematodes, and weeds, which affect in different developmental stages. The decreased production of a new crop cultivar is the result of biotic stress (Sanjay and Tiku 2009).

There are so many types of plant diseases that may include reduced quantity and quality of crop, threat to environment and human health, increased cost of production, less remunerative alternatives, and deterioration of natural resources (Eilenberg et al. 2001). After seeing the present scenario, there are many varieties of natural metabolites produced by biocontrol agents that are available. In numerous decade ago, biological metabolites are used for the control of various plant diseases. So, the purpose of living microbes to control the growth of plant pathogen is an important part of success (Kumar and Saxena 2009). Thus, it is a challenging task to overcome the damage caused by plant pathogens. It is compulsory to devise strategies to combat these problems. Although there are multiple pesticides, biocontrols or chemicals are available in the market, but to reduce the growth of plant pathogens, these compounds may have certain drawbacks; for example, the little amount of these compounds present in fruits or grains could harm human being and may be harmful to the environment, which can cause water and soil pollution, animal and food contamination, and beneficial microbe elimination. Keeping in view the hazardous effect of pesticides, natural products seem to be an eco-friendly approach to manage the disease and reduce the toxic effects of pesticides and fungicides to environment and human health.

1.2 Role of Natural Plant Products in Disease Management

The accessibility of nutritious and safe food could be increased by increasing the sustainability of agricultural production system, which is the global challenge for food security. The use of bioactive natural products isolated from natural products for plant protection and sustainability may be incorporated in plant production. In modern agriculture, the use of botanical and microbial natural products may lead to the development of natural inducers and antimicrobial agents of plant host defense system. In much previous research, there are so many plants, such as neem, citrus, garlic, moringa, etc., which have been used for the management of several bacterial and fungal diseases. On chemical basis, many plant metabolites act as antifungal agents like flavones, flavonoids, quinines, tannins, terpenes, and saponins. All these always played important role as antifungal agents. These are all produced by plants against microbial infections and found to be effective substances against harmful microorganisms and pathogens (Ciocan and Bara 2007). Flavonoids are allelopathic compounds, phytoalexins, and antimicrobial and detoxifying agents that always act as signal molecules and as protector from various stresses. Flavonoids also play an important role in abiotic stresses like temperature and drought tolerance, frost hardiness, and freezing tolerance (Iwashina 2003).

Isoflavonoids are involved in the defense mechanism against plant pathogen, and these can be characterized by migration of phenyl ring. In species *Mangifera indica*, the major phytoalexin is vestitol, belonging to class isoflavones (Lanot and Morris 2005). In several previous kinds of researches, scientist investigated the presence of flavonoid phytoalexins in cucumber and reported that silicon is involved in the defense mechanism against fungal proteins (Fawe et al. 2001). Flavonoids function as detoxifying agents, justified by flavonoid peroxidase that plays an important role in H_2O_2 scavenging (Yamasaki et al. 1997). Flavonoids are not only able to detoxify the reactive oxygen species but also chelate with heavy metals, resulting in divergence of molecular structures. Flavonoids also play an important role against plant viruses. Galangin (3,5,7-trihydroxyflavone), a flavonoid isolated from *Helichrysum aureonitens*, creates the defense mechanism against the viruses and gram-negative bacteria and fungi (Cowan 1999).

1.2.1 Coumarins

Coumarin is a colorless, crystalline solid that is bitter in taste. It was first screened from tonka beans in 1820 by A. Vogel (Munich). Basically, coumarins are phenolic substances containing various secondary metabolites, composed of pyrone rings and fused benzene. Coumarins are inducible antifungal chemicals that function as antifungal agents. Many coumarins, like scopolin, scopoletin, and umbelliferone, are formed in tissues of plant in response to fungus like *Fusarium oxysporum*, which attack potato roots. When roots of parsnip and celery were inoculated to *Sclerotinia sclerotiorum*, the level of furanocoumarins was decreased (Rahman 2000). Coumarins are also known as phytoalexin, and in much previous study, it was reported as antifungal agents that develop the defense mechanism against fungal bodies (Brooker et al. 2007). In few works, it was also regarded as antibacterial and anti-insecticidal agents. It has antibacterial effects on animal and pathogenic bacteria (Razavi et al. 2009). In other previous studies, coumarin and murraxocin acts as strong insecticidal agents having mortality on larva and eggs of insects (Sharma et al. 2006). Coumarin is also reported to act as an ovicide agent (Tasleem et al. 2012). There are many agents like carbazole derivatives; clausenidin, dentatin, and clauszoline extracted from *Clausena excavate* exhibit antimycotic activity. In several in vitro studies, coumarins showed antibacterial and antifungal activities (Nakajima and Kawazu 1980). Synthetic coumarins and angelicin derivatives were found effective against *A. niger*, *C. albicans*, *S. cerevisiae*, and *C. neoformans*.

1.2.2 Tannins or Gallotannin

Tannin is a polyphenolic biomolecule and is responsible for the flavoring and astringency in tea. Normally, tannin is found in the bark, wood, roots, leaves, and fruits of plants. The molecular weight of tannin ranges from 500 to 3000 kDa, and it is divided into two categories: (i) hydrolysable and (ii) condensed tannins. Hydrolyzable tannins are derived from flavonoid monomers and produce ellagic or gallic acids upon heating with sulfuric and hydrochloric acids. Hydrolyzable tannins are extracted from vegetable plants such as oak wood (*Quercus petraea*, *Quercus alba*, and *Quercus robur*), chestnut wood (*Castanea sativa*), tara pods (*Caesalpinia spinosa*), myrobalan, and gallnut. There is a carbohydrate, usually D-glucose, in the center of tannin molecule as multiple esters. Certain examples of tannins are gallic acids, which are esters of glucose found in leaves and bark of many plants. They are also formed by the condensation reaction of flavan derivatives (Tasleem et al. 2012). The examples of condensed tannins are polyflavonoid tannins, proanthocyanidins, pyrocatecollic-type tannins, catechol-type tannins, etc., formed by condensation of flavans. Condensed tannin is also found in grape, commonly called as procyanidins, consisting of 2–50 polymer (or more) catechin units linked by carbon-carbon bonds. Tannins create the defense mechanisms by the binding of dietary proteins of digestive enzymes. Many types of tannin act as insecticidal agents and could have a negative effect on insects because they were having basic gut pH and tannins don't act on those proteins (Barbehenn and Constabel 2011).

1.2.3 Alkaloids

Alkaloids usually contain more than nitrogen atoms in their heterocyclic ring. These compounds were having weak and neutral acidic properties (Manske 1965; Lewis 1998). The function of alkaloids in the plant is uncertain. There is no such importance of alkaloids, although they are regarded only as by-products of plant metabolism. Sometimes, they may act as reservoirs for protein synthesis. In previous research, it was reported that generally alkaloids are therapeutically significant plant substances (Tasleem et al. 2012). Alkaloids are produced by large group of various organisms like plant fungi, animals, and bacteria (Roberts 1998). Alkaloids have antimalarial, antibacterial, analgesic, and anticancer properties. The first reported alkaloid was morphine, derived from *Papaver somniferum* (opium poppy) in 1805.

The alkaloid like 2-(3-4-dimethyl-2-5-dihydro1H-pyrrol-2-yl)-1-methylethylpentanoate derived from *Datura* shows antifungal activities against *Candida* and *Aspergillus* species. Fragulanine is a cyclic peptide quinoline alkaloid isolated from *Melochia odorata* that showed antifungal activity against the pathogenic fungi. Other alkaloid,

3-methoxysampangine, derived from *Cleistopholis patens* were reported to have antifungal activity against *C. neoformans*, *C. albicans*, and *A. fumigatus*. N-Desmethylcycleanine, cycleanine, and cocsoline from *Albertisia villosa* are the antifungal alkaloids that were reported from higher plants (Tasleem et al. 2012).

1.2.4 Terpenoids

Terpenoids are commonly called as isoprenoids. Basically, these are the assorted class isolated from terpenes. Terpenes are large and diverse class of hydrocarbons produced by varieties of plants. They have strong odor and may protect the plants by deterring herbivores and by parasites of herbivores. Terpenoids act as antioxidants and execute various functions in plants and animals (e.g., carotenoids function as essential pigments for light extracting and provide photo protection and plant pigmentation). Plant terpenoids are used as herbal remedies against various pathogens and diseases. Terpenoids give odor to eucalyptus, flavors to ginger cinnamon and cloves, red color in tomatoes, and the yellow pigmentation in sunflowers (Specter 2009). Terpenes (diterpenes, triterpenes, tetraterpenes, hemiterpenes, and sesquiterpenes (C15)) contain oxygen in their side chain and are termed as terpenoids. Some common examples are artemisinin (sesquiterpenoids), farnesol, camphor (monoterpenes), and methanol.

In several research studies, it was reported that terpenes or terpenoids work against bacteria, viruses, and protozoa. Capsaicin enhances the growth and development of *Candida albicans*, which inhibit the growth of various types of bacteria (Cowan 1999). Terpenes are also considered as antifungal agents. Monoterpenoids are involved in the innate immunity against various plant pathogens. The major components of oils, that is, 1,8-cineole, p-cymene cineole, carvacrol, thymol, and geranial, exhibit antifungal activity. The extracts from *Agastache rugosa* (essential oil) were reported to have antifungal activity. The essential oil extracts from the leaves of *Litsea cubeba* contain n-transnerolidol 3, 7-dimethyl-1, 6-octadien-3-ol, and cis-ocimene that manifest antifungal activities. Tri-terpenoid glycosides obtained from *Bellis perennis* and *Solidago virgaurea* inhibit the growth and development of human pathogenic yeasts (*Cryptococcus* and *Candida* species) (Tasleem et al. 2012).

1.3 Plant Growth-Promoting Rhizobacteria (PGPR)

The root system of the plant system is surrounded by a narrow zone of soil called rhizosphere (Walker et al. 2003). The bacterial community colonizing this narrow environmental zone is term as "rhizobacteria" (Kloepper et al. 1991). The bacterial communities that colonize the roots of the plant and support their growths are called plant growth-promoting rhizobacteria (PGPR) (Beneduzi et al. 2012). These bacteria are one of the most effective and eco-friendly methods for the management of plant disease (Compant et al. 2005). PGPR as biocontrol agents have several benefits over chemical practices, because PGPR are nontoxic and they are naturally occurring microorganism having endurable applications.

The application of PGPR as a cost-effective control method of pest management in roots has been reported by several workers (Lucy et al. 2005; Whipps 2001). Different bacterial strains have shown to have ability for development as biocontrol agents on cereals. Bacterial isolates from the plant root, such as *Bacillus*, *Pseudomonas*, and *Azotobacter*, showed antagonistic activity to check the plant pathogen and act as disease management agent (Berg and Smalla 2009).

The biocontrol potential of *Pseudomonas* sp. and *Bacillus* spp. as important biocontrol agents to strive against root and soilborne microbial pathogens has been reported in several crops like wheat, tomato, potato and chickpea (Hussain and Khan 2020; Dashti et al. 2012; Perez-Montano et al. 2014). Several species of *Bacillus* such as *B. licheniformis*, *B. cereus*, *and B. thuringiensis* were reported to be potential biocontrol agents. *Bacillus* spp. screened from the chickpea rhizosphere have shown to reduce the pathogenic activity of fungus called *Fusarium oxysporum*, which is reported to cause *Fusarium* wilt disease. In a greenhouse experiments, *Bacillus* strains that was isolated and reported from the sorghum rhizosphere in Ethiopia and wild grass sp. in South Africa have antagonistic affects against the root rot disease caused by *F. oxysporum* and crown rot pathogens by *Pythium ultimum* (Idris et al. 2007).

The plant growth-promoting rhizobacteria exerted different mechanisms such as antibiosis, secreting toxin surface compound (bio-surfactants) and volatiles, chitinase cell wall-degrading enzymes, and α-1,3-glucanase and also induce systemic resistance in plants to deplete the soilborne pathogens (Perez-Montano et al. 2014; Haas and Défago 2005; Compant et al. 2005; Van Loon 2007, Whipps 2001). The earlier reported mechanism of biocontrol is the secretion of siderophore ligands that efficiently confiscate iron and inhibit the growth of pathogen (Raaijmakers et al. 2002).

1.4 Fungi as Biocontrol Agents

Nowadays, fungi biological control is considered to be a rapid and effective developing natural phenomenon with wide applications in industrial sector like food production and food yield. Harman et al. (2013) reported that *Trichoderma* species allows for the development of biocontrol strategies against economically important plant pathogen. Antagonistic effect of *Trichoderma* spp. is due to the secretion of secondary metabolites against *Pythium ultimum* and *Rhizoctonia solani* (Harman et al. 2004). Several other fungi like *Pochonia chlamydosporia* isolates have been evaluated as biocontrol agents against the root-knot nematodes (RKN) with different crops and experimental conditions (Shurf et al. 2014). Endophytic colonization of the root by *P. chlamydosporia* suppresses the growth of the pathogens and enhances the growth and development of the plant (Maciá-Vicente et al. 2009)

1.5 Role of Mycorrhizae in Disease Management

To develop effective and durable protection to root system, the role of mycorrhiza against pathogenic population is well established (Thakur et al. 2005). Mycorrhizal fungi are most common fungal association with roots of majority of plants.

The mycorrhiza (fungus root) is defined as an association between fungi and plants that establishes on tissue of root system during the time of active plant growth and makes unfavorable environment for pathogens. Mycorrhizal fungus acts as biocontrol agent against various plant pathogens, which is a relatively new and eco-friendly technique. Several studies have proved that tree seedlings with mycorrhizal associations exhibit more resistance to feeder roots against pathogenic fungi/bacteria/nematodes than non-mycorrhizal roots (Al-Karaki 2000).

Ectomycorrhizae multiply on the root surface producing a netlike structure called the Hartig net. They reduce the development of disease by exerting different mechanisms including antibiosis, by synthesizing antifungal compound and developing a barrier around the root of the plant (Duchesne 1994). Root rot disease of red pine caused by *F. oxysporum* and *F. moniliforme* is effectively controlled by ectomycorrhizal fungi like *Paxillus involutus*.

Vesicular arbuscular mycorrhizae (VAM) are another significant part of the microbial soil community that provide vital benefit for plant growth (Sukhada et al. 2010; Yinsuo et al. 2004). The VAM fungi are not only beneficial for plant development, but it also develop the resistance in the host plant against various soilborne plant pathogens (Ziedan et al. 2011; Upadhyaya et al. 2000). Among the VAM fungi, the genus *Glomus* is very common with species like *G. fasciculatum*, *G. mosseae*, *G. constrictum*, *G. monosporum*, and *G. macrosporum.* During the establishment on the root system, they prevent root infections by reducing the entry sites of pathogens and encouraging host defense. They have also been found to decrease the infection of root-knot nematode in different plant systems (Linderman 1994). The effect of *Pseudomonas syringae* on tomato is significantly decreased when the host plants are inoculated with mycorrhizae fungi (García-Garrido and Ocampo 2002). The mode of action involved in these interactions includes indirect effects, chemical interactions, and physical protection (Fitter and Garbaye 1994). The other strategy adopted by VAM fungi includes improved nutrition in the plant; increase lignification on the root system; and chemical composition of antifungal isoflavonoids, chitinase, etc. (Morris and Ward 1992).

Arbuscular mycorrhizal fungi (AMF) are reported to manage a number of crop diseases, especially root diseases (Xavier and Boyetchko 2004). They can affect the pathogens and suppress the diseases through the development of systemic resistance in the host (Jung et al. 2012; Pozo and Azcon-Aguilar 2007; Pineda et al. 2010). It has been found that mycorrhizal induce resistance (MIR) is the result of active depletion of components in the Salicylic acid (SA)-dependent defense pathway, which causes systemic priming of jasmonic acid-dependent defenses (Pozo and Azcon-Aguilar 2007; Hause et al. 2007).

1.6 Management of Plant Diseases Caused by Plant Parasitic Nematodes

Agriculture production in India has sustained losses of millions of dollars due to various factors. Among them, one of the major constraints is pest diseases, which are the limiting factor in the cultivation of crops. Among them, diseases caused by plant parasitic nematodes (PPN) are one of the constraints in reducing both the quality and yield of the crops. They cause 21.3% crop losses amounting to ₹102039.79 million ($1577 million) annually; the losses in 19 horticultural crops were assessed at ₹50224.98 million, while for 11 field crops it was estimated at ₹51814.81 million. Rice-rot nematode *Meloidogyne graminicola* was economically most important causing yield loss of ₹23272.32 million in rice. Citrus (₹9828.22 million) and banana (₹9710.46 million) among fruit crop and tomato (₹6035.2 million), brinjal (₹3499.12 million), and okra (₹2480.86 million) among the vegetable crops suffered comparatively more losses (Walia and Chakrabarty 2018; Khan et al. 2010). It has been estimated that overall losses amount to $78 billion globally due to RKN (Chen et al. 2004).

Plant parasitic nematodes are regarded as biggest enemy of crops due to their small size and natural habitats where they established a feeding site. They lead major structural changes in root system and metabolism. Root-knot nematodes (RKN), such as *Meloidogyne* spp., are the most dangerous nematodes worldwide. They act on more than 5000 plant species and cause economic losses in many horticultural and agricultural crops (Ntalli et al. 2010).

One or more nematode pests are always associated with every crop, which cause economic loss to crop, and their control is the major requirements for increasing the crop productivity. Pesticides are currently being used to manage these nematode pests leading to environmental and health concerns and resulting into the suppression of other naturally occurring biocontrol agents as well as resistance in their nematodes.

Many soil-inhabiting microorganisms, such as fungi, bacteria, protozoans, viruses, turbellarians, enchytraeids, mites, predatory nematodes, collembolans, and tardigrades, are parasites, predators, or antagonistic to plant parasitic nematodes. These microorganisms have been exploited as biocontrol agents for the management of PPN in several agricultural and horticultural crops. The increasing thrust toward sustainable agriculture and integrated pest management has led to biological control emerging from their status as a fringe sector to being viewed as an intrinsic part of crop protection. A brief description on the most promising organisms is furnished below.

1.6.1 *Paecilomyces lilacinus*

This is an opportunistic fungus prevalent in many soils. This fungus parasitizes the egg of root-knot nematodes and suppresses nematode hatching. Inoculation of this fungus at root zone could significantly reduce root loss due to nematode infestation and improve plant growth.

1.6.2 *Trichoderma* spp.

These are hypomycetous fungi widely used against several disease-causing fungi. Several species of *Trichoderma* colonize egg masses of nematodes. Although direct parasitization was not observed with any of these isolates, distortion of nematode eggs was frequently seen with *T. harzianum* and *T. viride*. Culture of these isolates showed antagonistic effect against nematodes. *T. harzianum* suppress root knot nematodes and improve growth of the plant.

1.6.3 *Pochonia chlamydosporia*

This fungus is known as biocontrol agent of root knot and cyst nematodes. Isolate of this fungus parasitizes root-knot nematode eggs and suppresses their hatching by more than 55% within 24 h. It is a promising biocontrol agent to controlled nematode problem and significantly increased yield.

1.6.4 Mycorrhizal Fungi

These fungi are obligate dependents on plants for nourishment. Their symbiotic association with roots increases the plant's ability to absorb water, phosphorus, and other elements. Mycorrhizae are known to increase host tolerance against nematode infestation due to enhanced "P" status of the host between fungus and the nematode. Significant increase in growth of plant and reduction in the root knot nematode populations were observed when challenged with such fungi like *Glomus mosseae*, *G. fasciculatum*, *Acaulospora laevis*, and *Gigaspora margarita.*

1.6.5 *Pasteuria penetrans*

This bacterium inoculation reduced nematode population and thereby improved the vegetative growth of the plants. However, mass multiplication of this bacterium is yet to be standardized.

1.6.6 *Pseudomonas fluorescens*

P. fluorescens is a rhizosphere bacterium, which is commonly found in soils. This bacterium is quite popular among the farmers to control fungal diseases. Some of the isolates of this bacterium possess inhibitory effect on nematode and enhance root growth.

1.6.7 Endophytic Bacteria

Endophytic bacteria like *Pseudomonas aeruginosa*, *Bacillus megaterium*, and *Curtobacterium luteum* offer excellent nematicidal properties. Their endophytic nature is an added advantage for effectively preventing the nematode entry into roots.

1.6.8 Entomopathogenic Nematodes

Entomopathogenic nematodes (EPNs) are emerging as a potent biocontrol agents. They have a great potential as biological control agents because they have wide host range, are easy to handle, have short life cycle, are economically produced at large scale, and are environmentally safe. They are symbiotically associated with bacteria, which played a significant role in suppressing nematode population.

1.7 Future Prospects

In the present scenario of crop production, the natural metabolites produced by biocontrol are of utmost importance; however, its application needs to be fully exploited. The research in this area is still confined to the laboratory, and very little attention has been given to produce the profit-making formulations of the bioagents. Moreover, cost-effective products have not been used conveniently by the farmers owing to the limited available information regarding its use. Therefore, to popularize the concept of biological control extension at basic research levels in this direction needs to be improved.

Most bioagents fail to perform infield trial experiments but perform well in the laboratory conditions. There are many reasons out of which probably blame to the ecological and physiological constraints that limit the efficacy of metabolites. To improve the selection and characterization of metabolites, biotechnology and other molecular tools are gaining importance to potentially solve the problem in near future. Increasing the efficacy of BCA, different methods such as mutation or protoplasm fusion could be a good idea. To better understand the mechanism of bioagents and to evaluate their environmental factors that favor the hasty growth of biological agents, there is an urgent need of efficient bioagents in cost-effective manner for market entrance.

1.8 Conclusion

There are currently more than 30 bacterial and fungal products for the management of soilborne diseases of agricultural crops. In the recent era, in India, there is a new progressive research and development (on applied level) on new biological agents at field levels that have find a new place on active application to disease-infected parts of the crops, but other problems have also emerged, which are new obstacle in the development of biological agents like lack of constancy on its effectiveness. When these biological agents are applied to control the diseases, they show some inconsistency or they are less effective at ground/base level, which may be due to that before any integration between abiotic and biotic factors that can provide and help to increase the efficacy of biocontrol activity. Biological agents are more valuable section of integrated pest management systems; while using other types of pest management strategies like chemical, mechanical, and organic, there should be more coordination between each other. Nowadays, people are highly aware of organic production and are very interested and active in participating in activities likes home gardening and terrace farming in urban populations and cities, as well as reducing the usages of harmful pesticides, fungicides, and insecticides, which are creating a market of the natural metabolite products. Probably, in spite of these metabolites, more improvement can be done on isolation, formulation, and their application processes, especially at farmer's field level. The twenty-first century will be the age of biotechnology that will contribute an important role in the field of plant pathology and lead to the development of new strategies for natural metabolites.

Conflict of Interest The authors declare that they have no conflict of interest.

References

Al-Karaki GN (2000) Growth of mycorrhizal tomato and mineral acquisition under soil stress. Mycorrhiza 10:51–54

Barbehenn RV, Constabel CP (2011) Tannins in plant–herbivore interactions. Phytochemistry 72:1551–1565

Beneduzi A, Ambrosini A, Passaglia LMP (2012) Plant growth-promoting rhizobacteria (PGPR): their potential as antagonists and biocontrol agents. Genet Mol Biol 35(4):1044–1051

Berg G, Smalla K (2009) Plant species and soil type cooperatively shape the structure and function of microbial communities in the rhizosphere. FEMS Microbiol Ecol 68:1–13

Brooker NL, Kuzimichev Y, Laas J, Pavlis L (2007) Evaluation of coumarin derivatives as antifungal agents soil-borne fungal pathogens. Commun Agric Appl Biol Sci 72:785–793

Chen ZX, Chen SY, Dickson DW (2004) Nematology advance and perspectives, vol 2. CAB International, Wallingford

Ciocan ID, Bara I (2007) Plant products and antimicrobial agents. Universitatii ale ªtiintifice Analele Alexandru Ioan Cuza, Tom VIII

Compant S, Duffy B, Nowak J, Clement C, EA BI (2005) Use of plant growth promoting bacteria for biocontrol of plant diseases: principles, mechanisms of action, and future prospects. Appl Environ Microbiol 71:4951–4959

Cowan MM (1999) Plant products as antimicrobial agents. Clin Microbiol Rev 12:564–582

Dashti NH, Ali NY, Cherian VM, Montasser MS (2012) Application of plant growth-promoting rhizobacteria (PGPR) in combination with a mild strain of *Cucumber mosaic virus* (CMV) associated with viral satellite RNAs to enhance growth and protection against a virulent strain of CMV in tomato. Can J Plant Pathol 34:177–186

Duchesne LC (1994) Role of ectomycorrhizal fungi in biocontrol. In: Pfleger FL, Linderman RG (eds) Mycorrhizae and plant health. APS Press, St. Paul, pp 27–45

Eilenberg J, Hajek A, Lomer C (2001) Suggestions for unifying the terminology in biological control. BioControl 46:387–400

Fawe A, Menzies J, Cherif M, Belanger R (2001) Silicon and disease resistance in dicotyledons. Plant Sci 8:159–169. https://doi.org/10.1016/S0928-3420(01)80013-6

Fitter AH, Garbaye J (1994) Interactions between mycorrhizal fungi and other soil organisms. Plant Soil 159:123–132

García-Garrido JM, Ocampo JA (2002) Regulation of the plant defence response in arbuscular mycorrhizal symbiosis. J Exp Bot 53(373):1377–1386. https://doi.org/10.1093/jexbot/53.373.1377

Haas D, Défago G (2005) Biological control of soil-borne pathogens by fluorescent pseudomonads. Nat Rev Microbiol 3:307–319

Harman GE, Howell CR, Viterbo A, Chet I, Lorito M (2004) *Trichoderma* species opportunistic, avirulent plant symbionts. Nat Rev Microbiol 2:43–56

Harman GE, Howell CR, Viterbo A, Chet I, Lorito M (2013) Trichoderma species – opportunistic, avirulent plant symbionts. Nat Rev Microbiol 2:43–56

Hause B et al (2007) Jasmonates in arbuscular mycorrhizal interactions. Phytochemistry 68:101–110

Hussain T, Khan AA (2020) Bacillus subtilis HussainT-AMU and its antifungal activity against potato black scurf caused by Rhizoctonia solani. Biocatal Agric Biotechnol 23:101433

Idris EES, Iglesias DJ, Talon M, Borriss R (2007) Tryptophan-dependent production of indole-3-acticacid (IAA) affects level of plant growth-promotion by *Bacillus amyloliquefaciens* FZB42. Mol Plant-Microbe Interact 20:619–626

Iwashina T (2003) Flavonoid function and activity to plants and other organisms. Biol Sci Space 17(1):24–44

Jung SC et al (2012) Mycorrhiza induced resistance and priming of plant defenses. J Chem Ecol 38:651–664

Khan MR, Jain RK, Singh RV, Pramanik A (2010) Economically important plant parasitic nematodes distribution ATLAS. Directorate of Information and Publications of Agriculture, New Delhi, p 137

Kloepper JW, Zablotowick RM, Tipping EM, Lifshitz R (1991) Plant growth promotion mediated by bacterial rhizosphere colonizers. In: Keister DL, Cregan PB (eds) The rhizosphere and plant growth. Kluwer Academic Publishers, Dordrecht, pp 315–326

Kumar J, Saxena SC (2009) Proceedings of the 21st training on recent advances in plant disease management, GBPUA&T Pantnagar, pp 1–3

Lanot A, Morris P (2005) Elicitation of isoflavan phytoalexins. In: Marquez AJ (ed) *Lotus japonicus* handbook. Springer, Amsterdam, pp 355–361

Lewis RA (1998) Lewis' dictionary of toxicology. CRC Press, Boca Raton, FL, p 51. ISBN 1–56670–223-2

Linderman RG (1994) Role of VAM fungi in biocontrol. In: Pfleger FL, Linderman RG (eds) Mycorrhizae and plant health. The American Phytopathological Society, St. Paul, MN, pp 1–27. ISBN 0–89054–158-2

Lucy M, Reed E, Glick BR (2005) Application of free living plant growth-promoting rhizobacteria. Antonie Van Leeuwenhoek 86:1–25

Maciá-Vicente JG, Rosso LC, Ciancio A, Jansson HB, Lopez-Llorca LV (2009) Colonisation of barley roots by endophytic *Fusarium equiseti* and *Pochonia chlamydosporia*: effects on plant growth and disease. Ann Appl Biol 155:391–401

Manske RHF (1965) The alkaloids. Chemistry and physiology, vol VIII. Academic, New York, p 673

Morris DL, Ward JB Jr (1992) Coumarin inhibits micronuclei formation induced by benzo(a) pyrene in male but not female ICR mice environ. Mol Mutagen 19:132–138

Ntalli NG, Menkissoglu-Spiroudi U, Giannakou I (2010) Nematicidal activity of powder and extracts of *Melia azedarach* fruits against *Meloidogyne incognita*. Ann Appl Biol 156:309–317

Pérez-Montaño F, Alías-Villegas C, Bellogín RA, del Cerro P, Espuny MR, Jiménez- Guerrero I, López-Baena FJ, Ollero FJ, Cubo T (2014) Plant growth promotion in cereal and leguminous agricultural important plants: from microorganism capacities to crop production. Microbiol Res 169:325–336

Nakajima S, Kawazu K (1980) Coumarin and euponin, two inhibitors of insect development from leaves of *Eupatorium japonicum*. Agric Biol Chem 44:2893–2899

Pineda A, Zheng S-J, van Loon JJA, Pieterse CMJ, Dicke M (2010) Helping plants to deal with insects: the role of beneficial soil-borne microbes. Trends Plant Sci 15(9):507–514

Pozo MJ, Azcon-Aguilar C (2007) Unraveling mycorrhiza-induced resistance. Curr Opin Plant Biol 10:393–398

Rahman AU (2000) Studies in natural product chemistry, vol 24. Elsevier, Amesterdam, pp 860–861. ISBN: 0-444-50643-8

Raaijmakers JM, Vlami M, de Souza JT (2002) Antibiotic production by bacterial biocontrol agents. Antonie Van Leeuwenhoek 81:537–547

Razavi SM, Ghasemiyan A, Salehi S, Zahri F (2009) Screening of biological activity of *Zosima absinthifoliafruits* extracts. Eur Asia J Biosci 4:25–28

Sanjay G, Tiku AK (2009) Botanicals in pest management: current status and future perspectives. Biomed Life Scipp, pp 317

Sharf R, Shiekh H, Syed A, Akhtar A, Robab MI (2014) Interaction between *Meloidogyne incognita* and *Pochonia chlamydosporia* and their effects on the growth of *Phaseolus vulgaris*. Arch Phytopathol Plant Protect 47(5):622–630

Sharma R, Negi DS, Shiu WK, Gibbons S (2006) Characterization of an insecticidal coumarin from *Boenninghausenia albiflora*. Phytother Res 20:607–609

Specter M (2009) A life of its own. The New Yorker, pp 56–65

Sukhada M, Manjula R, Rawal RD, Lakshmikantha HC, Saikat C, Ramachandra YL (2010) Evaluation of arbuscular mycorrhiza and other biocontrol agents in managing *Fusarium oxysporum* f.sp. *cubense* infection in nbanana cv. Neypoovan. Biocontrol Sci Tech 20:165–181

Tasleem A, Bhosalea JD, Kumara N, Mandala TK, Bendreb RS, Lavekara GS, Dabu R (2012) Natural products – antifungal agents derived from plants. J Asian Nat Prod Res 11(7):621–638

Thakur JS, Sharma YP, Lakhanpal TN (2005) Effect of ectomycorrhizal on the development of powdery mildew (*Podosphaera leucotricha*) of apple (*Malus domestica Borkh*) seedling. J Myc Pl Pathol 35:275–276

Upadhyaya RK, Mukerji KG, Chamola BP (2000) Sustainable management of arbuscular mycorrhizal fungi in the biocontrol of soil-borne plant diseases. In: Sharma MP, Adholeya A (eds) Biocontrol potential and its exploitation in sustainable agriculture, Vol. 1: crop diseases, weeds and nematodes. Kluwer Academic Publishers, New York, pp 117–138

van Loon LC (2007) Plant responses to plant growth-promoting rhizobacteria. Eur J Plant Pathol 119:243–254

Vogel A (1820) De l'existence de l'acide benzoïque dans la fève de tonka et dans les fleurs de mélilot [On the existence of benzoic acid in the tonka bean and in the flowers of melilot]. Journal de Pharmacie (in French) 6:305–309

Walker TS, Bais HP, Grotewold E, Vivanco JM (2003) Root exudation and rhizosphere biology. Plant Physiol 1320:44

Walia RK, Chakrabarty PK (2018) Nematode problems of crops in India; A comparative volume on four decade of AICRP (Nematode), ICAR-All India Coordinated Research Project on Nematodes in agriculture. M. S. Printers, New Delhi, pp 400

Whipps JM (2001) Microbial interactions and biocontrol in the rhizosphere. J Exp Bot 52:487–511

Xavier L, Boyetchko S (2004) Arbuscular mycorrhizal fungi in plant disease control. In: Arora D, Bridge P, Bhatnagar D (eds) Fungal biotechnology in agricultural, food, and environmental applications. Marcel Dekker, Inc, New York

Yamasaki H, Sakihama Y, Ikehara N (1997) Flavonoid-peroxidase reaction as a detoxification mechanism of plant cells against H202. Plant Physiol 115:1405–1412

Yinsuo J, Vincent MG, Colin JS (2004) The influence of *Rhizobium* and arbuscular mycorrhizal fungi on nitrogen and phosphorus accumulation by *Vicia faba*. Ann Bot 94:251–258

Ziedan EH, Elewa IS, Mostafa MH, Sahab AF (2011) Applications of mycorrhizae for controlling root rot diseases of sesame. J Plant Protect Res 51:354–361

Plant Disease Management by Bioactive Natural Products

2

Simranjeet Singh, Vijay Kumar, Shivika Datta, Daljeet Singh Dhanjal, and Joginder Singh

Abstract

The expansion in population and economies leads to development of new methods of agriculture to increase the productivity and meet the demand. Chemical fertilizers and pesticides are used in large scale to increase the production, but these pose environmental impacts on different ecosystems of life. Bioactive natural products derived from microorganisms and plants serve to minimize these impact. Various compounds isolated from plant sources, algal sources, microbial sources, marine sources, etc., act as elicitors for plant defense against various sources. Majority of them are chitosan, salicylic acid (SA), benzoic acid, benzothiadiazole, alkaloids, flavonoids, terpenes, proteins, peptides, blasticidin, mildiomycin, polyoxins, phenolic compounds, etc., which act against various plant diseases caused by bacteria, fungi, and viruses. The present chapter discusses the usages of bioactive natural products as tool agonists pant disease management. These derived bioactive compounds minimize the use of chemical agents and contribute to eco-friendly sustainable agriculture practices.

Simranjeet Singh and Vijay Kumar have equally contributed to this chapter.

S. Singh
Department of Biotechnology, Lovely Professional University, Jalandhar, Punjab, India

Punjab Biotechnology Incubators, Mohali, Punjab, India

Regional Advanced Water Testing Laboratory, Mohali, Punjab, India

V. Kumar
Department of Chemistry, Regional Ayurveda Research Institute for Drug Development, Gwalior, Madhya Pradesh, India

S. Datta
Department of Zoology, Doaba College, Jalandhar, Punjab, India

D. S. Dhanjal · J. Singh (✉)
Department of Biotechnology, Lovely Professional University, Jalandhar, Punjab, India
e-mail: joginder.15005@lpu.co.in

J. Singh, A. N. Yadav (eds.), *Natural Bioactive Products in Sustainable Agriculture*, https://doi.org/10.1007/978-981-15-3024-1_2

Keywords
Plant diseases · Defense · Metabolites · Phenol · Fungi · Algae · Marine source

2.1 Introduction

Exponential rise in global population contributes to new aspects of food production to overcome the growing needs (Mancosu et al. 2015; McKenzie and Williams 2015). Worldwide, out of seven persons, one is chronically undernourished. Therefore, overcoming food challenge stands as a major issue globally (McCarthy et al. 2018; D'Odorico et al. 2018). There are various challenges including weed and pest management and other phytopathogens which form an important basis for regulation of crop production and development (Damalas and Eleftherohorinos 2011; Grasswitz 2019). As population and economies expand, new methods of agriculture are developed and implemented to increase the productivity (Thornton 2010). Chemical fertilizers and pesticides are used in large scale to increase the production, but these pose various environmental impacts on different ecosystems of life (Aktar et al. 2009; Zhang et al. 2018). Most of the developed countries are strict in or ban the use of toxic chemical fertilizers and pesticides and have initiated various programs on integrated pest management (Rijal et al. 2018). Along with this, their focus remains on the development of natural product–based pesticides from the plant extracts and essential oils (George et al. 2014; Walia et al. 2017). Development of bioactive natural products for the development of pesticides is an eco-friendly and cost-effective process with no metabolite formation after action (Atanasov et al. 2015; Chaudhary et al. 2017). In modern agriculture, bioactive natural products derived from microorganisms and plants serve to minimize the impact of phytopathogens (Strobel and Daisy 2003). It is recommended to use the bioactive natural products in combination with other products for effective control and to minimize the tolerance by the pathogens (Hintz et al. 2015). The US Environmental Protection Agency defines biopesticides as compounds derived from natural compounds and categorized biopesticides as either microbial biopesticides or biochemical biopesticides (Chandler et al. 2011; Marrone 2019). Biopesticides are nontoxic to nontarget pests and have minimal toxicity to the environment and humans (Quarcoo et al. 2014). The phytochemicals derived from natural products cover wide range of various chemical groups such as peptides, proteins, saponins, vitamins, terpenes, polysaccharide, and polyphenols which act as natural inducers in host defense systems (Mujeeb et al. 2014; Shad et al. 2014).

Bioactive natural products play an important role in disease prevention and healthcare systems (Sofowora et al. 2013). The characterization of bioactive compounds and natural products has utilized various chemical methods and spectroscopic techniques to determine their structures (Altemimi et al. 2017; Tyśkiewicz et al. 2019). The use of infrared spectroscopy and ultraviolet spectrophotometry, mass spectroscopy, nuclear magnetic resonance, high-performance liquid chromatography, circular dichroism, and polarimetry provides additional information about

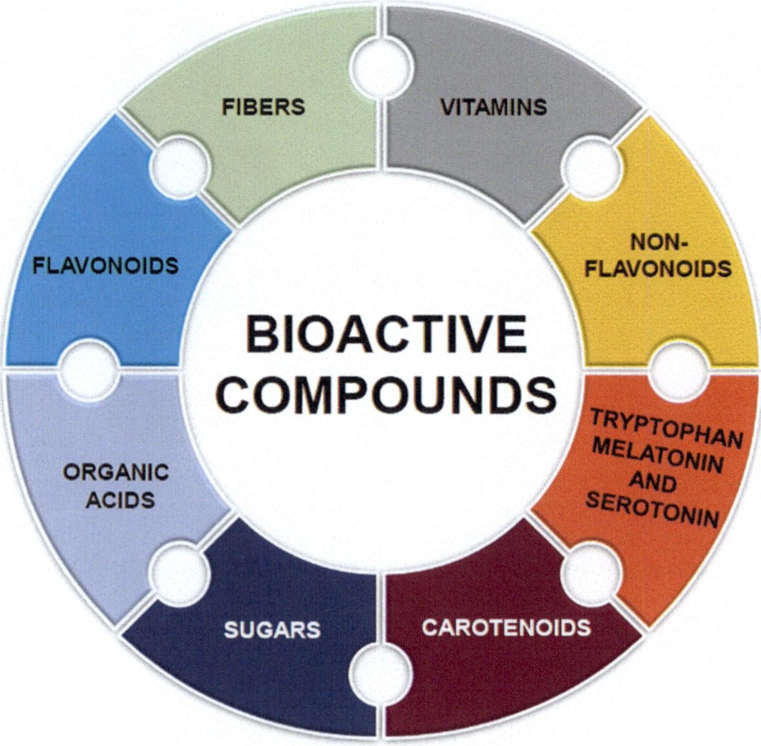

Fig. 2.1 Various bioactive molecules isolated from different natural sources

the various functional groups present in the bioactive compounds (Dias et al. 2016; Abu Khalaf et al. 2019). This chapter is designed to cover various bioactive natural products, their sources and types, and their mode of action. Efforts have also been made to summarize the role of bioactive natural compounds in plant defense mechanism in both indirect and direct methods. Various bioactive molecules isolated from different natural sources are well represented in Fig. 2.1.

2.2 Types of Bioactive Natural Products

Most of the bioactive natural products originate from microbes, animals, or plants, or they have prebiotic origin (Davies 2013; Davani-Davari et al. 2019). Microorganism and plants have proven to be good source of natural products such as bioinsecticides, antimicrobial agents, and antineoplastic agents (Cowan 1999; Hintz et al. 2015). Bioactive natural products derived from these sources are also reported to have various biological activities related to human health including biofilm inhibitory, anti-inflammatory, immunosuppressive, anticancer, antifungal, and antibiotic activities (Xu 2015; Pham et al. 2019). These products also possess selectivity,

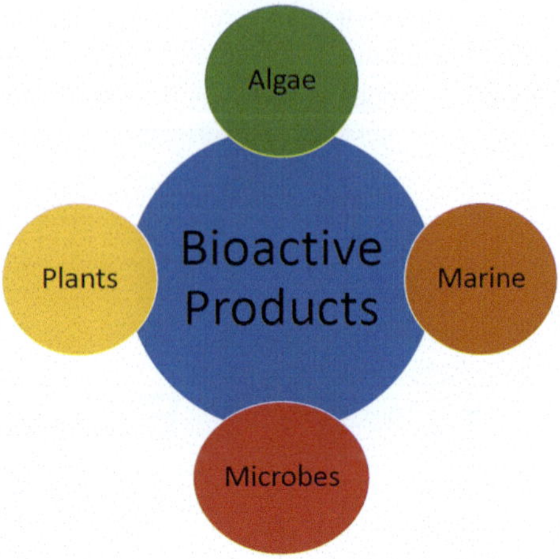

Fig. 2.2 Different sources of bioactive natural products for plant diseases

pharmacokinetic, and biopotency traits which act as drug traits (Guo 2017). Depending upon their type of source, they are categorized into microbial-, plant-, algal-, and marine-derived natural products (Fig. 2.2) (El Gamal 2010; Malve 2016).

2.2.1 Microbial-Derived Natural Products

Discovery of penicillin in 1929 opened the gates for microbial populations as a potent drug candidate (Tan and Tatsumura 2015). Till now, various microbial species have been screened and isolated for drug discovery which serves as antibacterial, antidiabetic, and anticancer agents, etc. (Raja et al. 2010; Cragg and Newman 2013). Genus *Streptomycetes* of the family Actinobacteria has been reported for the majority of naturally producing antibiotic agents (Barka et al. 2016; Sivalingam et al. 2019). Most of the antibiotics isolated from *Streptomycetes* act as inhibitors of protein synthesis, while some disrupt cell wall biosynthesis or integrity of cell membrane (Walsh and Wencewicz 2014; Louzoun Zada et al. 2018). There are also some antibiotics which induce the defense mechanism of host plants. Some bioactive natural products act on the protein coat of viruses, while some induce tolerance to plant against pathogen (Gupta and Birdi 2017; Köhl et al. 2019). In Canada, European Union, and the United States, uses of antibiotics in controlling plant disease are strictly prohibited (Kumar et al. 2019).

Species of the genus *Streptomyces* are also reported to produce proteins and peptides (blasticidin, mildiomycin, and polyoxins) which are used as a defense against pathogen (Sathya et al. 2017; Harir et al. 2018). Only few antimicrobial peptides have been registered and commercialized to protect crops because they can easily oxidize and hydrolyze and have less physical and chemical stability (Rai et al. 2016; Kumar et al. 2018). Antimicrobial peptides are used to develop various infection control agents and also used in cancer and diabetes treatment (Pfaltzgraff et al. 2018). Bacteriocins are reported to control spoilage and foodborne pathogens (Da Costa et al. 2019).

The extracts or metabolites isolated from microorganisms can be used as crude extracts, antibiotics, harpin, and proteins that prevent or reduce the diseases (Harish et al. 2008; Cheesman et al. 2017). Harpin is a heat-stable and glycine-rich protein, produced by pest *Erwinia amylovora,* which is effective against various pests on different hosts (Li et al. 2010; Zhou et al. 2019). Harpin is also reported to be effective against fungal pathogens during postharvesting (Alkan and Fortes 2015). It is also reported for accelerated growth, enhancing germination and flowering, stimulation of resistant responses, etc. (Yuan et al. 2017; Lymperopoulos et al. 2018). Four active regions have been reported in molecule of harpin, that is, HrpW (source: *E. amylovora*), HrpN (source: *E. amylovora*), PopA (source: *Ralstonia solanacearum*), and HrpZ (source: *Pseudomonas syringae*) (Kim and Beer 1998; Jin et al. 2001). These active regions are reported to likely increase the efficacy in growth stimulation and various other processes (Kurutas 2016). Another bioactive product, yeast extract hydrolysate, was reported to be isolated from *Saccharomyces cerevisiae* that also induces tolerance responses. Strobilurin isolated from fungus *Strobilurus tenacellus* is reported to the electron transport chain by binding with Qo site of cytochrome b (Ding et al. 2008; Song et al. 2015). Seventeen fungicides derived from strobilurin and its derivatives are commercialized (Bartlett et al. 2002).

2.2.2 Plant-Derived Natural Products

From earlier times, plant and plant-derived products have been used as good source of bioactive compounds (Amit Koparde et al. 2019). Various plant extracts, leaves, stems, and even roots are reported worldwide not only as medicines but also as biopesticides (Koma 2012).

Five families of the plant kingdom, Apocynaceae, Flacourtiaceae, Fabaceae, Lamiaceae, and Asteraceae, are reported as potent sources of biopesticides and bioactive natural products (Gakuya et al. 2013; Céspedes et al. 2014). To facilitate the development of biopesticides from plant sources, the extraction processes should be standardized, plant material should grow under natural conditions, and plant products should not affect nontarget species (Grzywacz et al. 2014; Tembo et al. 2018). Peels of pomegranate and citrus fruits and pomace from olives and grapes are also reported to inhibit various pathogens (Fourati et al. 2019).

The resins of conifers are also used against microbial pathogens and are good source of bioactive natural products (Termentzi et al. 2011; Gouda et al. 2016). Advancement in tissue culture techniques aids in the development of cell cultures of various plants to accelerate the production of bioactive natural products (Hussain et al. 2012; Cardoso et al. 2019). Salicylic acid and its derivatives isolated from *Reynoutria sachalinensis* of family Polygonaceae were also reported to have resistance to phytopathogens and play an important role in acquiring systemic resistance (Paul and Sharma 2002). It is phytotoxic in nature and used as preventative, trunk injection, seed treatment, and as anti-infectious agent (Lin et al. 2009). Salicylic acid also reduces size of lesion and virus titers (Park et al. 2009; Künstler et al. 2019). Anthraquinone compound isolated from extract of knotweed contains terpene in their functional group which is reported to act against phytopathogens and induce defense response of plants (Shan et al. 2008). Simple phenolics (salicylates) act as antifeedant to insect herbivores such as *Operophtera brumata* (L.) in *Salix* leaves, and there is a negative correlation between the salicylate levels and the larval growth; however, salicylic acid (SA) is much more important as phytohormone than as deterrent (War et al. 2012).

Biopesticides are also reported to be isolated from *Lupinus albus doce* of family Fabaceae (Gwinn 2018). It has a BLAD oligomer having chitinolytic and lectin-binding activities (Ruiz-López et al. 2010). It is friendly to honey bees and protects protein of legumes from allergic responses (Lucas et al. 2009). Alkaloids and benzophenanthridine alkaloids were also reported to be isolated from extracts of *Macleaya cordata* (family: Papaveraceae) which is effective against *Erysiphe graminis* and *Botrytis cinerea* (Kosina et al. 2010; Yu et al. 2014; Ge et al. 2015).

Phenolic compounds contribute to host plant resistance against insects and herbivores (Fürstenberg-Hägg et al. 2013). They are also reported to act as defensive mechanism against weeds, microbial infections, and herbivores (Kortbeek et al. 2019). Lignin, a phenolic heteropolymer, plays a central role in plant defense against insects and pathogens (Barakat et al. 2010). Phenols also play an important role in cyclic reduction of reactive oxygen species (ROS) such as superoxide anion and hydroxide radicals, H_2O_2, and singlet oxygen, which in turn activate a cascade of reactions leading to the activation of defensive enzymes. It limits the entry of pathogens by blocking it physically or increasing the leaf toughness that reduces the feeding by herbivores and also decreases the nutritional content of the leaf (Sharma et al. 2012; Nita and Grzybowski 2016). Lignin synthesis induced by herbivory or pathogen attack and its rapid deposition reduce further growth of the pathogen or herbivore fecundity (Miedes et al. 2014; Xie et al. 2018). Oxidation of phenols catalyzed by polyphenol oxidase (PPO) and peroxidase (POD) is a potential defense mechanism in plants against herbivorous insects. Quinones formed by oxidation of phenols bind covalently to leaf proteins and inhibit the protein digestion in herbivores (Felton and Gatehouse 1996; Haruta et al. 2001). In addition, quinones also exhibit direct toxicity to insects (Sugumaran and Bolton 1995). Alkylation of amino acids reduces the nutritional value of plant proteins for insects, which in turn negatively affects insect growth and development (Chen et al. 2005; Zhao et al. 2019). Plants are also reported to produce flavonoids which protect the plant against insect pests by influencing the behavior, growth, and development of insects (Mierziak et al. 2014). Tannins are also reported to act on phytophagous insects by affecting the growth of phytopathogens. They bind with proteins and midgut lesions and reduce the nutritional absorption efficiency of phytophagous insects (Barbehenn and Peter Constabel 2011; Vandenborre et al. 2011; Martinez et al. 2016). Various bioactive compounds isolated from plant sources are represented in Fig. 2.3.

Flavonoids contain anthocyanins, flavones, flavonols, flavanones, dihydroflavonols, chalcones, aurones, flavan, and proanthocyanidins (Iwashina 2000). More than 5000 flavonoids have been reported in plants (Hossain et al. 2016). Flavones such as flavonols, flavones, proanthocyanidins, flavan-3-ols, flavonones, flavans, and isoflavonoids function as feeding deterrents against many insect pests (Mierziak et al.

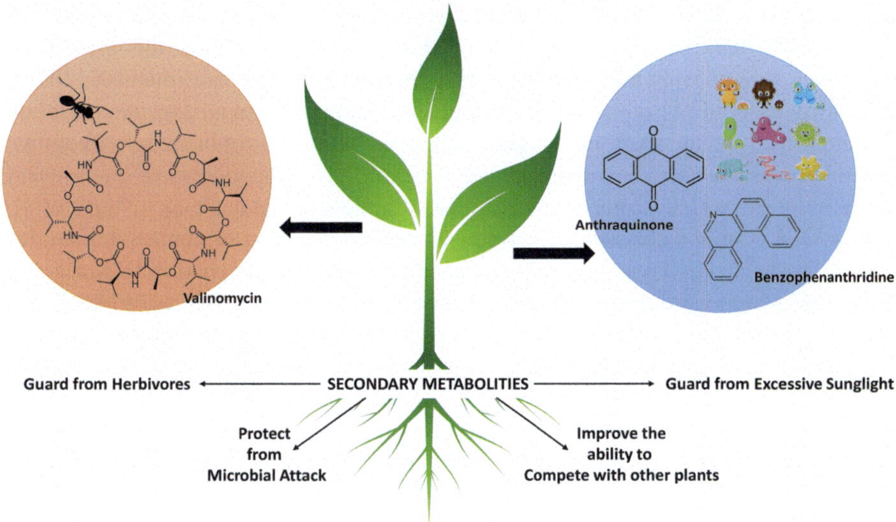

Fig. 2.3 Various bioactive compounds isolated from plant sources as defense for plant diseases

2014; Panche et al. 2016). Flavonoids such as flavones 5-hydroxyisoderricin, 7-methoxy-8-(3-methylbutadienyl)-flavanone, and 5-methoxyisoronchocarpin isolated from *Tephrosia villosa* (L.), *T. purpurea* (L.), and *T. vogelii* Hook, respectively, function as feeding deterrents against *Spodoptera exempta* (Walk.) and *Spodoptera littoralis* (Bios) (Chen et al. 2014; Samuel et al. 2019). Angustone A, licoisoflavone B, angustone B, angustone C, isoflavones, licoisoflavone A, luteone, licoisoflavone B, and wighteone are feeding deterrents to insects and also have antifungal activity against *Colletotrichum gloeosporioides* (Penz.) and *Cladosporium cladosporioides* (Fres.) (Wang et al. 2013).

Essential oils isolated from plant sources are used as treatment of various diseases in vegetables and fruits. They are either used singly or with other products as edible coating or film (Sánchez-González et al. 2011; Ju et al. 2019). These oils are major component for formulation of biopesticides which include *Allium sativum* (garlic, family *Amaryllidaceae*), *Syzygium aromaticum* (clove, family Myrtaceae), *Cinnamon cassia* and *Cinnamomum zeylanicum* (cinnamon of family Lauraceae), *Mentha piperita,* and *Thymus vulgaris* (peppermint and thyme of family Lamiaceae) (Pohlit et al. 2011). These essential oils are mixture of phenols and terpenes (Dhifi et al. 2016; Morsy 2017). The active component in thyme oil is thymol, menthol in peppermint, methyl eugenol in clove oil, cinnamaldehyde in cinnamon oil, and sulfur compounds in garlic oils (Nazzaro et al. 2013). Thyme oil is used to control cucurbit disease and bacterial spot in turnip (Sivakumar and Bautista-Baños 2014). Garlic is reported to be effective against *Alternaria* blight on mustard. Clove oil is used to control fungal infections (Liu et al. 2017).

Bioactive oil isolated from *Simmondsia chinensis* species of Simmondsiaceae family and is enriched in wax esters, which is primarily used to block the oxygen uptake of mycelium (Perez-Rosales et al. 2018). Hydrophobic and azadirachtin extract of oil isolated from *Azadirachta indica* (neem) is reported to act as biocidal agents with disrupts normal growth of pests. It also inhibited the growth of *Chaetomium globosum* (soft wood fungus) (Selim et al. 2014; Fatima et al. 2016). To control early blight of tomato, cotton seed oil isolated from *Gossypium hirsutum* of family Malvaceae was used which act as growth inhibitor of bacterial infections. Oil of species *Chenopodium ambrosioides* of family Chenopodiaceae is a renowned antihelminthic which protects storage foods from microbial attack (Naqvi et al. 2017). Oils contains carvacrol or ascaridole which are used to control nematodes and soilborne fungi (Oka et al. 2000). Oil of tea plants is enriched in terpenes and terpinenic which were having antimicrobial activities and also used as biopesticides against viruses and plant fungal infections by infecting membrane integrity (Guimarães et al. 2019). Oils isolated from barley (*Blumeria graminis*) are used as plant protection agents and as fungicides. Isoflavonoids (judaicin, judaicin-7-*O*-glucoside, 2-methoxyjudaicin, and maackiain) isolated from the wild relatives of chickpea act as antifeedant against *Helicoverpa armigera* (Hubner) at 100 ppm (Simmonds and Stevenson 2001). Judaicin and maackiain are deterrent to *S. littoralis* and *S. frugiperda,* respectively. Cyanopropenyl glycoside and alliarinoside strongly inhibit feeding by the native American butterfly, *Pieris napi oleracea* L., while a flavone glycoside, isovitexin-6″-β-D-glucopyranoside, acts as a direct feeding deterrent to the late instars (Anthoni et al. 2001).

2.2.3 Algal-Derived Natural Products

Biofuels obtained from algal bodies are also used as biopesticides as they hinder various metabolic processes and enzyme activity of pathogens (Hannon et al. 2010; Costa et al. 2019). Laminarin is another bioactive natural product isolated from brown algae which is used for preserving food items (Kadam et al. 2015; Ciko et al. 2018). It is an oligosaccharide having mannitol or glucose chain in its terminal position. The chains are composed of b(1,6) branches and b-(1,3)-D-glucans (Legentil et al. 2015). Laminarin is a well-known biopesticide, but it also acts as bio-stimulant. Saponins are isolated from *C. quinoa* that are reported to suppress various plant diseases (Yoon et al. 2013; Kregiel et al. 2017). They act as fungicides and resistance inducers, and have low mammalian toxicity (Dias 2012; Thakur and Sohal 2013; Lamichhane et al. 2018).

2.2.4 Marine-Derived Natural Products

Most of the marine-derived natural products are potent source of bioactive compounds which act to treat different ailments (Nair et al. 2015).The first bioactive compounds were reported in the 1950s which were isolated from *Cryptotheca crypta* (Caribbean sponge) that includes *spongothymidine* and *spongouridine* (Ruiz-Torres et al. 2017). These two compounds were reported as potent antiviral and anticancer agents. Other bioactive products like *discodermolide* isolated from

Discodermia dissoluta are reported to having strong anticancer activity (Khalifa et al. 2019; Ha et al. 2019). Microalgal species contain various bioactive compounds which are oleic acid, palmitoleic acid, linolenic acid, cyanovirin, vitamins E and B12, β-carotene, lutein, zeaxanthin, and phycocyanin functional groups which act as various antimicrobial agents having potential to prevent the diseases (De Morais et al. 2015; Jena and Subudhi 2019). Algal bodies contain heterocysts which fix atmospheric nitrogen and induce plant growth by synthesizing amino acids (Issa et al. 2014; Magnuson 2019).

2.3 Conclusion and Future Prospects

Disease control of plants is often regulated by chemical fertilizers and pesticides. These chemicals not only kill the pests but also increase the production of crops. However, all these chemical agents are hazardous to both aquatic and terrestrial ecosystems; besides this, various pest species have developed tolerance to wide range of these pesticides, insecticides, fungicides, etc. Bioactive compounds isolated from various sources have defense mechanisms against various stress and microbial pathogens. Various compounds isolated from plant sources, algal sources, microbial sources, marine sources, etc., act as elicitors for plant defense against various sources. These include chitosan, salicylic acid, benzoic acid, benzothiadiazole, alkaloids, flavonoids, terpenes, proteins, peptides, blasticidin, mildiomycin, polyoxins, and phenolic compounds which act against various plant diseases caused by bacteria, fungi, and viruses. Formulations of bioactive natural products are critical to overcome agricultural challenges such as being nutritious, safe, and adequate for supply to all human populations. These bioactive compounds minimize the use of chemical agents and contribute to eco-friendly sustainable agriculture practices.

Development of new gene cassettes using metabolic engineering into these sources will not only increase the efficiency of these bioactive compounds but also help in the production of secondary metabolites which act as valuable defense system.

References

Abu Khalaf R, Alhusban AA, Al-Shalabi E et al (2019) Isolation and structure elucidation of bioactive polyphenols. In: Studies in natural products chemistry. Elsevier B.V., Amsterdam, pp 267–337

Aktar W, Sengupta D, Chowdhury A (2009) Impact of pesticides use in agriculture: their benefits and hazards. Interdiscip Toxicol 2:1–12. https://doi.org/10.2478/v10102-009-0001-7

Alkan N, Fortes AM (2015) Insights into molecular and metabolic events associated with fruit response to post-harvest fungal pathogens. Front Plant Sci 6:889

Altemimi A, Lakhssassi N, Baharlouei A et al (2017) Phytochemicals: extraction, isolation, and identification of bioactive compounds from plant extracts. Plan Theory 6(4):pii: E42

Amit Koparde A, Chandrashekar Doijad R, Shripal Magdum C (2019) Natural products in drug discovery. In: Pharmacognosy – medicinal plants. IntechOpen, Rijeka, Croatia

Anthoni U, Christophersen C, Nielsen PH (2001) O-benzylation of polyphenolics. Preparation of 1,2,4-tribenzyloxybenzene. Synth Commun 31:2223–2229. https://doi.org/10.1081/SCC-100104477

Atanasov AG, Waltenberger B, Pferschy-Wenzig EM et al (2015) Discovery and resupply of pharmacologically active plant-derived natural products: a review. Biotechnol Adv 33:1582–1614

Barakat A, Bagniewska-Zadworna A, Frost CJ, Carlson JE (2010) Phylogeny and expression profiling of CAD and CAD-like genes in hybrid Populus (P. deltoides × P. nigra): evidence from herbivore damage for subfunctionalization and functional divergence. BMC Plant Biol 10:100. https://doi.org/10.1186/1471-2229-10-100

Barbehenn RV, Peter Constabel C (2011) Tannins in plant-herbivore interactions. Phytochemistry 72:1551–1565

Barka EA, Vatsa P, Sanchez L et al (2016) Taxonomy, physiology, and natural products of actinobacteria. Microbiol Mol Biol Rev 80:1–43. https://doi.org/10.1128/mmbr.00019-15

Bartlett DW, Clough JM, Godwin JR et al (2002) The strobilurin fungicides. Pest Manag Sci 58:649–662. https://doi.org/10.1002/ps.520

Cardoso JC, de Oliveira MEB, Cardoso F de C (2019) Advances and challenges on the in vitro production of secondary metabolites from medicinal plants. Hortic Bras 37:124–132. https://doi.org/10.1590/s0102-053620190201

Céspedes CL, Salazar JR, Ariza-Castolo A et al (2014) Biopesticides from plants: Calceolaria integrifolia s.l. Environ Res 132:391–406. https://doi.org/10.1016/j.envres.2014.04.003

Chandler D, Bailey AS, Mark Tatchell G et al (2011) The development, regulation and use of biopesticides for integrated pest management. Philos Trans R Soc B Biol Sci 366:1987–1998. https://doi.org/10.1098/rstb.2010.0390

Chaudhary S, Kanwar RK, Sehgal A et al (2017) Progress on Azadirachta indica based biopesticides in replacing synthetic toxic pesticides. Front Plant Sci 8:610

Cheesman MJ, Ilanko A, Blonk B, Cock IE (2017) Developing new antimicrobial therapies: are synergistic combinations of plant extracts/compounds with conventional antibiotics the solution? Pharmacogn Rev 11:57–72

Chen H, Wilkerson CG, Kuchar JA et al (2005) Jasmonate-inducible plant enzymes degrade essential amino acids in the herbivore midgut. Proc Natl Acad Sci U S A 102:19237–19242. https://doi.org/10.1073/pnas.0509026102

Chen Y, Yan T, Gao C et al (2014) Natural products from the genus tephrosia. Molecules 19:1432–1458. https://doi.org/10.3390/molecules19021432

Ciko AM, Jokić S, Šubarić D, Jerković I (2018) Overview on the application of modern methods for the extraction of bioactive compounds from marine macroalgae. Mar Drugs 16(10):pii: E348. https://doi.org/10.3390/md16100348

Costa JAV, Freitas BCB, Cruz CG et al (2019) Potential of microalgae as biopesticides to contribute to sustainable agriculture and environmental development. J Environ Sci Health – Part B Pestic Food Contam Agric Wastes 54:366–375

Cowan MM (1999) Plant products as antimicrobial agents. Clin Microbiol Rev 12:564–582

Cragg GM, Newman DJ (2013) Natural products: a continuing source of novel drug leads. Biochim Biophys Acta, Gen Subj 1830:3670–3695

D'Odorico P, Davis KF, Rosa L et al (2018) The global food-energy-water nexus. Rev Geophys 56:456–531

Da Costa RJ, Voloski FLS, Mondadori RG et al (2019) Preservation of meat products with bacteriocins produced by lactic acid bacteria isolated from meat. J Food Qual 2019:1–12

Damalas CA, Eleftherohorinos IG (2011) Pesticide exposure, safety issues, and risk assessment indicators. Int J Environ Res Public Health 8:1402–1419

Davani-Davari D, Negahdaripour M, Karimzadeh I et al (2019) Prebiotics: definition, types, sources, mechanisms, and clinical applications. Foods 8(3):pii: E92

Davies J (2013) Specialized microbial metabolites: functions and origins. J Antibiot (Tokyo) 66:361–364

De Morais MG, Vaz BDS, De Morais EG, Costa JAV (2015) Biologically active metabolites synthesized by microalgae. Biomed Res Int 2015:835761

Dhifi W, Bellili S, Jazi S et al (2016) Essential oils' chemical characterization and investigation of some biological activities: a critical review. Medicines 3:25. https://doi.org/10.3390/medicines3040025

Dias MC (2012) Phytotoxicity: an overview of the physiological responses of plants exposed to fungicides. J Bot 2012:1–4. https://doi.org/10.1155/2012/135479

Dias DA, Jones OAH, Beale DJ et al (2016) Current and future perspectives on the structural identification of small molecules in biological systems. Metabolites 6(4):46

Ding MG, Butler CA, Saracco SA et al (2008) Introduction of cytochrome b mutations in Saccharomyces cerevisiae by a method that allows selection for both functional and non-functional cytochrome b proteins. Biochim Biophys Acta Bioenerg 1777:1147–1156. https://doi.org/10.1016/j.bbabio.2008.04.029

El Gamal AA (2010) Biological importance of marine algae. Saudi Pharm J 18:1–25

Fatima N, Mukhtar U, Ihsan-Ul-Haq et al (2016) Biological evaluation of endophytic fungus chaetomium sp. NF15 of justicia adhatoda l.: a potential candidate for drug discovery. Jundishapur J Microbiol 9:e29978. https://doi.org/10.5812/jjm.29978

Felton GW, Gatehouse JA (1996) Antinutritive plant defence mechanisms. In: Biology of the insect midgut. Springer Netherlands, Dordrecht, pp 373–416

Fourati M, Smaoui S, Ennouri K et al (2019) Multiresponse optimization of pomegranate peel extraction by statistical versus artificial intelligence: predictive approach for foodborne bacterial pathogen inactivation. Evid Based Complement Alternat Med 2019:1–18. https://doi.org/10.1155/2019/1542615

Fürstenberg-Hägg J, Zagrobelny M, Bak S (2013) Plant defense against insect herbivores. Int J Mol Sci 14:10242–10297

Gakuya DW, Itonga SM, Mbaria JM et al (2013) Ethnobotanical survey of biopesticides and other medicinal plants traditionally used in Meru central district of Kenya. J Ethnopharmacol 145:547–553. https://doi.org/10.1016/j.jep.2012.11.028

Ge Y, Liu P, Yang R et al (2015) Insecticidal constituents and activity of alkaloids from Cynanchum mongolicum. Molecules 20:17483–17492. https://doi.org/10.3390/molecules200917483

George DR, Finn RD, Graham KM, Sparagano OA (2014) Present and future potential of plant-derived products to control arthropods of veterinary and medical significance. Parasit Vectors 7:28

Gouda S, Das G, Sen SK et al (2016) Endophytes: a treasure house of bioactive compounds of medicinal importance. Front Microbiol 7:1538

Grasswitz TR (2019) Integrated pest management (IPM) for small-scale farms in developed economies: challenges and opportunities. Insects 10:179. https://doi.org/10.3390/insects10060179

Grzywacz D, Stevenson PC, Mushobozi WL et al (2014) The use of indigenous ecological resources for pest control in Africa. Food Secur 6:71–86. https://doi.org/10.1007/s12571-013-0313-5

Guimarães AC, Meireles LM, Lemos MF et al (2019) Antibacterial activity of terpenes and terpenoids present in essential oils. Molecules 24(13):pii: E2471. https://doi.org/10.3390/molecules24132471

Guo Z (2017) The modification of natural products for medical use. Acta Pharm Sin B 7:119–136

Gupta PD, Birdi TJ (2017) Development of botanicals to combat antibiotic resistance. J Ayurveda Integr Med 8:266–275

Gwinn KD (2018) Bioactive natural products in plant disease control. In: Studies in natural products chemistry. Elsevier B.V., Amsterdam, pp 229–246

Ha MW, Song BR, Chung HJ, Paek S-M (2019) Design and synthesis of anti-cancer chimera molecules based on marine natural products. Mar Drugs 17:500. https://doi.org/10.3390/md17090500

Hannon M, Gimpel J, Tran M et al (2010) Biofuels from algae: challenges and potential. Biofuels 1:763–784. https://doi.org/10.4155/bfs.10.44

Harir M, Bendif H, Bellahcene M, Fortas Z, Rebecca Pogni Z (2018) Streptomyces secondary metabolites. In: Basic biology and applications of actinobacteria. IntechOpen, Rijeka, Croatia

Harish S, Saravanakumar D, Radjacommare R et al (2008) Use of plant extracts and biocontrol agents for the management of brown spot disease in rice. BioControl 53:555–567. https://doi.org/10.1007/s10526-007-9098-9
Haruta M, Pedersen JA, Constabel CP (2001) Polyphenol oxidase and herbivore defense in trembling aspen (Populus tremuloides): cDNA cloning, expression, and potential substrates. Physiol Plant 112:552–558
Hintz T, Matthews KK, Di R (2015) The use of plant antimicrobial compounds for food preservation. Biomed Res Int 2015:246264
Hossain MK, Dayem AA, Han J et al (2016) Molecular mechanisms of the anti-obesity and antidiabetic properties of flavonoids. Int J Mol Sci 17:569
Hussain A, Ahmed I, Nazir H, Ullah I (2012) Plant tissue culture: current status and opportunities. In: Recent advances in plant in vitro culture. InTech, Rijeka, Croatia
Issa AA, Hemida M, Ohyam T (2014) Nitrogen fixing cyanobacteria: future prospect. In: Advances in biology and ecology of nitrogen fixation. InTech, Rijeka, Croatia
Iwashina T (2000) The structure and distribution of the flavonoids in plants. J Plant Res 113:287–299
Jena J, Subudhi E (2019) Microalgae: an untapped resource for natural antimicrobials. In: The role of microalgae in wastewater treatment. Springer, Singapore, pp 99–114
Jin Q, Hu W, Brown I et al (2001) Visualization of secreted Hrp and Avr proteins along the Hrp pilus during type III secretion in Erwinia amylovora and Pseudomonas syringae. Mol Microbiol 40:1129–1139. https://doi.org/10.1046/j.1365-2958.2001.02455.x
Ju J, Xie Y, Guo Y et al (2019) Application of edible coating with essential oil in food preservation. Crit Rev Food Sci Nutr 59:2467–2480. https://doi.org/10.1080/10408398.2018.1456402
Kadam SU, Tiwari BK, O'Donnell CP (2015) Extraction, structure and biofunctional activities of laminarin from brown algae. Int J Food Sci Technol 50:24–31. https://doi.org/10.1111/ijfs.12692
Khalifa SAM, Elias N, Farag MA et al (2019) Marine natural products: a source of novel anticancer drugs. Mar Drugs 17:491. https://doi.org/10.3390/md17090491
Kim JF, Beer SV (1998) HrpW of Erwinia amylovora, a new harpin that contains a domain homologous to pectate lyases of a distinct class. J Bacteriol 180(19):5203–5210
Köhl J, Kolnaar R, Ravensberg WJ (2019) Mode of action of microbial biological control agents against plant diseases: relevance beyond efficacy. Front Plant Sci 10:845. https://doi.org/10.3389/fpls.2019.00845
Koma S (2012) Plants as potential sources of pesticidal agents: a review. In: Pesticides – advances in chemical and botanical pesticides. InTech, Rijeka, Croatia
Kortbeek RWJ, van der Gragt M, Bleeker PM (2019) Endogenous plant metabolites against insects. Eur J Plant Pathol 154:67–90. https://doi.org/10.1007/s10658-018-1540-6
Kosina P, Gregorova J, Gruz J et al (2010) Phytochemical and antimicrobial characterization of Macleaya cordata herb. Fitoterapia 81:1006–1012. https://doi.org/10.1016/j.fitote.2010.06.020
Kregiel D, Berlowska J, Witonska I et al (2017) Saponin-based, biological-active surfactants from plants. In: Application and characterization of surfactants. InTech, Rijeka, Croatia
Kumar P, Kizhakkedathu JN, Straus SK (2018) Antimicrobial peptides: diversity, mechanism of action and strategies to improve the activity and biocompatibility in vivo. Biomol Ther 8(1):pii: E4
Kumar M, Jaiswal S, Sodhi KK et al (2019) Antibiotics bioremediation: perspectives on its ecotoxicity and resistance. Environ Int 124:448–461
Künstler A, Király L, Kátay G et al (2019) Glutathione can compensate for salicylic acid deficiency in tobacco to maintain resistance to tobacco mosaic virus. Front Plant Sci 10:1115. https://doi.org/10.3389/fpls.2019.01115
Kurutas EB (2016) The importance of antioxidants which play the role in cellular response against oxidative/nitrosative stress: current state. Nutr J 15(1):71
Lamichhane JR, Osdaghi E, Behlau F et al (2018) Thirteen decades of antimicrobial copper compounds applied in agriculture. A review. Agron Sustain Dev 38(3):28
Legentil L, Paris F, Ballet C et al (2015) Molecular interactions of β-(1→3)-glucans with their receptors. Molecules 20:9745–9766. https://doi.org/10.3390/molecules20069745

Li JG, Liu HX, Cao J et al (2010) PopW of Ralstonia solanacearum, a new two-domain harpin targeting the plant cell wall. Mol Plant Pathol 11:371–381. https://doi.org/10.1111/j.1364-3703.2010.00610.x

Lin T-C, Ishizaka M, Ishii H (2009) Acibenzolar-S-methyl-induced systemic resistance against anthracnose and powdery mildew diseases on cucumber plants without accumulation of phytoalexins. J Phytopathol 157:40–50. https://doi.org/10.1111/j.1439-0434.2008.01448.x

Liu Q, Meng X, Li Y et al (2017) Antibacterial and antifungal activities of spices. Int J Mol Sci 18(6):pii: E1283

Louzoun Zada S, Green KD, Shrestha SK et al (2018) Derivatives of ribosome-inhibiting antibiotic chloramphenicol inhibit the biosynthesis of bacterial cell wall. ACS Infect Dis 4:1121–1129. https://doi.org/10.1021/acsinfecdis.8b00078

Lucas JA, Ramos Solano B, Montes F et al (2009) Use of two PGPR strains in the integrated management of blast disease in rice (Oryza sativa) in Southern Spain. Field Crop Res 114:404–410. https://doi.org/10.1016/j.fcr.2009.09.013

Lymperopoulos P, Msanne J, Rabara R (2018) Phytochrome and phytohormones: working in tandem for plant growth and development. Front Plant Sci 9:1037

Magnuson A (2019) Heterocyst thylakoid bioenergetics. Life 9(1):pii: E13. https://doi.org/10.3390/life9010013

Malve H (2016) Exploring the ocean for new drug developments: marine pharmacology. J Pharm Bioallied Sci 8:83–91

Mancosu N, Snyder R, Kyriakakis G, Spano D (2015) Water scarcity and future challenges for food production. Water 7:975–992. https://doi.org/10.3390/w7030975

Marrone PG (2019) Pesticidal natural products – status and future potential. Pest Manag Sci 75:2325–2340. https://doi.org/10.1002/ps.5433

Martinez M, Santamaria ME, Diaz-Mendoza M et al (2016) Phytocystatins: defense proteins against phytophagous insects and acari. Int J Mol Sci 17(10):pii: E1747

Mc Carthy U, Uysal I, Badia-Melis R et al (2018) Global food security – issues, challenges and technological solutions. Trends Food Sci Technol 77:11–20

McKenzie FC, Williams J (2015) Sustainable food production: constraints, challenges and choices by 2050. Food Secur 7:221–233. https://doi.org/10.1007/s12571-015-0441-1

Miedes E, Vanholme R, Boerjan W, Molina A (2014) The role of the secondary cell wall in plant resistance to pathogens. Front Plant Sci 5:358

Mierziak J, Kostyn K, Kulma A (2014) Flavonoids as important molecules of plant interactions with the environment. Molecules 19:16240–16265. https://doi.org/10.3390/molecules191016240

Morsy NFS (2017) Chemical structure, quality indices and bioactivity of essential oil constituents. In: Active ingredients from aromatic and medicinal plants. InTech, London

Mujeeb F, Bajpai P, Pathak N (2014) Phytochemical evaluation, antimicrobial activity, and determination of bioactive components from leaves of aegle marmelos. Biomed Res Int 2014:1. https://doi.org/10.1155/2014/497606

Nair DG, Weiskirchen R, Al-Musharafi SK (2015) The use of marine-derived bioactive compounds as potential hepatoprotective agents. Acta Pharmacol Sin 36:158–170

Naqvi RZ, Zaidi SSEA, Akhtar KP et al (2017) Transcriptomics reveals multiple resistance mechanisms against cotton leaf curl disease in a naturally immune cotton species, Gossypium arboreum. Sci Rep 7:15880. https://doi.org/10.1038/s41598-017-15963-9

Nazzaro F, Fratianni F, De Martino L et al (2013) Effect of essential oils on pathogenic bacteria. Pharmaceuticals 6:1451–1474

Nita M, Grzybowski A (2016) The role of the reactive oxygen species and oxidative stress in the pathomechanism of the age-related ocular diseases and other pathologies of the anterior and posterior eye segments in adults. Oxidative Med Cell Longev 2016:1

Oka Y, Nacar S, Putievsky E et al (2000) Nematicidal activity of essential oils and their components against the root-knot nematode. Phytopathology 90:710–715. https://doi.org/10.1094/PHYTO.2000.90.7.710

Panche AN, Diwan AD, Chandra SR (2016) Flavonoids: an overview. J Nutr Sci 5:e47

Park SW, Liu PP, Forouhar F et al (2009) Use of a synthetic salicylic acid analog to investigate the roles of methyl salicylate and its esterases in plant disease resistance. J Biol Chem 284:7307–7317. https://doi.org/10.1074/jbc.M807968200

Paul P, Sharma P (2002) Azadirachta indica leaf extract induces resistance in barley against leaf stripe disease. Physiol Mol Plant Pathol 61:3–13. https://doi.org/10.1006/pmpp.2002.0412

Perez-Rosales E, Alcaraz-Meléndez L, Puente ME et al (2018) Endophytic bacteria isolated from wild jojoba [Simmondsia chinensis L. (Schneider)] roots improve in vitro propagation. Plant Cell Tissue Organ Cult 135:515–522. https://doi.org/10.1007/s11240-018-1483-9

Pfaltzgraff A, Brandenburg K, Weindl G (2018) Antimicrobial peptides and their therapeutic potential for bacterial skin infections and wounds. Front Pharmacol 9:281

Pham JV, Yilma MA, Feliz A et al (2019) A review of the microbial production of bioactive natural products and biologics. Front Microbiol 10:1404. https://doi.org/10.3389/fmicb.2019.01404

Pohlit AM, Lopes NP, Gama RA et al (2011) Patent literature on mosquito repellent inventions which contain plant essential oils – a review. Planta Med 77:598–617

Quarcoo F, Bonsi C, Tackie N (2014) Pesticides, the environment, and human health. In: Pesticides – toxic aspects. InTech, Rijeka, Croatia

Rai M, Pandit R, Gaikwad S, Kövics G (2016) Antimicrobial peptides as natural bio-preservative to enhance the shelf-life of food. J Food Sci Technol 53:3381–3394

Raja A, Gajalakshmi P, Mohamed Mahroop Raja M (2010) Drugs from the natural bio sources for human disease. Int J Pharmacol 6:511–514. https://doi.org/10.3923/ijp.2010.360.363

Rijal J, Regmi R, Ghimire R et al (2018) Farmers' knowledge on pesticide safety and pest management practices: a case study of vegetable growers in Chitwan, Nepal. Agriculture 8:16. https://doi.org/10.3390/agriculture8010016

Ruiz-López MA, García-López PM, Rodríguez-Macías R et al (2010) Mexican wild lupines as a source of quinolizidine alkaloids of economic potential. México

Ruiz-Torres V, Encinar JA, Herranz-López M et al (2017) An updated review on marine anticancer compounds: the use of virtual screening for the discovery of small-molecule cancer drugs. Molecules 22:22. https://doi.org/10.3390/molecules22071037

Samuel VJ, Mahesh AR, Murugan V (2019) Phytochemical and pharmacological aspects of Tephrosia genus: a brief review. J Appl Pharm Sci 9:117–125. https://doi.org/10.7324/JAPS.2019.90317

Sánchez-González L, Vargas M, González-Martínez C et al (2011) Use of essential oils in bioactive edible coatings: a review. Food Eng Rev 3:1–16

Sathya A, Vijayabharathi R, Gopalakrishnan S (2017) Plant growth-promoting actinobacteria: a new strategy for enhancing sustainable production and protection of grain legumes. 3 Biotech 7:102

Selim KA, El-Beih AA, Abdel-Rahman TM, El-Diwany AI (2014) Biological evaluation of endophytic fungus, Chaetomium globosum JN711454, as potential candidate for improving drug discovery. Cell Biochem Biophys 68:67–82. https://doi.org/10.1007/s12013-013-9695-4

Shad AA, Ahmad S, Ullah R et al (2014) Phytochemical and biological activities of four wild medicinal plants. Sci World J 2014:1. https://doi.org/10.1155/2014/857363

Shan B, Cai YZ, Brooks JD, Corke H (2008) Antibacterial properties of Polygonum cuspidatum roots and their major bioactive constituents. Food Chem 109:530–537. https://doi.org/10.1016/j.foodchem.2007.12.064

Sharma P, Jha AB, Dubey RS, Pessarakli M (2012) Reactive oxygen species, oxidative damage, and antioxidative defense mechanism in plants under stressful conditions. J Bot 2012:1–26. https://doi.org/10.1155/2012/217037

Simmonds MS, Stevenson PC (2001) Effects of isoflavonoids from Cicer on larvae of Helicoverpa armigera. J Chem Ecol 27:965–977. https://doi.org/10.1023/a:1010339104206

Sivakumar D, Bautista-Baños S (2014) A review on the use of essential oils for postharvest decay control and maintenance of fruit quality during storage. Crop Prot 64:27–37

Sivalingam P, Hong K, Pote J, Prabakar K (2019) Extreme environment streptomyces: potential sources for new antibacterial and anticancer drug leads? Int J Microbiol 2019:1

Sofowora A, Ogunbodede E, Onayade A (2013) The role and place of medicinal plants in the strategies for disease prevention. Afr J Tradit Complement Altern Med 10:210–229
Song Z, Clain J, Iorga BI et al (2015) Saccharomyces cerevisiae-based mutational analysis of the bc1 complex Qo site residue 279 to study the trade-off between atovaquone resistance and function. Antimicrob Agents Chemother 59:4053–4058. https://doi.org/10.1128/AAC.00710-15
Strobel G, Daisy B (2003) Bioprospecting for microbial endophytes and their natural products. Microbiol Mol Biol Rev 67:491–502. https://doi.org/10.1128/mmbr.67.4.491-502.2003
Sugumaran M, Bolton J (1995) Direct evidence for quinone-quinone methide tautomerism during tyrosinase catalyzed oxidation of 4-allylcatechol. Biochem Biophys Res Commun 213:469–474. https://doi.org/10.1006/bbrc.1995.2155
Tan SY, Tatsumura Y (2015) Alexander Fleming (1881–1955): discoverer of penicillin. Singap Med J 56:366–367
Tembo Y, Mkindi AG, Mkenda PA et al (2018) Pesticidal plant extracts improve yield and reduce insect pests on legume crops without harming beneficial arthropods. Front Plant Sci 9:1425. https://doi.org/10.3389/fpls.2018.01425
Termentzi A, Fokialakis N, Leandros Skaltsounis A (2011) Natural resins and bioactive natural products thereof as potential antimicrobial agents. Curr Pharm Des 17:1267–1290. https://doi.org/10.2174/138161211795703807
Thakur M, Sohal BS (2013) Role of elicitors in inducing resistance in plants against pathogen infection: a review. ISRN Biochem 2013:1–10. https://doi.org/10.1155/2013/762412
Thornton PK (2010) Livestock production: recent trends, future prospects. Philos Trans R Soc B Biol Sci 365:2853–2867
Tyśkiewicz K, Konkol M, Kowalski R et al (2019) Characterization of bioactive compounds in the biomass of black locust, poplar and willow. Trees – Struct Funct 33:1235
Vandenborre G, Smagghe G, Van Damme EJM (2011) Plant lectins as defense proteins against phytophagous insects. Phytochemistry 72:1538–1550
Walia S, Saha S, Tripathi V, Sharma KK (2017) Phytochemical biopesticides: some recent developments. Phytochem Rev 16:989–1007
Walsh CT, Wencewicz TA (2014) Prospects for new antibiotics: a molecule-centered perspective. J Antibiot (Tokyo) 67:7–22
Wang X, Radwan MM, Taráwneh AH et al (2013) Antifungal activity against plant pathogens of metabolites from the endophytic fungus Cladosporium cladosporioides. J Agric Food Chem 61:4551–4555. https://doi.org/10.1021/jf400212y
War AR, Paulraj MG, Ahmad T et al (2012) Mechanisms of plant defense against insect herbivores. Plant Signal Behav 7:1306
Xie M, Zhang J, Tschaplinski TJ et al (2018) Regulation of lignin biosynthesis and its role in growth-defense tradeoffs. Front Plant Sci 9:1427. https://doi.org/10.3389/fpls.2018.01427
Xu J (2015) Bioactive natural products derived from mangrove-associated microbes. RSC Adv 5:841–892
Yoon MY, Cha B, Kim JC (2013) Recent trends in studies on botanical fungicides in agriculture. Plant Pathol J 29:1–9. https://doi.org/10.5423/PPJ.RW.05.2012.0072
Yu X, Gao X, Zhu Z et al (2014) Alkaloids from the tribe Bocconieae (Papaveraceae): a chemical and biological review. Molecules 19:13042–13060. https://doi.org/10.3390/molecules190913042
Yuan HY, Saha S, Vandenberg A, Bett KE (2017) Flowering and growth responses of cultivated lentil and wild lens germplasm toward the differences in red to far-red ratio and photosynthetically active radiation. Front Plant Sci 8:386. https://doi.org/10.3389/fpls.2017.00386
Zhang L, Yan C, Guo Q et al (2018) The impact of agricultural chemical inputs on environment: global evidence from informetrics analysis and visualization. Int J Low-Carbon Technol 13:338–352. https://doi.org/10.1093/ijlct/cty039
Zhao H, Ullah H, McNeill MR et al (2019) Inhibitory effects of plant trypsin inhibitors Msti-94 and Msti-16 on Therioaphis trifolii (Monell) (Homoptera: Aphididae) in Alfalfa. Insects 10:154. https://doi.org/10.3390/insects10060154
Zhou X, Liu Y, Huang J et al (2019) High temperatures affect the hypersensitive reaction, disease resistance and gene expression induced by a novel harpin HpaG-Xcm. Sci Rep 9:990. https://doi.org/10.1038/s41598-018-37886-9

Biological Control Agents: Diversity, Ecological Significances, and Biotechnological Applications

3

Simranjeet Singh, Vijay Kumar, Daljeet Singh Dhanjal, and Joginder Singh

Abstract

Plants and pathogens interact with a wide range of organisms throughout their life cycle. These interactions play an important role in affecting plant health. At multiple levels of scale, interaction between microbes and plants leads to biological control of phytopathogens. Biological control agent refers to the use of natural or modified organisms, gene products, to reduce the effects of undesirable organisms and to favor desirable organisms such as crops, beneficial insects, and microorganisms. These agents induce enhanced resistance against a pathogen, compete for space and nutrients with pathogens, interact by antibiosis mechanism, secrete antimicrobial agents against pathogens, and kill and invade pathogen spores as well as mycelium, cell, and endospores of pathogen. These agents become suitable alternatives for plant disease management. Biological control agents are an alternative technique to chemical fertilizer and pesticides to control the various pest diseases. The biotechnological applications like genetic manipulation allow us to amend the properties of these biological agents to survive under stress and nutrient-limited conditions, enhance the production of anti-

Simranjeet Singh, Vijay Kumar, and Joginder Singh have equally contributed to this chapter.

S. Singh
Department of Biotechnology, Lovely Professional University, Jalandhar, Punjab, India

Punjab Biotechnology Incubators, Mohali, Punjab, India

Regional Advanced Water Testing Laboratory, Mohali, Punjab, India

V. Kumar
Department of Chemistry, Regional Ayurveda Research Institute for Drug Development, Gwalior, Madhya Pradesh, India

D. S. Dhanjal · J. Singh (✉)
Department of Biotechnology, Lovely Professional University, Jalandhar, Punjab, India
e-mail: joginder.15005@lpu.co.in

J. Singh, A. N. Yadav (eds.), *Natural Bioactive Products in Sustainable Agriculture*, https://doi.org/10.1007/978-981-15-3024-1_3

bacterial and antifungal compounds, grown under varied pH and temperature conditions, as compared to the original strain. Advancement in molecular approaches has escalated the potential for improving the properties of these biocontrol agents.

Keywords
Pesticides · Environmental pollution · Biological agents · Genetic engineering

3.1 Introduction

Various diseases of plants should be controlled in order to maintain the quality of food in agricultural sector (Sidhu et al. 2019; Singh et al. 2019a). Various approaches and treatment methods are used worldwide to control, mitigate, and prevent plant diseases (O'Brien 2017; Singh et al. 2019b). Beyond good agronomic and horticultural practices, farmers mostly rely on pesticides and chemical fertilizers. Such approaches to agricultural sector put a remarkable improvement in crop quality and productivity in the last ten decades (Patibanda and Ranganathswamy 2018). Excessiveness and misuse of these chemical fertilizers and pesticides cause various environmental health hazards and disturb the ecological balance of biotic components in the ecosystem (Bale et al. 2008; Singh et al. 2019c). Pesticides have high mammalian, chronic, and acute toxicity level. Additionally, a wide range of nerve and muscular disorders is seen in many cases when exposed to different pesticides (Jayaraj et al. 2016; Singh et al. 2019d). Regular contact with these chemical agents may cause headache, weakness in muscles, sweating, and protracted exposure to pesticides damage the liver, kidney, CNS, TSH, and bladder (Colnot and Dekant 2017; Singh et al. 2018). They have an adverse effect on cellular respiration. Long-term exposure causes damage to the immune system, skin sensitization, and cancer (Anderson and Meade 2014; Singh et al. 2019e). Higher dose of pesticides decreases the potential of beneficial microbial population in soil to induce the growth-promoting mechanism and production of enzymes (Ahmad et al. 2018). The permeability of the plant cell and transcuticular diffusion is affected by spraying pesticides on them (Wang and Liu 2007). Recent studies on the effect of pesticides on plant growth emphasize its effect on delay in seed germination experiments. To overcome this, various organization and governing bodies have framed strict regulations and amendments on its usage today. Most of the hazardous pesticides are banned in various countries due to their high toxicity, and others are in vain (Shakir et al. 2016).

Scientists have developed various alternative inputs to these pesticides for controlling various diseases and pests. Out of these alternatives, one of them is "biological controls" (Bale et al. 2008). In 1919, "Harry Scott Smith" was the first to use this term and was brought widespread by Paul H. DeBach. "Biocontrols" are commonly used in the fields of plant pathology and entomology (Cumagun 2014). In plant pathology, the biocontrol is used against the microbial population to suppress the growth of phytopathogens, and in entomology "biocontrol" is used to suppress growth of different pest insects against microbial pathogens, entomopathogenic nematodes, or predatory insects (Wratten 2008). In both these fields, the microbes which suppress or kill the pathogen or pest are known as "biological control agent" (Usta 2013). In addition to this, "biological control" also refers to the use of natural products fermented or extracted from various sources. These extracts are either used pure or in mixture forms with specific activities having multiple effects on the target pathogen or pests (Siegwart et al. 2015). According to the US Research Council, biological control refers to "the use of natural or modified organisms, genes, or gene products, to reduce the effects of undesirable organisms and to favour desirable organisms such as crops, beneficial insects, and microorganisms" (Sandhu et al. 2012). Other definitions of biocontrol solely depend on the target of suppression which includes number, sources, type, degree, and human intervention. Biological control suppresses the damaging activities of pathogens commonly known as natural enemies (Heydari and Pessarakli 2010).

From recent advancement in the field of genetic engineering technology, transgenic approach has been introduced in accelerating the effectiveness of biological control to confer various properties of phytopathogens (O'Connell et al. 2012). The prospect of using transgenic approach in biocontrol agents depends on the accessibility of various plasmid sequences, which improve or modulate the behavior of biocontrol agents against phytopathogens (Whipps 2001). This chapter will address various types of biological control agents, various mechanisms involved for biological control, etc. Efforts have also been made to summarize the ecological significances of biological control agents. Recent advancement in genetic engineering to accelerate the function of biocontrol agents have also been incorporated in this chapter.

3.2 Types of Interactions Contributing to Biological Control

Pathogens and plants interact with a wide range of organisms throughout their life cycle. These interactions play an important role in affecting plant health. There are various direct and indirect approaches in which organisms interact (Köhl et al. 2019). When species and plant involve interference or consumption competition, it is referred as direct, while indirect is mediated by a third party and doesn't involve processes such as induced resistance, apparent competition, and resource competition (Hibbing et al. 2010). These types of interactions can be found both at the microscopic and macroscopic level. At multiple levels of scale, interaction between microbes and plants leads to biological control of phytopathogens (Passera et al. 2019). Direct interactions include herbivory, parasitism, antiobiosis, and predation, while indirect includes predation, amensalism, competition, neutralism, commensalism, protocooperation, mutualism, etc. (Abramovitch et al. 2006). The strength of both the interactions not only depends on the individual species involved but also on suitability of abiotic and biotic environmental factors (Filizola et al. 2019). However, in lab conditions, these interactions show promising result, but in field studies, these interactions proved ineffective due to various environmental constraints.

Using Odum's concept various interactions are defined on the basis of the outcomes for each other (Mwakinyali et al. 2019). Mutualism is an interaction between species in which both derive benefit. Sometimes, these interactions involve both biochemical and physical contact. A perfect example of such interaction is seen between mycorrhizal fungi and plants (Zeilinger et al. 2016). Species of *Rhizobium* genus either reproduce in the soil or in legume plants forming a mutualistic association. These mutualistic characteristics fortify the host plant with improved nutrition by stimulating defense mechanism of the host (Maróti and Kondorosi 2014).

Another interaction, protocooperation, is also a type of mutualism in which both organisms depend on each other for their survival. They are also considered as facultative mutualists as their survival depends on disease suppression and host organism and environmental conditions (Zohair et al. 2018). Further, commensalism is defined in which one species benefitted and the second one neither benefitted nor benefitted (Sharma et al. 2017).

Most of the microbial population are reported to form commensalism relation with the host species or plant as their presence frequently results in negative or positive consequences to the host plant species. Sometimes, their presence raises various challenges for infecting pathogens, as there is decrease in disease severity, or pathogen infection is a characteristic of commensal interactions (Baumler and Fang

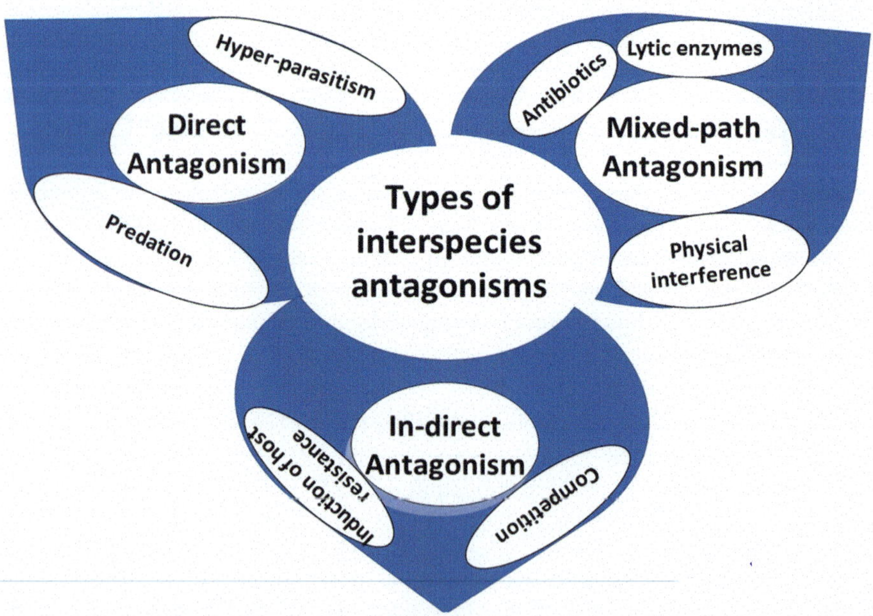

Fig. 3.1 Different types of interspecies antagonisms; their mechanisms in biological control of pathogens

2013). Different types of interspecies antagonisms and their mechanisms in biological control of pathogens are depicted in Fig. 3.1.

Neutralism interaction is described as when one species has no effect on other. Competition between and within species is often related to decreased activity, growth, or fecundity of both the interacting species (Svenning et al. 2014). Biocontrol occurs when pathogen species competes with nonpathogen species for nutrient around and inside the plant. Parasitism is a type of interaction in which two different organisms coexist for a long time and smaller benefits (parasite) and the other is harmed (host) to some extent (Syed Ab Rahman et al. 2018). Biological control results in different degrees from all these interactions depending upon environmental conditions in which they occur (Kidd and Amarasekare 2012).

3.3 Types of Biological Control Agents (BCA)

Based on whether "biological control agents" is indigenous or exotic to the field where they are introduced or applied, they are divided as classical or augmentation BCAs (Lacey et al. 2015). In augmentation BCA, the agent may or may not present already in the utilization area, whereas in classical BCA, an exotic natural pest is introduced to manage the pest organism. However, augmentation is inoculative in which an introduced organism proliferates (Sundh and Goettel 2013).

For augmentative BCA, scientists use microbial biological control agents (MBCA) that are registered as plant protection units developed by biocontrol industries. Sometimes, secondary metabolites produced by various microbes act as antimicrobial agents and were used in the product formation (Montesinos 2003). These compounds are regarded as chemical actives in the European Union. In 2017, a total of 101 microbial biological control agents have been registered in the United States, New Zealand, Japan, Europe, Canada, Brazil, Australia, etc. for disease control (Woo et al. 2014).

3.4 Mechanisms Involved for Biological Control

MBCA protect agricultural products from damage caused by different diseases through various modes of action. They induce prime-enhanced resistance mechanism against various infection caused by a pathogen (Latz et al. 2020). MBCA worked without the direct antagonistic contact with the pathogen species. They compete for space availability and nutrients with pathogens in indirect approach (Franken et al. 2019).

In direct approach, they interact by antibiosis or hyperparasitism mechanism with the pathogen species. Hyperparasites kill and invade spores, mycelium, cell of bacterial pathogens, and endospores of fungal pathogens (Navarro et al. 2019). Another direct-mode approach is the production of antimicrobial agents having inhibiting affects against pathogens (Droby et al. 2016). Mechanisms involved for biological control are well illustrated in Fig. 3.2.

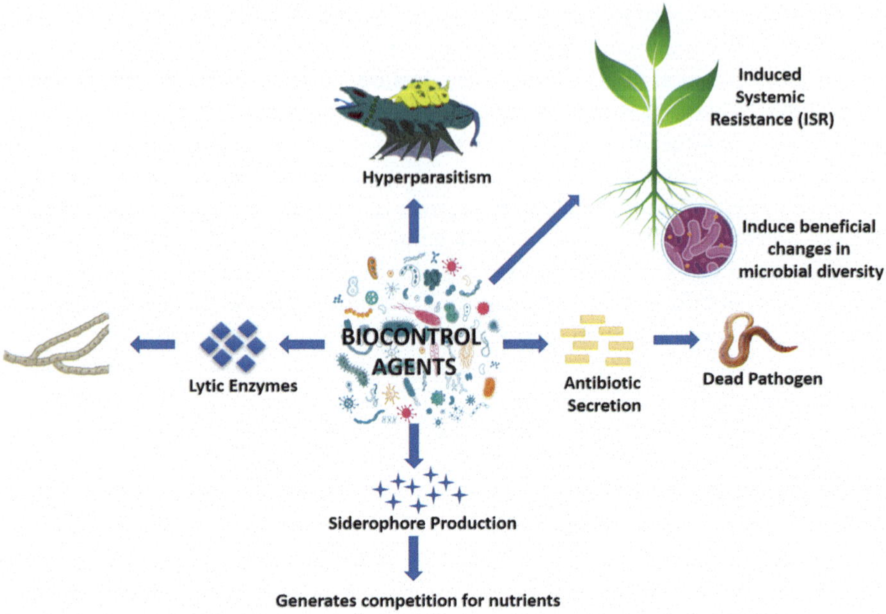

Fig. 3.2 Mechanistic actions of biological control agents against different pathogens

Biocontrol agents are used against a wide range of target species such as weeds, nematodes, mites, insects, micro-pathogens (fungi or bacteria), and even vertebrates. Some phytopathogens are so toxic as they invade their host to death (Baron et al. 2019). Many MBCAs that control bacterial or fungal plant pathogens depend on one or a combination of various mechanisms. Competitive displacement describes the competition for food or space, while antibiosis mechanism involves production of antimicrobial enzymes or certain metabolites. Conservative biocontrol mechanism is linked to indirect stimulation of organisms (Singh 2014).

Plant protects themselves with wide range of chemical and physical mechanism against various pathogens. Resistance or enhanced tolerance is one of the potential agronomic strategies to protect biological losses in agricultural sector. In constitutive mechanism, plant cuticles are induced by specific resistance mechanisms which are triggered by recognition receptors (PAMPs) in response to stimuli (Peterson et al. 2016). Resistance can be developed in the site of pathogen functioning or signals to host plant resulting in systemic acquired resistance (SAR) which responds directly to necrotizing pathogens. SAR is activated after the plant is being exposed to elicitors from nonpathogen, artificial stimulants (Walters et al. 2005).

The second type is induced systemic resistance (ISR) in which the pathogens are induced by plant growth-promoting microbes. In the presence of stimulus, both these induced resistances (SAR and ISR) increase (Choudhary et al. 2007). Priming of plants leads to sensitization for increased defense to long-lasting system of stronger defense in the presence of a stimulus. Microorganisms produce resistance-induced stimuli called microbe pathogen-associated molecular patterns (MAMPs) which are recognized by pattern recognition receptor (PRRs) (Hwang et al. 2015).

ISR is also reported to be involved in the production of phytoalexins, pathogenesis-related proteins, phenolic compounds, generation of reactive oxygen species, etc. They also aid in the formation of various physical barriers like modification of cuticles and cell plants by the induced plant (Mandal and Ray 2011). These processes are energy-dependent which help plants to maintain these defense mechanisms to be active. While in direct induced resistance mechanism, defense priming allows reaction in a robust and fast manner with low-energy costs (Mauch-Mani et al. 2017). Plant are also exposed to such stimuli from other sources such as herbivore, bacteria or fungi, abiotic stress, etc. Plant defense mechanism or reactions are stimulated by microbiological control agents (MBCAs) which depend on plant genotype (Fatima and Anjum 2017). Production of siderophores, biosurfactants, VOCs, antibiotics, pyocyanin, and DAP acts as elicitors in induced systemic resistance. Mode of action of MBCAs depends on the release of signal components which induce defense mechanism to respond against pathogen. However, biocontrol mechanism depends on the adaptability and survivability of the MBCAs on the plant (Aznar and Dellagi 2015).

3.5 Ecological Significances of Biological Control Agents

Biocontrol of weeds is now driven by ecological perspective as it undergoes a threatening era of self-examination. Weed control by biological agents has always been ruled by some ecological principles because the target and tools of biological control agents are living organisms (Suckling and Sforza 2014). Despite progress in the field of science and technology, there is still a need to develop the capability of biocontrol agents to work efficiently in their new environments. One ecological approach is to improve the predictability to enhance existing theoretical operations to control weed programs using biocontrol agents (Brodeur 2012).

Population ecology provides a sight to understand the concepts for weed biocontrol strategies which can manipulate both host plant and pathogenic number. Finally, the biocontrol agents should be released in a manner that an outbreak is generated which will minimize the density of weed to threshold level without damaging the population dynamics of nontarget plants or organisms (McEvoy 2018).

Various researchers have confirmed the mechanism underlying outbreaks and suggest that it is based on natural insect plant systems and information related to pest species. Better predictability, increased impact, and high rates of establishment are the major advantages while linking biological control systems to ecological theories (Arnold et al. 2019).

In future aspects of weed biocontrol, incorporation of ecological data system into the framework should be done for checking out and making new policies for effective biological control. For this ecological data should be required for both to release a decision related to impacts of weeds on native communities and their potential and nontarget impressions of biological control agents (Chattopadhyay et al. 2017).

3.6 Recent Advances and Genetic Engineering in the Biological Control Applications

Biological control agents have become the viable alternative for plant disease management. Moreover, advancement in molecular approaches has escalated the potential for improving the properties of these biocontrol agents (Syed Ab Rahman et al. 2018). The genetic manipulation techniques allow us to amend the properties like their ability to survive under stress condition, survive under nutrient-limited conditions, and enhance the production of antibacterial and antifungal compounds, grown under varied pH and temperature conditions as compared to the original strain (Romano et al. 2018). The different approaches used to enhance the ability of biocontrol agents via genetic manipulations are as follows.

3.6.1 Genetic Modification for Developing Resistance Against Fungicides

When plants are highly prone to disease, the fungal biological control becomes ineffective. Thus, in order to control disease, there is need for the involvement of other approaches (Bardin et al. 2015). Hence, amending the genetic material and incorporation of fungicide tolerance gene in the biocontrol agent improve its activity (Brunner et al. 2005). In a study, UV rays were used to induce benomyl-resistant in *Fusarium* isolate that are antagonistic to *Fusarium* wilt (Bubici et al. 2019).

3.6.2 Genetic Modification for Developing Hypovirulence

Hypo-virulent strains have been found to be responsible for spreading diseases in chestnut and greenhouse crop (Zhong et al. 2016). Nowadays, researchers are focusing on genetically engineered *Cryphonectria parasitica* (Murrill) Barr. A different approach is the insertion of synthetic transcripts of virus or induction of mutations in mitochondria or nuclear DNA. This approach can uplift and protect the greenhouse crops (Hoegger et al. 2003; Lan et al. 2008).

3.6.3 Enhancement in Bacteriocins Synthesis

The disease in various plants is controlled by *Agrobacterium radiobacter*, which synthesizes bacteriocin "agrocin 84" that regulates the growth of phytopathogens. The group researchers used plasmid pAgK84 containing agrocin 84-encoding gene, bacteriocin immunity gene, and conjugated transfer gene. The transfer of pAgK84 plasmid to *Agrobacterium radiobacter* resulted in the development of insensitive strain of *A. radiobacter* to agrocin 84 (Kim et al. 2006). An alternative approach is to enhance the efficiency of bacteriocin-synthesizing biological control agents via increasing the secretion of bacteriocins or either by constructing biocontrol strain effective in synthesizing bacteriocin (Yang et al. 2014).

3.6.4 Enhancement in Siderophore Synthesis

For many years, siderophore production has been extensively explored. Siderophores are iron-chelating compounds. The molecular evidence obtained from fungal gene showed no sign of siderophore gene (Kurth et al. 2016). Recently, scientists found that *Pseudomonas* sp. M114 contains a receptor for pseudobactin MT3A, a siderophore which is not synthesized by this bacterium. This suggest that by using heterologous siderophores, the gene will offer competitive advantage against rhizospheric bacteria (Ahmed and Holmström 2014).

3.6.5 Enhancement in Antibiotic Synthesis

Recently, genes encoding for producing the antibiotics have been identified which will be effective during the biocontrol the phytopathogens. The availability of antibiotic genes can aid in improving the biocontrol agents (Olanrewaju et al. 2017). In a study, in which cosmid pME3090 was inserted in *P. fluorescens*, CHA0 resulted in the increase production of 2,4-diacetylphloroglucinol and pyoluteorin antibiotics (Schnider-Keel et al. 2000). Moreover, genetically modified *Pseudomonas fluorescens* BL915 strain showed the increased production of pyrrolnitrin antibiotic (Souza and Raaijmakers 2003).

3.6.6 Enhancement in Lytic Enzyme Synthesis

Mycoparasitism is the mechanism used for biocontrol of phytopathogenic fungi. Chitinase, glucanase, and protease are the lytic enzymes synthesized by bacterial as well as fungal agents to degrade the cell wall of fungi causing their cessation and inhibition of their growth (Markovich and Kononova 2003). In the past few decades, extensive research is being done on cloning and characterization of the gene encoding for lytic enzyme. The gathered information from these techniques will aid us in improving the quality of biocontrol agents (Gajera et al. 2012). Genes encoding for chitinase enzyme have been cloned in diverse microbes like *Enterobacter agglomerans*, *Serratia marcescens*, and *Trichoderma harzianum*. These modified strains of microbes were found to be antagonistic to phytopathogenic fungi (Swiontek Brzezinska et al. 2014).

3.6.7 Enhanced Root Colonization Ability

Few microbes have been found to be beneficial to plants without their symbiotic relation with plants known as plant growth-promoting rhizobacteria (PGPR). The growth of the plant is improved by inhibiting the growth of phytopathogenic microbes (Gouda et al. 2018). Recently, genetically modified PGPR are employed for biocontrol of these plant pathogens and to improve their ability to grow in nutrient-limited conditions. These amendments enhance the root colonization ability of the PGPR (Backer et al. 2018). Genetic changes transform the bacterial strains and provide them the ability to degrade the salicylate and naphthalene. Additionally, the insertion of antifreezing gene in the PGPR has enabled them to survive and proliferate in cold temperature and perform the biocontrol activity (Glick and Bashan 1997; Sanghera et al. 2011).

3.7 Conclusion and Future Prospects

Uses of biological control agents are an alternative technique to chemical fertilizer and pesticides to control various pest diseases. Biological control agents not only kill the plant pathogens but also have no toxic effect on ecosystem. Biological control agents effectively reduce the crop damage by pests and disease development in plants. Biological control agents also have an important role in integrated pest management systems. However, biological control should be applied when they meet the criteria for their proper functioning. Ecological factors also play an important role in activity and performance of biological control agents. New application strategies should be developed to enhance the effectiveness of these agents. It is also important to isolate novel and biocontrol microorganisms having high effectiveness against most of the phytopathogens.

References

Abramovitch RB, Anderson JC, Martin GB (2006) Bacterial elicitation and evasion of plant innate immunity. Nat Rev Mol Cell Biol 7:601–611

Ahmad M, Pataczek L, Hilger TH et al (2018) Perspectives of microbial inoculation for sustainable development and environmental management. Front Microbiol 9. https://doi.org/10.3389/fmicb.2018.02992

Ahmed E, Holmström SJM (2014) Siderophores in environmental research: roles and applications. Microb Biotechnol 7:196–208. https://doi.org/10.1111/1751-7915.12117

Anderson SE, Meade BJ (2014) Potential health effects associated with dermal exposure to occupational chemicals. Environ Health Insights 8s1:EHI.S15258. https://doi.org/10.4137/ehi.s15258

Arnold JE, Egerer M, Daane KM (2019) Local and landscape effects to biological controls in urban agriculture – a review. Insects 10:215. https://doi.org/10.3390/insects10070215

Aznar A, Dellagi A (2015) New insights into the role of siderophores as triggers of plant immunity: what can we learn from animals? J Exp Bot 66:3001–3010

Backer R, Rokem JS, Ilangumaran G et al (2018) Plant growth-promoting rhizobacteria: context, mechanisms of action, and roadmap to commercialization of biostimulants for sustainable agriculture. Front Plant Sci 9:871

Bale JS, Van Lenteren JC, Bigler F (2008) Biological control and sustainable food production. Philos Trans R Soc B Biol Sci 363:761–776

Bardin M, Ajouz S, Comby M et al (2015) Is the efficacy of biological control against plant diseases likely to be more durable than that of chemical pesticides? Front Plant Sci 6. https://doi.org/10.3389/fpls.2015.00566

Baron NC, Rigobelo EC, Zied DC (2019) Filamentous fungi in biological control: current status and future perspectives. Chilean J Agric Res 79:307–315. https://doi.org/10.4067/s0718-58392019000200307

Baumler A, Fang FC (2013) Host specificity of bacterial pathogens. Cold Spring Harb Perspect Med 3. https://doi.org/10.1101/cshperspect.a010041

Brodeur J (2012) Host specificity in biological control: insights from opportunistic pathogens. Evol Appl 5:470–480. https://doi.org/10.1111/j.1752-4571.2012.00273.x

Brunner K, Zeilinger S, Ciliento R et al (2005) Improvement of the fungal biocontrol agent Trichoderma atroviride to enhance both antagonism and induction of plant systemic disease resistance. Appl Environ Microbiol 71:3959–3965. https://doi.org/10.1128/AEM.71.7.3959-3965.2005

Bubici G, Kaushal M, Prigigallo MI et al (2019) Biological control agents against Fusarium wilt of banana. Front Microbiol 10:616

Chattopadhyay P, Banerjee G, Mukherjee S (2017) Recent trends of modern bacterial insecticides for pest control practice in integrated crop management system. 3 Biotech 7:60

Choudhary DK, Prakash A, Johri BN (2007) Induced systemic resistance (ISR) in plants: mechanism of action. Indian J Microbiol 47:289–297. https://doi.org/10.1007/s12088-007-0054-2

Colnot T, Dekant W (2017) Approaches for grouping of pesticides into cumulative assessment groups for risk assessment of pesticide residues in food. Regul Toxicol Pharmacol 83:89–99. https://doi.org/10.1016/j.yrtph.2016.12.004

Cumagun CJR (2014) Advances in formulation of trichoderma for biocontrol. In: Biotechnology and biology of trichoderma. Elsevier B.V, Amsterdam, pp 527–531

Droby S, Wisniewski M, Teixidó N et al (2016) The science, development, and commercialization of postharvest biocontrol products. Postharvest Biol Technol 122:22–29. https://doi.org/10.1016/j.postharvbio.2016.04.006

Fatima S, Anjum T (2017) Identification of a potential ISR determinant from *Pseudomonas aeruginosa* PM12 against Fusarium wilt in tomato. Front Plant Sci 8. https://doi.org/10.3389/fpls.2017.00848

Filizola PRB, Luna MAC, de Souza AF et al (2019) Biodiversity and phylogeny of novel Trichoderma isolates from mangrove sediments and potential of biocontrol against Fusarium strains. Microb Cell Factories 18:89. https://doi.org/10.1186/s12934-019-1108-y

Franken P, Takken FLW, Rep M (2019) Transcript accumulation in a trifold interaction gives insight into mechanisms of biocontrol. New Phytol 224:547–549. https://doi.org/10.1111/nph.16141

Gajera HP, Bambharolia RP, Patel SV et al (2012) Antagonism of trichoderma spp. against Macrophomina phaseolina: evaluation of coiling and cell wall degrading enzymatic activities. J Plant Pathol Microbiol 3:1000149. https://doi.org/10.4172/2157-7471.1000149

Glick BR, Bashan Y (1997) Genetic manipulation of plant growth-promoting bacteria to enhance biocontrol of phytopathogens. Biotechnol Adv 15:353–378

Gouda S, Kerry RG, Das G et al (2018) Revitalization of plant growth promoting rhizobacteria for sustainable development in agriculture. Microbiol Res 206:131–140

Heydari A, Pessarakli M (2010) A review on biological control of fungal plant pathogens using microbial antagonists. J Biol Sci 10:273–290

Hibbing ME, Fuqua C, Parsek MR, Peterson SB (2010) Bacterial competition: surviving and thriving in the microbial jungle. Nat Rev Microbiol 8:15–25

Hoegger PJ, Heiniger U, Holdenrieder O, Rigling D (2003) Differential transfer and dissemination of hypovirus and nuclear and mitochondrial genomes of a hypovirus-infected *Cryphonectria parasitica* strain after introduction into a natural population. Appl Environ Microbiol 69:3767–3771. https://doi.org/10.1128/AEM.69.7.3767-3771.2003

Hwang EE, Wang MB, Bravo JE, Banta LM (2015) Unmasking host and microbial strategies in the agrobacterium-plant defense tango. Front Plant Sci 6:1–17. https://doi.org/10.3389/fpls.2015.00200

Jayaraj R, Megha P, Sreedev P (2016) Review article. Organochlorine pesticides, their toxic effects on living organisms and their fate in the environment. Interdiscip Toxicol 9:90–100

Kidd D, Amarasekare P (2012) The role of transient dynamics in biological pest control: insights from a host-parasitoid community. J Anim Ecol 81:47–57. https://doi.org/10.1111/j.1365-2656.2011.01881.x

Kim JG, Park BK, Kim SU et al (2006) Bases of biocontrol: sequence predicts synthesis and mode of action of agrocin 84, the Trojan horse antibiotic that controls crown gall. Proc Natl Acad Sci U S A 103:8846–8851. https://doi.org/10.1073/pnas.0602965103

Köhl J, Kolnaar R, Ravensberg WJ (2019) Mode of action of microbial biological control agents against plant diseases: relevance beyond efficacy. Front Plant Sci 10. https://doi.org/10.3389/fpls.2019.00845

Kurth C, Kage H, Nett M (2016) Siderophores as molecular tools in medical and environmental applications. Org Biomol Chem 14:8212–8227

Lacey LA, Grzywacz D, Shapiro-Ilan DI et al (2015) Insect pathogens as biological control agents: Back to the future. J Invertebr Pathol 132:1–41. https://doi.org/10.1016/j.jip.2015.07.009

Lan X, Yao Z, Zhou Y et al (2008) Deletion of the cpku80 gene in the chestnut blight fungus, Cryphonectria parasitica, enhances gene disruption efficiency. Curr Genet 53:59–66. https://doi.org/10.1007/s00294-007-0162-x

Latz MAC, Jensen B, Collinge DB, Lyngs Jørgensen HJ (2020) Identification of two endophytic fungi that control Septoria tritici blotch in the field, using a structured screening approach. Biol Control 141:104128. https://doi.org/10.1016/j.biocontrol.2019.104128

Mandal S, Ray RC (2011) Induced systemic resistance in biocontrol of plant diseases. In: Bioaugmentation, biostimulation and biocontrol. Springer, Berlin, pp 241–260

Markovich NA, Kononova GL (2003) Lytic enzymes of Trichoderma and their role in protecting plants from fungal diseases. Prikl Biokhim Mikrobiol 39:389–400

Maróti G, Kondorosi É (2014) Nitrogen-fixing rhizobium-legume symbiosis: are polyploidy and host peptide-governed symbiont differentiation general principles of endosymbiosis? Front Microbiol 5. https://doi.org/10.3389/fmicb.2014.00326

Mauch-Mani B, Baccelli I, Luna E, Flors V (2017) Defense priming: an adaptive part of induced resistance. Annu Rev Plant Biol 68:485–512. https://doi.org/10.1146/annurev-arplant-042916-041132

McEvoy PB (2018) Theoretical contributions to biological control success. BioControl 63:87–103

Montesinos E (2003) Development, registration and commercialization of microbial pesticides for plant protection. Int Microbiol 6:245–252

Mwakinyali SE, Ding X, Ming Z et al (2019) Recent development of aflatoxin contamination biocontrol in agricultural products. Biol Control 128:31–39

Navarro MOP, Piva ACM, Simionato AS et al (2019) Bioactive compounds produced by biocontrol agents driving plant health. In: Microbiome in plant health and disease. Springer, Singapore, pp 337–374

O'Brien PA (2017) Biological control of plant diseases. Australas Plant Pathol 46:293–304

O'Connell DM, Wratten SD, Pugh AR, Barnes AM (2012) "New species association" biological control? Two coccinellid species and an invasive psyllid pest in New Zealand. Biol Control 62:86–92. https://doi.org/10.1016/j.biocontrol.2012.03.011

Olanrewaju OS, Glick BR, Babalola OO (2017) Mechanisms of action of plant growth promoting bacteria. World J Microbiol Biotechnol 33(11):197

Passera A, Compant S, Casati P et al (2019) Not just a pathogen? Description of a plant-beneficial Pseudomonas syringae strain. Front Microbiol 10. https://doi.org/10.3389/fmicb.2019.01409

Patibanda AK, Ranganathswamy M (2018) Effect of agrichemicals on biocontrol agents of plant disease control. In: Microorganisms for green revolution, pp 1–21. https://doi.org/10.1007/978-981-10-7146-1_1

Peterson JA, Ode PJ, Oliveira-Hofman C, Harwood JD (2016) Integration of plant defense traits with biological control of arthropod pests: challenges and opportunities. Front Plant Sci 7:1794

Romano S, Jackson SA, Patry S, Dobson ADW (2018) Extending the "one strain many compounds" (OSMAC) principle to marine microorganisms. Mar Drugs 16(7):244

Sandhu SS, Sharma AK, Beniwal V et al (2012) Myco-biocontrol of insect pests: factors involved, mechanism, and regulation. J Pathog 2012:1–10. https://doi.org/10.1155/2012/126819

Sanghera GS, Wani SH, Hussain W, Singh NB (2011) Engineering cold stress tolerance in crop plants. Curr Genomics 12:30–43. https://doi.org/10.2174/138920211794520178

Schnider-Keel U, Seematter A, Maurhofer M et al (2000) Autoinduction of 2,4-diacetylphloroglucinol biosynthesis in the biocontrol agent Pseudomonas fluorescens CHA0 and repression by the bacterial metabolites salicylate and pyoluteorin. J Bacteriol 182:1215–1225. https://doi.org/10.1128/JB.182.5.1215-1225.2000

Shakir SK, Kanwal M, Murad W et al (2016) Effect of some commonly used pesticides on seed germination, biomass production and photosynthetic pigments in tomato (*Lycopersicon esculentum*). Ecotoxicology 25:329–341. https://doi.org/10.1007/s10646-015-1591-9

Sharma M, Tarafdar A, Ghosh R, Gopalakrishanan S (2017) Biological control as a tool for eco-friendly management of plant pathogens. In: Advances in soil microbiology: recent trends and future prospects. Springer, Singapore, pp 153–188

Sidhu GK, Singh S, Kumar V, Dhanjal DS, Datta S, Singh J (2019) Toxicity, monitoring and biodegradation of organophosphate pesticides: a review. Crit Rev Environ Sci Technol 53:1–53

Siegwart M, Graillot B, Lopez CB et al (2015) Resistance to bio-insecticides or how to enhance their sustainability: a review. Front Plant Sci 6:1–19. https://doi.org/10.3389/fpls.2015.00381

Singh HB (2014) Management of Plant Pathogens with microorganisms. Proc Indian Natn Sci Acad 80:443–454. https://doi.org/10.16943/ptinsa/2014/v80i2/55120

Singh S, Kumar V, Chauhan A, Datta S, Wani AB, Singh N, Singh J (2018) Toxicity, degradation and analysis of the herbicide atrazine. Environ Chem Lett 16(1):211–237

Singh S, Kumar V, Kapoor D, Kumar S, Singh S, Dhanjal DS, Datta S, Samuel J, Dey P, Wang S, Prasad R (2019a) Revealing on hydrogen sulfide and nitric oxide signals co-ordination for plant growth under stress conditions. Physiol Plant. https://doi.org/10.1111/ppl.13002

Singh S, Kumar V, Singh S, Singh J (2019b) Influence of humic acid, iron and copper on microbial degradation of fungicide Carbendazim. Biocatal Agric Biotechnol:101196. https://doi.org/10.1016/j.bcab.2019.101196

Singh S, Kumar V, Singh J (2019c) Kinetic study of the biodegradation of glyphosate by indigenous soil bacterial isolates in presence of humic acid, Fe (III) and Cu (II) ions. J Environ Chem Eng 7:103098

Singh S, Kumar V, Sidhu GK, Datta S, Dhanjal DS, Koul B, Singh J (2019d) Plant growth promoting rhizobacteria from heavy metal contaminated soil promote growth attributes of *Pisum sativum* L. Biocatal Agric Biotechnol 17:665–671

Singh S, Kumar V, Romero R, Sharma K, Singh J (2019e) Applications of nanoparticles in wastewater treatment. In: Nanobiotechnology in bioformulations. Springer, Cham, pp 395–418

Souza JT, Raaijmakers JM (2003) Polymorphisms within the prnD and pltC genes from pyrrolnitrin and pyoluteorin-producing Pseudomonas and Burkholderia spp. FEMS Microbiol Ecol 43:21–34. https://doi.org/10.1111/j.1574-6941.2003.tb01042.x

Suckling DM, Sforza RFH (2014) What magnitude are observed non-target impacts from weed biocontrol? PLoS One 9:e84847

Sundh I, Goettel MS (2013) Regulating biocontrol agents: a historical perspective and a critical examination comparing microbial and macrobial agents. BioControl 58:575–593. https://doi.org/10.1007/s10526-012-9498-3

Svenning JC, Gravel D, Holt RD et al (2014) The influence of interspecific interactions on species range expansion rates. Ecography (Cop) 37:1198–1209. https://doi.org/10.1111/j.1600-0587.2013.00574.x

Swiontek Brzezinska M, Jankiewicz U, Burkowska A, Walczak M (2014) Chitinolytic microorganisms and their possible application in environmental protection. Curr Microbiol 68:71–81. https://doi.org/10.1007/s00284-013-0440-4

Syed Ab Rahman SF, Singh E, Pieterse CMJ, Schenk PM (2018) Emerging microbial biocontrol strategies for plant pathogens. Plant Sci 267:102–111

Usta C (2013) Microorganisms in biological pest control — a review (Bacterial toxin application and effect of environmental factors). In: Current progress in biological research. InTech, Rijeka

Walters D, Walsh D, Newton A, Lyon G (2005) Induced resistance for plant disease control: maximizing the efficacy of resistance elicitors. Phytopathology 95:1368–1373

Wang CJ, Liu ZQ (2007) Foliar uptake of pesticides-present status and future challenge. Pestic Biochem Physiol 87:1–8

Whipps JM (2001) Microbial interactions and biocontrol in the rhizosphere. J Exp Bot 52:487–511. https://doi.org/10.1093/jxb/52.suppl_1.487

Woo SL, Ruocco M, Vinale F et al (2014) Trichoderma-based products and their widespread use in agriculture. Open Mycol J 8:71–126

Wratten SD (2008) Conservation biological control and biopesticides in agricultural. In: Encyclopedia of ecology. Elsevier, Amsterdam, pp 744–747

Yang SC, Lin CH, Sung CT, Fang JY (2014) Antibacterial activities of bacteriocins: application in foods and pharmaceuticals. Front Microbiol 5:241

Zeilinger S, Gupta VK, Dahms TES et al (2016) Friends or foes? Emerging insights from fungal interactions with plants. FEMS Microbiol Rev 40:182–207. https://doi.org/10.1093/femsre/fuv045

Zhong J, Chen D, Zhu HJ et al (2016) Hypovirulence of Sclerotium rolfsii caused by associated RNA mycovirus. Front Microbiol 7. https://doi.org/10.3389/fmicb.2016.01798

Zohair MM, El-Beih AA, Sadik MW et al (2018) Promising biocontrol agents isolated from medicinal plants rhizosphere against root-rot fungi. Biocatal Agric Biotechnol 15:11–18. https://doi.org/10.1016/j.bcab.2018.04.015

Role of the Potent Microbial Based Bioagents and Their Emerging Strategies for the Ecofriendly Management of Agricultural Phytopathogens

4

Touseef Hussain, Nadeem Akthar, Raghavendra Aminedi, Mohd. Danish, Yasar Nishat, and Seema Patel

Abstract

Food security is a global concern, and it is a substantial challenge to feed the ever-increasing population. The anthropological operations, abiotic and biotic stresses, have limited the crop productivity to a great extent. Phytopathogens are the major biotic constraints and pose a significant threat to food production. The extensive utility of hazardous chemicals for pathogen control is unhealthy to mankind and environment as well. In this context alternative strategies or agents that are operative in terms of cost-effectiveness, feasibility, and practicality for sustainable agricultural production are imperative. Biocontrol agents comprising of bacteria, fungi, raw plant materials, and vermicompost have become attractive in terms of pathogen control and improved crop productivity. This chapter describes the immense role of biocontrol agents in pathogen suppression and sustainable crop production. Various strategies such as production of bioactive compounds and mechanisms adopted by biocontrol agents to fight pathogens with convincing examples are discussed. Furthermore, emerging biocontrol strategies covering both conventional and biotechnological approaches that are in infancy and their emphasis as a need for improved crop production are also

T. Hussain (✉) · M. Danish · Y. Nishat
Department of Botany, Aligarh Muslim University, Aligarh, Uttar Pradesh, India

N. Akthar
Department of Animal Biosciences, University of Guelph, Guelph, ON, Canada

R. Aminedi
Laboratory of Plant-Microbe Interactions, Regional Centre for Biotechnology, NCR Biotech Science Cluster, Faridabad, Haryana, India

S. Patel
Bioinformatics and Medical Informatics Research Center, San Diego State University, San Diego, CA, USA

J. Singh, A. N. Yadav (eds.), *Natural Bioactive Products in Sustainable Agriculture*, https://doi.org/10.1007/978-981-15-3024-1_4

discussed. The major challenge is to develop cost-effective spray bio-formulations and new application methods feasible for practical applications against a broad range of phytopathogens. Undoubtedly, exploitation of biocontrol agents/strategies offers a promising ray to address food security, in particular when well optimized for a particular plant and/or soil type.

Keywords
Biological control agents · Phytopathogens · Plant root diseases · Crop losses · Rhizosphere

4.1 Introduction

The substantial pressure on land for agricultural farming due to increased urbanization, rapid changes in the agroclimatic conditions, and rising population led to the emergence of alarming food security globally. In addition, yield losses in agricultural crops caused by plant pathogens/pests such as viruses, bacteria, fungi, oomycetes, nematodes, insects, and mollusks pose a significant risk. Estimates reveal that the crop loss due to plant pathogens in Western countries is approximately 25% while that in developing countries is as high as 50% (Dubey et al. 2015; reviewed by Aboutorabi 2018). Annually it is estimated that plant diseases, either directly or indirectly, account for losses worth 40 billion dollar worldwide (Roberts et al. 2006). These plant pathogens are major biotic constraints of food production and have led to severe disasters in the past significantly impacting human history, for instance, the Irish potato famine resultant of *Phytophthora infestans* (1840s), the Great Bengal rice famine resultant of *Cochliobolus miyabeanus* (1943), the Bengal brown leaf spot disease of rice due to *Helminthosporium oryzae* (1940s), and the USA epidemic of southern corn leaf blight, caused by *Bipolaris maydis* (1970–1971), among others. While effective management stratagems such as the development of resistant varieties and genetically engineered cultivars and usage of agrochemicals were employed to tackle these diseases, they do have loopholes. Developing a disease-resistant variety through conventional breeding is time-consuming, and the resistant trait is effective against only a few diseases. Likewise, crop plants engineered for resistance via genetic engineering face environmental risks and public concerns and have not been approved by national governmental policies. So far, the effective means of disease control had been achieved by the use of agrochemicals. However, they are cost-intensive, hazardous, and environmentally unfriendly as they severely affect soil fertility, microfauna, and human health (Aktar et al. 2009; Prashar and Shah 2016). The extensive usage of chemical pesticides results in their accumulation in plant tissues, inhibition of beneficial microbes, environmental contamination, and the emergence of disease-resistant pathovars. The reckless usage of chemicals is detrimental to organisms across all taxonomical hierarchies (Kawahara et al. 2005; Rastogi et al. 2010; Mnif et al. 2011; Schwartz et al. 2015). Some of the biotic catastrophes encompass algal bloom (Heisler et al. 2008), coral bleaching (Danovaro et al. 2008), mass death of bees (Kasiotis et al.

2014), food chain contamination (Chen et al. 2007), avian reproductive loss (Jagannath et al. 2008), and mad cow disease (O'Brien 2000). In India, the rising trend of farmer suicide is partly linked to the depressive effects of the pesticides (Chitra et al. 2006). Organophosphates have been consistently linked to cancers and neurological disorders (Rastogi et al. 2010; Neupane et al. 2014). In this regard, much attention has to be paid toward alternative disease management strategies that are free from environmental risks and other related issues. The recruitment of biological control agents (BCA) as control measures against phytopathogens seems to be a quite promising approach.

Among the various phytopathogens, root pathogens significantly affect the root system and interfere with nutrient utilization and water uptake of plants, thus causing severe yield losses. Generally, root systems secrete specific exudates to the soil which improve plant nutrient availability and encourage the interaction with the rhizospheric beneficial microbes (Broeckling et al. 2008; Carvalhais et al. 2011; Trivedi et al. 2017). The rhizospheric area is the nutrient-wealthy zone of the soil due to a variety of chemicals such as auxins, peptides, amino acids, and sugars, liberated by the roots (Carvalhais et al. 2011; Takahashi 2013). Several crops are vulnerable to root diseases caused by bacteria and fungi (Alstrom and Van Vurde 2001; Maheshwari 2012). A diverse range of microorganisms thrive in the rhizosphere. Bacteria compete for the rhizospheric niche due to its resistance to climatic and edaphic fluctuations, unlike phylloplane region (O'Callaghan et al. 2006). The bacteria residing in this habitat are known as rhizobacteria (Prashar et al. 2014; Alsohim et al. 2014; Ali et al. 2014).

Biological control is the technique which attempts to mitigate the plant disease by the use of a bacterium, virus, fungus, or a combination of them to the plant or the soil (D'aes et al. 2011; Maheshwari 2012; Guo et al. 2014). The biological control agents (BCA) inhibit the phytopathogen by various offense modalities (Pearson and Callaway 2003). The most important benefit of using BCAs is that they are specific for a pathogen and are likely to be inoffensive to nontarget species. Microbial antagonists which are used to suppress diseases in plants are termed as biocontrol agents (Maheshwari 2012). The plant diseases are generally suppressed by the antagonistic effect of bioagents which occupies the same rhizospheric zone and utilizes the same food as the pathogens (Berendsen et al. 2012; Prashar et al. 2014). In the last few decades, there have been many details of the wide range of applications of BCA for the plant root diseases management (Kloeppe et al. 1999; Kokalis-burelle 2002; Mavrodi et al. 2012). The data has been presented in Table 4.1. Eco-friendly and sustainable attributes of biocontrol have spurred intense research on the potential candidates.

For BCA to be marketable, apart from efficacy, two other key factors that must be fulfilled are safety and affordable cost. It is not easy to find such candidates, and biocontrol paradigms still lag behind in replacing the detrimental chemical-driven agro-market. Biological control can be achieved at several levels such as partial, *substantial*, and complete.

Crop plants have to deal with a gamut of pathogens, insects, and herbivores. Root diseases are major problems, which can sabotage the plant health. It is because the roots are vital for nutrition uptake and are in direct contact with an array of soil microorganisms. Soil-dwelling pathogens include *Rhizoctonia*, *Verticillium*, *Phytophthora*,

Table 4.1 Biocontrol agents (BCA), their target phytopathogens, and the host plants

Biocontrol Agents	Crop	Pathogen	References
Acremonium strictum, *Trichoderma harzianum*	(*Solanum lycopersicum*) Tomato	*Meloidogyne incognita*	Goswami et al. (2008)
Trichoderma viride, *T. harzianum*, *Pseudomonas fluorescens*	(*Arachis hypogaea*) Groundnut	*Macrophomina phaseolina*	Karthikeyan et al. (2007)
T. harzianum	(*Solanum lycopersicum*) Tomato	*Meloidogyne javanica*	Sahebani and Hadavi (2008)
Verticillium chlamydosporium, *Photorhabdus luminescens*	(*Cucumis sativus*) Cucumber	*Meloidogyne incognita*	Zakaria et al. (2013)
Microsphaeropsis sp.	(*Malus domestica*) Apple	*Phytophthora cactorum*	Alexandar and Stewart (2001)
Pochonia chlamydosporia	(*Solanum tuberosum*) Potato, (*Solanum lycopersicum*) Tomato	*M. incognita*	Sellittoa et al. (2016)
Trichoderma asperellum	(*Xanthosoma sagittifolium*) Cocoyam	*Pythium myriotylum*	Mbarga et al. (2012)
Trichoderma viride	(*Glycine max*) Soybean	*Fusarium oxysporum f.* sp. *adzuki*, *Pythium arrenomanes*	John et al. (2010)
T. harzianum, *Pseudomonas fluorescens*, and *Bacillus subtilis*	(*Carthamus tinctorius*) Safflower	*Macrophomina phaseolina* (*root rot disease*)	Govindappa et al. (2010)
Paecilomyces lilacinus	(*Solanum lycopersicum*) Tomato	*M. incognita*	Oclarit and Cumagun (2010)
B. amyloliquefaciens	(*Triticum aestivum*) Wheat	*Fusarium head blight*	Dunlap et al. (2013)
B. polymyxa	(*Oryza sativa*) Rice	*R. solani*, *Pyricularia grisea*	Kavitha et al. (2005)
B. cereus	*Arabidopsis thaliana*	*Pseudomonas syringae*	Chowdhary et al. (2015)
Bacillus spp.	Ginseng	*Fusarium* cf. *incarnatum*	Song et al. (2014)

Fusarium, *Pythium*, *Sclerotinia*, *Rosellinia*, etc. (Hussain and Khan 2020; Karima et al. 2012; Matny 2015; Mostert et al. 2017). How BCA might be used to control the root pathogens and to benefit farmers in developing countries is the focus of this article.

4.2 Merits and Demerits of Biological Control Agents (BCA)

BCA has two opposite facets like all remedies. While low-cost, mild, and eco-friendliness are its positive traits, its unpredictable efficacy is a major hurdle.

4.3 Mechanisms Adopted by Biological Control Agents

The BCA activity is mediated either directly by the antagonism of soilborne pathogens or indirectly by triggering plant defense responses (Pozo and Azcon-Aguilar 2007; Jamalizadeh et al. 2011). Direct antagonism results from physical contact with the pathogen or pathogen selectivity through the mechanisms expressed by the BCA. Elicitation of plant host defense system by BCA is considered as a direct form of antagonism. BCA employ several modalities to challenge the phytopathogens (Rahman et al. 2017). Such strategies include antibiosis through the production of antimicrobial compounds, competition for available nutrients and niches, disruption of pathogen signaling, and the elicitation of plant defenses (Sturz and Christie 2003; Bais et al. 2004; Compant et al. 2013). The schematic diagram below shows some of the pathogen-vanquishing modes of BCA (Fig. 4.1). The pathogen-disarming pathways have been described in the following sections.

4.3.1 Microbial Antagonisms

The microbes considered as ideal candidates for BCA have the ability to proliferate in the rhizospheres, a niche which defends the roots and acts hostile to pathogens. The advantageous microbes colonize the roots and secrete pathogen-antagonizing metabolites into the root system where they directly aid in the suppression of pathogenic bacterial growth (Shoda 2000). This microbial antagonism between beneficial microbes and pathogens is the most significant strategy of disease control, in which

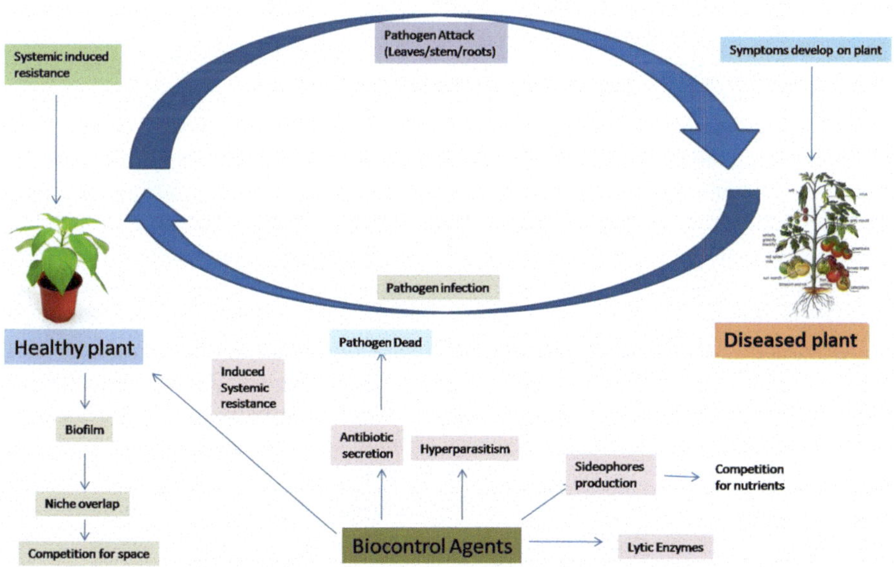

Fig. 4.1 The strategies exerted by the BCAs for the management of phytopathogens

the metabolically active populations of beneficial microbes offer protection either by direct antagonism or by priming of host plant defenses (Nihorimbere et al. 2011). It also involves antibiosis, where the secretion of diffusible antibiotics, volatile organic compounds, toxins, and extracellular cell wall-degrading enzymes (such as chitinase, β-1,3-glucanase, beta-xylosidase, pectin methylesterase) is key to pathogen control (Shoda 2000; Compant et al. 2005).

4.3.2 Parasitism

Parasitism, the concept of one organism living off another, is ubiquitous. BCA exploit parasitism to control pathogens. They produce cell wall-lysing enzymes such as chitinases and glucosaminidases to degrade the cell wall of fungal pathogens, followed by penetration and killing (Guigon-Lopez et al. 2015). *Trichoderma* has been discovered to parasitize *Macrophomina phaseolina* and *Pythium myriotylum*, among others (Kubicek et al. 2001). A nematophagous fungus *Pochonia chlamydosporia* uses its proteases to infect the eggs of nematodes. The fungal infectivity of the nematode is enhanced by chitosan (Escudero et al. 2016).

4.3.3 Competition

When two or more than two organisms demand the same nutrition from one source for survival, the interaction becomes competitive. BCA exploit this nexus to deter pathogens. They prevent the establishment of pathogen by utilizing the same nutrients which are needed for the development and infectivity of the pathogen. A microbe must be able to tap the accessible nutrients in the form of exudates, leachates, or senesced tissue in order to thrive in the phyllosphere or rhizosphere. The niche surrounding the rhizosphere is a significant source of carbon (Roviara 1965), where the photosynthate allocation can be as high as 40% (Degenhardt et al. 2003). The nutrient abundance around root surfaces attracts a large diversity of microbes, both favorable and pathogens. Effective BCA compete for these nutrients and protects the plants from phytopathogens (Duffy 2001). This competition approach has been proven to be successful particularly for soilborne pathogens such as *Fusarium* and *Pythium* that infect through mycelial contact as compared to the pathogens that directly germinate on plant aerial surfaces. For instance, *Enterobacter cloacae*, a BCA, reduces the growth of *Pythium ultimum* by enhancing the catabolism of nutrients (van Dijk and Nelson 2000; Kageyama and Nelson 2003). Migration of BCA to the root surface is governed by the chemical attractants of the root exudates such as organic acids, amino acids, and specific sugars (Nelson 1990; Welbaum et al. 2004; De Weert et al. 2002). For example, *Pseudomonas fluorescence* consumed iron required for the pathogen *Fusarium oxysporum*, and *Chryseobacterium* sp. WR21 restrained *Ralstonia solanacearum* by claiming root exudates for itself (Huang et al. 2017).

4.3.4 Production of Antimicrobial Compounds

4.3.4.1 Antibiotics

The role of antibiotics as BCA has been recognized. Among others, polymyxin, circulin, and colistin are antibiotics generally produced by a number of *Bacillus* species, Antibiotics such as phenazines, phloroglucinols, pyoluteorin, pyrrolnitrin, cyclic lipopeptides, and hydrogen cyanide (HCN) affect pathogenic fungi as well as Gram-positive and Gram-negative bacteria. For antibiotics to act as effective BCA, they must be produced in proximity to the pathogens and in adequate amount (Weller et al. 2007; Mavrodi et al. 2012). Some of the antibiotics documented to suppress plant pathogen growth include 2, 4-diacetyl phloroglucinol (DAPG) (against *Pythium* spp.) (Weller et al. 2007), agrocin 84 (against *Agrobacterium tumefaciens*) (Kerr 1980), iturin A (against *Botrytis cinerea* and *Rhizoctonia solani*) (Kloepper et al. 2004), and phenazines (against *Gaeumannomyces graminis* var. *tritici*) (Thomashow et al. 1990). Several species of the genus *Pseudomonas* elaborate antifungal metabolites phenazines, pyrrolnitrin, DAPG, and pyoluteorin (Bloemberg and Lugtenberg 2001). Phenazine has redox activity, which can suppress phytopathogens such as *F. oxysporum* and *Gaeumannomyces graminis* (Chin-A-Woeng et al. 1998). Pyrrolnitrin produced by *Pseudomonas chlororaphis* DF190 and PA23 inhibits the fungus *Leptosphaeria maculans*, which causes blackleg lesion in canola (Ramarathnam et al. 2011). *Trichoderma* has emerged as an epitome of antagonism (Verma et al. 2007). The efficacy of *Trichoderma virens*-produced agrocin K84 in the management of *Pythium* damping off of cotton seedlings has been widely observed. *Bacillus methylotrophicus* R2-2 and *Lysobacter antibioticus* 13-6 inhibited tomato root-knot-causing nematode *Meloidogyne incognita* (Zhou et al. 2016). Lipopeptide class of surfactants secreted by *Pseudomonas* and *Bacillus* species have been intended as BCA for their antagonistic effect on bacteria, fungi, oomycetes, protozoa, and nematodes, among other pests. The role of antibiotics as effective BCA has been demonstrated by the inability of soilborne root disease suppression by the mutant strains of biocontrol bacteria that fail to produce phenazines and ploroglucinols (Keel et al. 1992; Thomashow et al. 1990).

4.3.4.2 Iron-Chelating Siderophores

Bacteria and fungi-derived siderophores can inhibit plant pathogens by competing for iron, copper, zinc, and manganese (Leong and Expert 1989; Bloemberg and Lugtenberg 2003). Siderophores are iron-chelating compounds, and they aid in the transport of iron across cell membranes (Neilands 1981; Hider and Kong 2010). The ability of BCA to compete for nutrients can limit the growth of pathogens (Handelsman and Parke 1989; Nelson 1990; Harman and Nelson 1994). Iron being an essential growth element for all life forms, its deficit in soil niche and on plant surfaces creates a furious competition (Leong and Expert 1989; Loper and Henkels 1997). Iron deficiency induces BCAs to produce siderophores to acquire ferric ion (Whipps 2001). Siderophore as a mechanism of biological control was first demonstrated by plant growth-promoting strains of *Pseudomonas fluorescens* such as A1, BK1, TL3B1, and B10 against the pathogen *Erwinia carotovora* (Kloepper et al.

1980). Siderophores with higher affinity for iron and sequestration ability have been widely observed in fluorescent pseudomonads (Loper and Buyer 1991; Sullivan and Gara 1992). Significance of *P. fluorescens* siderophores in plant pathology suppression has been demonstrated in the past (Costa and Loper 1994; Leong and Expert 1989). Owing to their potential role in disease suppression, engineered bacterial strains with enhanced production of siderophores could be developed.

4.3.4.3 Biocidal Volatiles

Volatile organic compounds (VOCs) are lipophilic substances with high diffusion tendency through biological membranes (Pichersky et al. 2006). The VOCs emitted by soil bacteria act above the ground as well as within the soil. Some VOCs act as signaling molecules and play critical role in communications (Kai et al. 2009). Microbial VOCs as a weapon of defense against pathogenic fungi have received substantial attention in recent times (Mackie and Wheatley 1999; Strobel et al. 2001; Fernando et al. 2005; Gu et al. 2007; Zou et al. 2007; Liu et al. 2008; Wan et al. 2008; Arrebola et al. 2010). VOCs from soil bacteria can impede the growth of phytopathogenic fungi (Alström 2001; Wheatley 2002). For example, the VOCs produced by rhizobacteria inhibit the growth of pathogenic fungus *Sclerotiana sclerotiorum* (Giorgio et al. 2015). Volatiles produced by the bacterial strains, *Bacillus megaterium* KU143 and *Pseudomonas protegens* AS15, significantly inhibited the growth of *Aspergillus candidus*, *Aspergillus fumigatus*, *Penicillium fellutanum*, and *Penicillium islandicum* in stored rice grains (Mannaa and Kim 2018). Volatile compounds from *Trichoderma* was able to restrain plant pathogens (*Fusarium oxysporum*, *Rhizoctonia solani*, *Sclerotium rolfsii*, *Sclerotinia sclerotiorum*, and *Alternaria brassicicola*) (Amin et al. 2010; Meena et al. 2017). However, correlation of VOCs with BCAs, plant pathogens, and plants, along with biotic and abiotic, factors is explored to a limited level (Campos et al. 2010).

4.3.4.4 Lytic Enzymes (Chitinases and Glucanases)

The prevention of potential plant pathogen is usually done by the battery of hydrolytic enzymes produced by the microbes extracellularly. For example, chitinase and β-1, 3-glucanase cleave chitin and β-1, 3-glucan, the major components of fungal cell walls (Lam and Gaffney 1993), resulting in its weakening followed by death (Chernin and Chet 2002). The production of chitinase by *S. plymuthica*, *Serratia marcescens*, *Paenibacillus* sp., and *Streptomyces* sp. was found to be inhibitory against *Botrytis cinerea*, *Sclerotium rolfsii*, and *Fusarium oxysporum* f. sp. *cucumerinum* (Ordentlich et al. 1988; Frankowski et al. 2001). Similarly, laminarinase produced by *Pseudomonas stutzeri* digest and lyse mycelia of *F. solani* (Lim et al. 1991). The cell walls of *F. oxysporum*, *R. solani*, *S. rolfsii*, and *Pythium ultimum* are destroyed by the β-1, 3-glucanase synthesized by *Paenibacillus* and *B. cepacia* (Fridlender et al. 1993). The genetic evidence for the role of these enzymes in biocontrol was demonstrated by the genetic modification of *E. coli* with ChiA. The ChiA transformants showed disease in a number of incidences of southern blight of bean caused by *Sclerotium rolfsii* (Shapira et al. 1989). Similarly, ChiA from *S. marcescens* (Haran et al. 1993) transformed into *Trichoderma harzianum* was strongly able to inhibit *Sclerotium rolfsii* than the native strain. *Trichoderma*, a saprophytic fungus, produces a multitude of lysing enzymes and antagonizes pathogens by creating pressure on the available nutrients and space (Olmedo-Monfil and Casas-Flores 2014). A step forward in the paradigm of biocontrol was established by the generation of transgenic plants harboring gene for endochitinase from *T. harzianum* with increased resistance to plant pathogenic fungi (Lorito et al. 1993). These demonstrations suggest the importance of these enzymes in biocontrol and which could be enhanced using chitinolytic enzymes. In addition, search for other enzymes secreted by the biocontrol agents and their introduction into non-biocontrol microbes or via genetic engineering of plants against a broad range of phytopathogenic fungi could serve as an effective means of control measure.

4.3.4.5 Detoxification of Virulence Factors

The detox system revolves around the interaction of a protein produced by a BCA with another toxin produced by pathogen to decrease its virulence potential of the toxin on a temporary or permanent basis. For instance, *Alcaligenes denitrificans* and *P. dispersa* are involved in the detoxification of albicidin toxin produced by *Xanthomonas albilineans* (Zhang and Birch 1996, 1997; Walker et al. 1988; Basnayake and Birch 1995). Similarly, fusaric acid, a phytotoxin produced by various *Fusarium* species, is hydrolyzed by strains of *B. cepacia* and *Ralstonia solanacearum* (Toyoda and Utsumi 1991; Toyoda et al. 1988).

4.4 Priming and Induced Systemic Resistance

Resistance mediated by plants against various biotic stresses comprising both beneficial and non-beneficial and abiotic factors is distinct. Induced systemic resistance (ISR) in plants resembles systemic acquired resistance (SAR) if bacteria and pathogen remain separated. The complexity of ISR has been explored in a plant model *Arabidopsis* with three different pathways. Out of three, two involve in the release of pathogenesis-related (PR) proteins with an alternate route induction. In one of the pathways, PR proteins are produced in response to pathogen attack, while in the other pathway, they are produced in response to wounding or necrosis-inducing plant pathogens. In the pathway induced by pathogen, resistance is mediated by salicylic acid (SA) produced by plants, contrary to the wounding pathway that relies on jasmonic acid (JA) as the signaling molecule. Salicylic acid-mediated defense is triggered by pathogen infection that leads to the production of PR proteins such as PR-1, PR-2, chitinases, and some peroxidases (Kageyama and Nelson 2003; Park and Kloepper 2000; Ramamoorthy et al. 2001). These PR proteins can effectively destroy the invading cells and/or augment the cell membranes to withstand infections. On the contrary, the third pathway is rhizobacteria-induced systemic resistance (ISR), leads of systemic resistance, which is elicited by naturally present nonpathogenic bacteria associated with root. The induction of elicitors like volatile compound, protein, and antibiotic by biocontrol agents enhances the gene expression of jasmonic acid/salicylic acid/ethylene pathways (Hase et al. 2008). *Trichoderma*-emitted terpenes and compound such as 6-pentyl-2H-pyran-2-one-stimulated strains tended plant growth (Lee et al. 2016). BCA like *Pythium oligandrum* suppresses *Ralstonia solanacearum*-caused wilt disease in tomato by inducing an ethylene-dependent defense response (Hase et al. 2008). A number of compounds like polyacrylic acid, ethylene, and acetyl salicylic acid, various amino acid derivatives, the herbicide phosphinothricin, and harpin produced by *Erwinia amylovora* induce resistance of host plant against soilborne pathogens (Wei and Beer 1996). Application of biocontrol fungi, bacteria, bacteriophages, and compounds that induce ISR in the plant has been reported as an effective alternative tool for soilborne disease control.

4.5 Emerging Biocontrol Strategies

4.5.1 Usage of Plant Exudates to Attract Beneficial Biocontrol Microbes

The composition and function of rhizosphere microbial populations are significantly impacted by the exudate secreted by the root. Certain beneficial microbes are attracted by specific root exudates to meet the specific needs. The role of plant chemical exudates in favoring specific microbiomes has been well explored in the past (Rahman et al. 2017). For instance, specific nitrogen-fixing rhizobacteria are attracted by flavonoids released from legumes (Cooper 2007), and some beneficial rhizobacteria aid in activating plant defense responses to combat foliar diseases (Ryu et al. 2004). The application of soil microbiomes in agriculture has been extensively practiced as a strategy to enhance plant nutrition and disease resistance (Cao et al. 2011; Kavoo-Mwangi et al. 2013). The correlation between different exudate profiles and an attraction of different microbial populations was established in hormone-treated plants and defense signaling mutants (Carvalhais et al. 2013, 2015). Furthermore, signaling by strigolactone, a plant hormone, attracted mycorrhiza and other microbes that aid in phosphate solubilization, water supply, and defense (Rahman et al. 2017). The organic compounds malate, succinate, and fumarate aid in the attraction of a beneficial microbe *Pseudomonas fluorescens* and protect the plant from various pathogen attacks (Oku et al. 2014). Based on the successful demonstrated evidences, the implementation of plant exudates could be one of the viable approaches to attract beneficial microbes to control different plant diseases. In addition, the rhizosphere microbial population could be manipulated by simply spraying plants with signaling chemicals or altering the genotype via plant breeding to attract beneficial microbes (Carvalhais et al. 2015; Wintermans et al. 2016).

4.5.2 Use of Substrates to Maintain Beneficial Biocontrol Microbes

Substrates are the nutrients required for the growth and metabolism of microbes. Beneficial biocontrol microbes can be cultured using specific substrates. The majority of plant-associated microbes can be grown using systematic bacterial isolation approaches (Bai et al. 2015). This is advantageous to harness the beneficial microbiomes from the natural soil microbiota for the biocontrol of plant diseases by providing the right substrates. The nutritional flexibility of beneficial microbes, especially bacteria, renders them suitable for different types of environments.

4.5.3 Phyllosphere Biocontrol

Foliar diseases by fungal pathogens can significantly affect various crop plants (Madden and Nutter 1995; Dean et al. 2012). A better grasp of the role of foliar microbiomes could be a crucial step toward crop protection. Microbial biocontrol agents serve as an eco-friendly alternative to synthetic chemical control (Maksimov et al. 2011). Spray application of microbial biocontrol agents has been found to be effective on foliar diseases, including blights, leaf spots, and mildew (Heydari and Pessarakli 2010). The potential of microbial formulation in controlling stem rot pathogen of avocado plants has been tested (Demoz and Korsten 2006). Several bacteria could inhibit the growth of tomato plant bacterial stem rot pathogen *Erwinia chrysanthemi* under greenhouse condition (Aysan et al. 2003). Reduced fungal growth was noted on flowers of blueberries treated with the bacterial strain (*B. subtilis* QRD137) producing the biological products Serenade (Scherm et al. 2004). Plants defend themselves by producing antimicrobial compounds on the leaf surface or by enhancing the proliferation of beneficial microbes through the release of diverse phytochemicals (Vorholt 2012). It has been demonstrated in the past that leaf-colonizing microbes can prevent foliar disease progression in plants (Morris and Monier 2003). Niche occupation is believed to play an important role in the development of crop protection against pathogens (Lindow 1987). The implementation of innovative strategies such as profiling of phyllosphere microbiome (Vorholt 2012) and their interactions with the plant and microbe–microbe interactions can reveal new insights toward plant pathogen mitigation.

4.5.4 Breeding Microbe-Optimized Plants

Plants have evolved to interact with certain type of microbes. For instance, different ecotypes of *Arabidopsis* showed about fourfold increment in yield when inoculated with the bacterium *Pseudomonas simiae* WCS417r (Wintermans et al. 2016). This demonstrates that the outcome of the beneficial interaction is influenced by the genetic profile of the plant (Smith et al. 1999). Breeding of plants optimized to attract and maintain encouraging the colonization of beneficial microbes is the prime objective of this approach. The adoption of genetic engineering can lead to the generation of microbe-optimized plants capable of producing exudates attractive to the beneficial bacteria (Trivedi et al. 2017).

4.5.5 Engineering Microbiome, Plant-Optimized Microbes/ Microbiomes

This strategy involves engineering or breeding individual microbes or microbial consortia harboring beneficial microbes followed by their maintenance for application on crop plants meant for different soil types. This leads to the formulation of plant-/soil-optimized microbes and plant-/soil-optimized microbiomes that can serve as an inoculum for different crops in different soils. This strategy is rather new, but soil microbiomes have shown evidence of promoting plant–microbe interactions (Berendsen et al. 2012). Naturally occurring plant microbiomes play a protective role in the face of disease development (Bulgarelli et al. 2013).

4.5.6 Pairing Plant Seed with Optimal Microbiome and Soil Amendment Practices for Specific Soil Type

Efforts have been made and are still underway by the researchers to find the microbes that suit a particular crop to grow better. Smearing of seeds with promising microbes for specific soil type is one of the ideal strategies for optimizing plant–microbe interactions. The microbiomes coating the seeds assist the plants to absorb nutrients and as BCA offering protection against pathogens. In order to ensure that beneficial microbes are maintained, certain soil amendments may be essential.

Formulations of beneficial bacteria such as *Rhizobium* for legume seed treatment have already been in market. Apart from promoting the development of nitrogen-fixing nodules on leguminous plant roots, they also aid in the suppression of pathogens. The effectiveness of granular and aqueous extracts of vermicompost-based bioformulations enriched in microbial growth-promoting compounds has been demonstrated in the past (Kalra et al. 2010). Likewise, Rice et al. (1995) successfully cultured the phosphate-solubilizing fungus *Penicillium bilaii* with *Rhizobium*. Enhanced soybean nodulation was noted when co-inoculation of *Bradyrhizobium* and *Bacillus megaterium* was conducted (Liu and Sinclair 1990). Recently it was demonstrated that mixtures of rhizobacteria enhanced biological control of multiple plant diseases, promoting plant growth (Liu et al. 2018).

4.6 Current Scenario and the Need of Adopting of BCA in India

Food security is a topmost priority of all countries (Porter et al. 2014), and it is urgent for developing countries with rapid population growth. To meet the nutritional requirement of the ever-growing number of consumers, the production of food crops must be raised. But the excessive reliance on chemical fertilizers and pesticides is an unsustainable paracrine that comes with heavy price in terms of environment as well as organism health.

Despite the rising awareness of the environmental threats, in the developing countries like India, classical biocontrol programs are generally not used for agricultural practices. Therefore, it is important to extensively explore and evaluate their biocontrol potential against pathogens. Pilot program on the mass production of some BCA like *Trichoderma* can help in managing crop pathologies (Korolev et al. 2008; Cumagun 2014). *Trichoderma* can modulate the host plant signaling to combat cucumber mosaic virus (Vitti et al. 2015), *Fusarium* (Wang et al. 2005), and *Botrytis cinerea* (Elad et al. 1998), among other pathogens.

In Karnataka, the tomato plant wilt caused by *Ralstonia solanacearum* was abrogated by *Pseudomonas fluorescens* under greenhouse conditions (Vanitha et al. 2009). So, BCAs hold promises in promoting Indian agriculture, while reducing the dependency on the vicious chemical pesticides.

In order to assist the farmers in adopting new technologies, the government should take several approaches. The use of local languages to generalize literature and enhance financing to the biocontrol projects; if needed, the importing of BCA, promotion of research and development on biocontrol, and formulating manuals of BCA–plant host can be promising initiatives in this regard.

4.7 Future Prospects

In the present scenario of crop production, biocontrol is of utmost importance; however, its application needs to be optimized. Novel BCA can be screened from root microbiome (Lareen et al. 2016). Addition of resistance inducers like Bion (benzo(1,2,3) thiadiazole-7-carbothioic acid S-methyl ester) (BTH) and salicylic acid (SA) improved the efficiency of BCA such as *Trichoderma hamatum*, *Trichoderma harzianum*, and *Paecilomyces lilacinus* (Abo-Elyousr et al. 2009). BCA with synergistic function against phytopathogens might result in desirable outcome.

The research in this area is still confined to the laboratory, and very little attention has been given to formulate profitable bioagents. Moreover, the few cost-effective products have not been used conveniently by the farmers owing to the limited available information regarding their usage. Therefore, initiatives toward the popularization of biological control are required. BCAs which appear promising in laboratory settings often fail when implicated in the field. A myriad of physiological and ecological factors has been recognized causal for the futility. To improve the selection and characterization of BCA, biotechnology and other molecular tools are gaining importance to potentially solve the problem in the near future. For augmenting the efficacy of BCA, different methods such as mutation or protoplasm fusion could be a good intervention. A better understanding of the BCA mechanism and the evaluation of environmental factors is critical for viable marketability. Also, every technology leaves behind a trail of menace, so it is important to account for them, before things go out of control. For example, if BCA turn against the host plant, certain situations must be monitored. This concern is not without foundation, as *Pythium* touted as a BCA is a phytopathogen as well (Mavrodi et al. 2012; Alsohim et al. 2014). So, efficacy of the BCA mostly hinges on the plant host status, pathogens, and environmental conditions. The temperature, soil water level, moisture, oxygen, carbon, nitrogen, trace elements, indigenous microflora compositions, and the density of BCA have been found to be critical players in the pathogen control performance (Raaijmakers et al. 2002; Innocenti et al. 2015). The nematicidal effect of *Pseudomonas fluorescens* can be affected by the cyanobacterium *Calothrix parietina* (Hashem and Abo-Elyousr 2011). Alteration of relations from symbiosis to mutualism to parasitism, with changing milieu, has been reported (Redman et al. 2001). Introduction of alien species often poses environmental risks, so the impact must be assessed. Antibiotics as the driver of "drug resistance" are a grave problem. Similarly, BCA are capable of spreading perils, so they ought to be used in a controlled manner.

4.8 Conclusion

The importance of biocontrol agents in phytopathogen suppression has been known since decades back. However, the laboratory findings have not been successfully implicated in the field. In-depth understanding of the various strategies leading to positive plant-beneficial microbe association is imperative. The combination of microbial biofertilizers, biocontrol microbes, and soil amendments and the planting of microbe-optimized crops can forge positive plant–microbe interactions. In addition, potent bio-formulations and effective application methods need to be developed for successful implementation against a broad range of pathogens on a commercial scale in a cost-effective manner. This is an under-investigated area that deserves major research efforts, as it is paramount for enhancing crop yields and ensuring food security in a sustainable manner.

Conflict of Interest None of the Authors have conflict of interest.

References

Abo-Elyousr KAM, Hashem M, Ali EH (2009) Integrated control of cotton root rot disease by mixing fungal biocontrol agents and resistance inducers. Crop Prot 28:295–301. https://doi.org/10.1016/j.cropro.2008.11.004

Aboutorabi M (2018) A review on the biological control of plant diseases using various microorganisms. J Res Med Dent Sci 6:30–35

Aktar MW, Sengupta D, Chowdhury A (2009) Impact of pesticides use in agriculture: their benefits and hazards. Interdiscip Toxicol 2:1–12

Alexander BJR, Stewart A (2001) Glasshouse screening for biological control agents of *Phytophthora cactorum* on apple (*Malus domestica*). N Z J Crop Hortic Sci 29:159–169

Ali S, Hameed S, Imran A, Iqbal M, Lazarovits G (2014) Genetic, physiological and biochemical characterization of *Bacillus* sp. strain RMB7 exhibiting plant growth promoting and broad spectrum antifungal activities. Microb Cell Factories 13:144

Alsohim AS, Taylor TB, Barrett GA, Gallie J, Zhang XX, Altamirano-Junqueira AE, Johnson LJ, Rainey PB, Jackson RW (2014) The biosurfactant viscosin produced by *Pseudomonas fluorescens* SBW25 aids spreading motility and plant growth promotion. Environ Microbiol 16:2267–2281. https://doi.org/10.1111/1462-2920.12469

Alström S (2001) Characteristics of bacteria from oilseed rape in relation to their biocontrol activity against *Verticillium dahliae*. J Phytopathol 149(2):57–64

Alstrom S, Van Vurde JWL (2001) Endophytic bacteria and biocontrol of plant diseases. In: De Boer SH (ed) Plant pathogenic bacteria. Kluwer Academic Publishers, pp 60–67. https://doi.org/10.1007/978-94-010-0003-1_11

Amin F, Razdan VK, Mohiddin FA, Bhat KA, Sheikh PA (2010) Effect of volatile metabolites of *Trichoderma* species against seven fungal plant pathogens in-vitro. J Phytol 2:34–37

Arrebola E, Sivakumar D, Korsten L (2010) Effect of volatile compounds produced by *Bacillus* strains on postharvest decay in citrus. Biol Control 53:122–128. https://doi.org/10.1016/j.biocontrol.2009.11.010

Aysan Y, Karatas A, Cinar O (2003) Biological control of bacterial stem rot caused by *Erwinia chrysanthemi* on tomato. Crop Prot 22:807–811

Bai Y, Müller DB, Srinivas G, Garrido-Oter R, Potthoff E, Rott M, Dombrowski N, Münch PC, Spaepen S, Remus-Emsermann M (2015) Functional overlap of the *Arabidopsis* leaf and root microbiota. Nature 528:364

Bais HP, Fall R, Vivanco JM (2004) Biocontrol of *Bacillus subtilis* against infection of *Arabidopsis* roots by *Pseudomonas syringae* is facilitated by biofilm formation and surfactin production. Plant Physiol 134:307–319

Basnayake WVS, Birch RG (1995) A gene from *Alcaligenes denitrificans* that confers albicidin resistance by reversible antibiotic binding. Microbiology 141:551–560

Berendsen RL, Pieterse CMJ, Bakker PAHM (2012) The rhizosphere microbiome and plant health. Trends Plant Sci 17:478–486

Bloemberg GV, Lugtenberg BJ (2001) Molecular basis of plant growth promotion and biocontrol by rhizobacteria. Curr Opin Plant Biol 4:343–350

Bloemberg GV, Lugtenberg BJJ (2003) Phenazines and their role in biocontrol of *Pseudomonas* spp. New Phytol 157(3):503–523

Broeckling CD, Broz AK, Bergelson J, Manter DK, Vivanco JM (2008) Root exudates regulate soil fungal community composition and diversity. Appl Environ Microbiol 74:738–744

Bulgarelli D, Schlaeppi K, Spaepen S, van Themaat EVL, Schulze-Lefert P (2013) Structure and functions of the bacterial microbiota of plants. Annu Rev Plant Biol 64:807–838

Campos VP, Pinho RSC, Freire ES (2010) Volatiles produced by interacting microorganisms potentially useful for the control of plant pathogens. Ciênc Agrotec 34:525–535. https://doi.org/10.1590/S1413-70542010000300001

Cao Y, Zhang Z, Ling N, Yuan Y, Zheng X, Shen B, Shen Q (2011) *Bacillus subtilis* SQR 9 can control Fusarium wilt in cucumber by colonizing plant roots. Biol Fertil Soils 47:495–506

Carvalhais LC, Dennis PG, Fedoseyenko D, Hajirezaei MR, Borriss R, Wirén NV (2011) Root exudation of sugars, amino acids, and organic acids by maize as affected by nitrogen, phosphorus, potassium, and iron deficiency. J Plant Nutr Soil Sci 174:3–11

Carvalhais LC, Dennis PG, Badri DV, Tyson GW, Vivanco JM, Schenk PM (2013) Activation of the jasmonic acid plant defence pathway alters the composition of rhizosphere bacterial communities. PLoS One 8:e56457

Carvalhais LC, Dennis PG, Badri DV, Kidd BN, Vivanco JM, Schenk PM (2015) Linking jasmonic acid signaling, root exudates, and rhizosphere microbiomes. Mol Plant-Microbe Interact 28:1049–1058

Chen S, Shi L, Shan Z, Hu Q (2007) Determination of organochlorine pesticide residues in rice and human and fish fat by simplified two-dimensional gas chromatography. Food Chem 104:1315–1319

Chernin L, Chet I (2002) Enzymes in the environment: activity, ecology, and applications. Marcel Dekker, New York, pp 171–225

Chin-A-Woeng TF, Bloemberg GV, van der Bij AJ, van der Drift KM, Schripsema J, Kroon B, Scheffer RJ, Keel C, Bakker PA, Tichy HV, de Bruijn FJ (1998) Biocontrol by phenazine-1-carboxamide-producing Pseudomonas chlororaphis PCL1391 of tomato root rot caused by *Fusarium oxysporum* f. sp. *radicis-lycopersici*. Mol Plant-Microbe Interact 11:1069–1077. https://doi.org/10.1094/MPMI.1998.11.11.1069

Chitra GA, Muraleedharan VR, Swaminathan T, Veeraraghavan D (2006) Use of pesticides and its impact on health of farmers in South India. Int J Occup Environ Health 12:228–233

Compant S, Duffy B, Nowak J, Clément C, Barka EA (2005) Use of plant growth-promoting bacteria for biocontrol of plant diseases: principles, mechanisms of action, and future prospects. Appl Environ Microbiol 71:4951–4959. https://doi.org/10.1128/AEM.71.9.4951-4959.2005

Compant S, Duffy B, Nowak J, Clement C, Barka EA (2013) Use of plant growth promoting bacteria for biocontrol of plant diseases: principles, mechanisms of action, and future prospects. Appl Environ Microbiol 71:4951–4959

Cooper J (2007) Early interactions between legumes and rhizobia: disclosing complexity in a molecular dialogue. J Appl Microbiol 103:1355–1365

Costa JM, Loper JE (1994) Characterization of siderophore production by the biological-control agent *Enterobacter cloacae*. Mol Plant-Microbe Interact 7:440–448

Cumagun CJR (2014) Advances in formulation of *Trichoderma* for biocontrol. In: Gupta VK, Schmoll M, Herrera-Estrella A, Upadhyay RS, Druzhinina I, Tuohy M (eds) Biotechnology and biology of *Trichoderma*. Elsevier, Amsterdam, pp 527–531

D'aes J, Hua GK, De Maeyer K, Pannecoucque J, Forrez I, Ongena M, Dietrich LE, Thomashow LS, Mavrodi DV, Höfte M (2011) Biological control of *Rhizoctonia* root rot on bean by phenazine- and cyclic lipopeptide-producing *Pseudomonas* CMR12a. Phytopathology 101:996–1004

Danovaro R, Bongiorni L, Corinaldesi C, Giovannelli D, Damiani E, Astolfi P, Greci L, Pusceddu A (2008) Sunscreens cause coral bleaching by promoting viral infections. Environ Health Perspect 116:441–447

de Weert S, Vermeiren H, Mulders IH, Kuiper I, Hendrickx N, Bloemberg GV, Vanderleyden J, De Mot R, Lugtenberg BJ (2002) Flagella-driven chemotaxis towards exudate components is an important trait for tomato root colonization by *Pseudomonas fluorescens*. Mol Plant-Microbe Interact 15:1173–1180

Dean R, Van Kan JA, Pretorius ZA, Hammond-Kosack KE, Di Pietro A, Spanu PD, Rudd JJ, Dickman M, Kahmann R, Ellis J (2012) The top 10 fungal pathogens in molecular plant pathology. Mol Plant Pathol 13:414–430

Degenhardt J, Gershenzon J, Baldwin IT et al (2003) Attracting friends to feast on foes: engineering terpene emission to make crop plants more attractive to herbivore enemies. Curr Opin Biotechnol 14:169–176

Demoz BT, Korsten L (2006) *Bacillus subtillis* attachment, colonization, and survival on avocado flowers and its mode of action on stem-end rot pathogens. Biol Control 37:68–74

Dubey RK, Tripathi V, Abhilash PC (2015) Book review: principles of plant-microbe interactions: microbes for sustainable agriculture. Front Plant Sci 6:986

Duffy BK (2001) Encyclopedia of plant pathology. Wiley, New York, pp 243–244

Dunlap CA, Bowman MJ, Schisler DA (2013) Genomic analysis and secondary metabolite production in Bacillus amyloliquefaciens AS 43.3: a biocontrol antagonist of *Fusarium* head blight. Biol Control 64(2):166–175

Elad Y, Kirshner B, Yehuda N, Sztejnberg A (1998) Management of powdery mildew and gray mold of cucumber by *Trichoderma harzianum* T39 and *Ampelomyces quisqualis* AQ10. BioControl 43:241–251. https://doi.org/10.1023/A:1009919417481

Escudero N, Ferreira SR, Lopez-Moya F (2016) Chitosan enhances parasitism of *Meloidogyne javanica* eggs by the nematophagous fungus *Pochonia chlamydosporia*. Fungal Biol 120:572–585. https://doi.org/10.1016/j.funbio.2015.12.005

Fernando DWG, Ramarathnama R, Krishnamoorthy AS, Savchuk SC (2005) Identification and use of potential bacterial organic antifungal volatiles in biocontrol. Soil Biol Biochem 37:955–964. https://doi.org/10.1016/j.soilbio.2004.10.021

Frankowski J, Lorito M, Scala F et al (2001) Purification and properties of two chitinolytic enzymes of *Serratia plymuthica* HRO-C48. Arch Microbiol 176:421–426

Fridlender M, Inbar J, Chet I (1993) Biological control of soilborne plant pathogens by a β-1,3 glucanase-producing *Pseudomonas cepacia*. Soil Biol Biochem 25:1211–1221

Giorgio A, Lo Cantore P, Shanmugaiah V, Lamorte D, Iacobellis NS (2015) Rhizobacteria isolated from common bean in southern Italy as potential biocontrol agents against common bacterial blight. Eur J Plant Pathol 144(2):297–309. https://doi.org/10.1007/s10658-015-0767-8

Goswami J, Pandey RK, Tewari JP, Goswami BK (2008) Management of root knot nematode on tomato through application of fungal antagonists, *Acremonium strictum* and *Trichoderma harzianum*. Arch Phytopathol Plant Protect 43(3):237–240

Govindappa M, Lokesh S, Ravishankar RV, Rudra Naik V, Raju SG (2010) Induction of systemic resistance and management of safflower *Macrophomina phaseolina* root-rot disease by biocontrol agents. Arch Phytopathol Plant Protect 43(1):26–24

Gu YQ, Mo MH, Zhou JP, Zou CS, Zhang KQ (2007) Evaluation and identification of potential organic nematicidal volatiles from soil bacteria. Soil Biol Biochem 39:2567–2575. https://doi.org/10.1016/j.soilbio.2007.05.011

Guigon-Lopez C, Vargas-Albores F, Guerrero-Prieto V (2015) Changes in *Trichoderma asperellum* enzyme expression during parasitism of the cotton root rot pathogen *Phymatotrichopsis omnivora*. Fungal Biol 119:264–273. https://doi.org/10.1016/j.funbio.2014.12.013

Guo Q, Dong W, Li S, Lu X, Wang P, Zhang X, Wang Y, Ma P (2014) Fengycin produced by Bacillus subtilis NCD-2 plays a major role in biocontrol of cotton seedling damping-off disease. Microbiol Res 169:533–540

Handelsman J, Parke JL (1989) Plant-microbe interactions. In: Molecular and genetic perspectives, vol 3. McGraw-Hill, New York, pp 27–61

Haran S, Schickler H, Peer S, Logemann S, Oppenheim A, Chet I (1993) Increased constitutive chitinase activity in transformed *Trichoderma harzianum*. Biol Control 3:101–108

Harman GE, Nelson EB (1994) Seed treatment: progress and prospects. BCPC, Farnham, pp 283–292

Hase S, Takahashi S, Takenaka S (2008) Involvement of jasmonic acid signalling in bacterial wilt disease resistance induced by biocontrol agent *Pythium oligandrum* in tomato. Plant Pathol 57:870–876. https://doi.org/10.1111/j.1365-3059.2008.01858.x

Hashem M, Abo-Elyousr KA (2011) Management of the root-knot nematode Meloidogyne incognita on tomato with combinations of different biocontrol organisms. Crop Prot 30:285–292. https://doi.org/10.1016/j.cropro.2010.12.009

Heisler J, Glibert P, Burkholder J, Anderson D, Cochlan W, Dennison W, Gobler C, Dortch Q, Heil C, Humphries E, Lewitus A, Magnien R, Marshall H, Sellner K, Stockwell D, Stoecker D, Suddleson M (2008) Eutrophication and harmful algal blooms: a scientific consensus. Harmful Algae 8:3–13

Heydari A, Pessarakli M (2010) A review on biological control of fungal plant pathogens using microbial antagonists. J Biol Sci 10:273–290

Hider RC, Kong X (2010) Chemistry and biology of siderophores. Nat Prod Rep 27:637–657

Huang J, Wei Z, Hu J (2017) *Chryseobacterium nankingense* sp. nov. WR21 effectively suppresses *Ralstonia solanacearum* growth via intensive root exudates competition. BioControl 62:567–577. https://doi.org/10.1007/s10526-017-9812-1

Hussain T, Khan AA (2020) *Bacillus subtilis* HussainT-AMU and its antifungal activity against potato black scurf caused by *Rhizoctonia solani*. Biocatal Agric Biotechnol 23:101433

Innocenti G, Roberti R, Piattoni F (2015) Biocontrol ability of *Trichoderma harzianum* strain T22 against *Fusarium* wilt disease on water-stressed lettuce plants. BioControl 60:573–581. https://doi.org/10.1007/s10526-015-9662-7

Jagannath A, Shore RF, Walker LA, Ferns PN, Gosler AG (2008) Eggshell pigmentation indicates pesticide contamination. J Appl Ecol 45:133–140

Jamalizadeh M, Etebarian HR, Aminian H, Alizadeh A (2011) A review of mechanisms of action of biological control organisms against post-harvest fruit spoilage. EPPO Bull 41:65–71

John RP, Tyagi RD, Prévost D, Brar S, Pouleur S, Surampalli R (2010) Mycoparasitic *Trichoderma viride* as a biocontrol agent against *Fusarium oxysporum* f. sp. *adzuki* and *Pythium arrhenomanes* and as a growth promoter of soybean. Crop Prot 29:1452–1459. https://doi.org/10.1016/j.cropro.2010.08.004

Kageyama K, Nelson EB (2003) Differential inactivation of seed exudate stimulation of *Pythium ultimum* sporangium germination by *Enterobacter cloacae* influences biological control efficacy on different plant species. Appl Environ Microbiol 69(2003):1114–1120

Kai M, Haustein M, Molina F, Petri A, Scholz B, Piechulla B (2009) Bacterial volatiles and their action potential. Appl Microbiol Biotechnol 81:1001–1012. https://doi.org/10.1007/s00253-008-1760-3

Kalra A, Chandra M, Awasthi A, Singh AK, Khanuja SPS (2010) Natural compounds enhancing growth and survival of rhizobial inoculants in vermicompost based formulations. Biol Fertil Soils 46:521–524

Karima, Haggag HE, Nadia G EG (2012) In vitro study on *Fusarium solani* and *Rhizoctonia solani* isolates causing the damping off and root rot diseases in tomatoes. Nat Sci 10:16–25

Karthikeyan V, Sankaralingam A, Akkeeran S (2007) Biological control of groundnut stem rot caused by *Sclerotium rolfsii* (Sacc.). Arch Phytopathol Plant Protect 39(3):215–223

Kasiotis KM, Anagnostopoulos C, Anastasiadou P, Machera K (2014) Pesticide residues in honeybees, honey and bee pollen by LC-MS/MS screening: reported death incidents in honeybees. Sci Total Environ 485–486:633–642

Kavitha S, Senthilkumar S, Gnanamanickam S (2005) Isolation and partial characterization of antifungal protein from *Bacillus polymyxa* strain VLB16. Process Biochem 40(10):3236–3243

Kavoo-Mwangi A, Kahangi E, Ateka E, Onguso J, Mukhongo R, Mwangi E, Jefwa J (2013) Growth effects of microorganisms based commercial products inoculated to tissue cultured banana cultivated in three different soils in Kenya. Appl Soil Ecol 64:152–162

Kawahara J, Horikoshi R, Yamaguchi T, Kumagai K, Yanagisawa Y (2005) Air pollution and young children's inhalation exposure to organophosphorus pesticide in an agricultural community in Japan. Environ Int 31:1123–1132

Keel C, Schnider U, Maurhofer M (1992) Suppression of root diseases by Pseudomonas fluorescens CHAO: importance of the bacterial secondary metabolite 2,4-diacetylphloroglucinol. Mol Plant-Microbe Interact 5(1992):4–13

Kerr A (1980) Biological control of crown gall through production of agrocin 84. Plant Dis 64:25–30

Kloeppe JW, Rodríguez-Kábana R, Zehnder AW, Murphy JF, Sikora E, Fernández C (1999) Plant root-bacterial interactions in biological control of soilborne diseases and potential extension to systemic and foliar diseases. Australas Plant Pathol 28:21–26

Kloepper JW, Leong J, Teinize M, Schroth MN (1980) Enhanced plant growth by siderophores produced by plant growth promoting rhizobacteria. Nature 286:885–886

Kloepper JW, Ryu CM, Zhang S (2004) Induced systemic resistance and promotion of plant growth by *Bacillus* spp. Phytopathology 94:1259–1266

Kokalis A, Burelle N, Vavrina CS, Rosskopf EM, Shelby RA (2002) Field evaluation of plant growth-promoting rhizobacteria amended transplant mixes and soil solarization for tomato and pepper production in Florida. Plant Soil 238:257–266

Korolev N, Rav David D, Elad Y (2008) The role of phytohormones in basal resistance and *Trichoderma*-induced systemic resistance to *Botrytis cinerea* in *Arabidopsis thaliana*. BioControl 53:667–683. https://doi.org/10.1007/s10526-007-9103-3

Kubicek CP, Mach RL, Peterbauer CK, Lorito M (2001) *Trichoderma*: from genes to biocontrol. J Plant Pathol 83:11–23

Lam ST, Gaffney TD (1993) Biotechnology in plant disease control. Wiley, New York, pp 291–320

Lareen A, Burton F, Schäfer P (2016) Plant root-microbe communication in shaping root microbiomes. Plant Mol Biol 90:575–587

Lee S, Yap M, Behringer G, Hung R, Bennett JW (2016) Volatile organic compounds emitted by *Trichoderma* species mediate plant growth. Fungal Biol Biotechnol 3:7. https://doi.org/10.1186/s40694-016-0025-7

Leong SA, Expert D (1989) Plant-microbe interactions. In: Molecular and genetic perspectives, vol 3. McGraw-Hill, New York, pp 62–83

Lim HS, Kim YS, Kim SD (1991) *Pseudomonas stutzeri* YPL-1 genetic transformation and antifungal mechanism against *Fusarium solani*, an agent of plant root rot. Appl Environ Microbiol 57:510–516

Lindow SE (1987) Competitive exclusion of epiphytic bacteria by Ice-*Pseudomonas syringae* mutants. Appl Environ Microbiol 53:2520–2527

Liu Z, Sinclair J (1990) Enhanced soybean plant growth and nodulation by *Bradyrhizobium* in the presence of strains of *Bacillus megaterium*. Phytopathology 80:1024

Liu W, Wu W, Zhu B, Du Y, Liu F (2008) Antagonistic activities of volatiles from four strains of *Bacillus* spp. and *Paenibacillus* spp. against soil-borne plant pathogens. Agric Sci China 7:1104–1114. https://doi.org/10.1016/S1671-2927(08)60153-4

Liu K, McInroy JA, Hu CH, Kloepper JW (2018) Mixtures of plant-growth-promoting rhizobacteria enhance biological control of multiple plant diseases and plant-growth promotion in the presence of pathogens. Plant Dis 102:67–72

Loper JE, Buyer JS (1991) Siderophores in microbial interactions on plant surfaces. Mol Plant-Microbe Interact 4:5–13

Loper JE, Henkels MD (1997) Availability of iron to *Pseudomonas fluorescens* in rhizosphere and bulk soil evaluated with an ice nucleation reporter gene. Appl Environ Microbiol 63:99–105

Lorito M, Harman GE, Hayes CK, Broadway RM, Tronsmo A, Woo SL, Di Pietro A (1993) Chitinolytic enzymes produced by *Trichoderma harzianum*: antifungal activity of purified endochitinase and chitobiosidase. Phytopathol 83(3):302–307

Mackie AE, Wheatley RE (1999) Effects and incidence of volatile organic compound interactions between soil bacterial and fungal isolates. Soil Biol Biochem 31:375–385. https://doi.org/10.1016/S0038-0717(98)00140-0

Madden L, Nutter F Jr (1995) Modeling crop losses at the field scale. Can J Plant Pathol 17:124–137

Maheshwari DK (2012) Bacteria in agrobiology: disease management. Springer, Berlin

Maksimov I, Abizgil'Dina R, Pusenkova L (2011) Plant growth promoting rhizobacteria as alternative to chemical crop protectors from pathogens. Appl Biochem Microbiol 47:333–345

Mannaa M, Kim KD (2018) Biocontrol activity of volatile-producing *Bacillus megaterium* and *Pseudomonas protegens* against *Aspergillus* and *Penicillium* spp. predominant in stored rice grains: study II. Mycobiology 46:52–63. https://doi.org/10.1080/12298093.2018.1454015

Matny ON (2015) Fusarium head blight and crown rot on wheat & barley: losses and health risks. Adv Plants Agric Res 2:38–43

Mavrodi OV, Walter N, Elateek S, Taylor CG, Okubara PA (2012) Suppression of *Rhizoctonia* and *Pythium* root rot of wheat by new strains of *Pseudomonas*. Biol Control 62:93–102. https://doi.org/10.1016/j.biocontrol.2012.03.013

Mbarga JB, Ten Hoopen GM, Kuaté J, Adiobo A, Ngonkeu MEL, Ambang Z, Akoa A, Tondje PR, Begoude BAD (2012) *Trichoderma asperellum*: a potential biocontrol agent for *Pythium myriotylum*, causal agent of cocoyam (*Xanthosomas agittifolium*) root rot disease in Cameroon. Crop Prot 36:18–22

Meena M, Swapnil P, Zehra A, Dubey MK, Upadhyay RS (2017) Antagonistic assessment of *Trichoderma* spp. by producing volatile and non-volatile compounds against different fungal pathogens. Arch Phytopathol Plant Protect 50:629–648. https://doi.org/10.1080/03235408.2017.1357360

Mnif W, Hassine AI, Bouaziz A, Bartegi A, Thomas O, Roig B (2011) Effect of endocrine disruptor pesticides: a review. Int J Environ Res Public Health 8:2265–2303

Morris CE, Monier JM (2003) The ecological significance of biofilm formation by plant-associated bacteria. Annu Rev Phytopathol 41:429–453

Mostert D, Molina AB, Daniells J, Fourie G, Hermanto C, Chao C-P, Fabregar E, Sinohin VG, Masdek N, Thangavelu R, Li C, Yi G, Mostert L, Viljoen A (2017) The distribution and host range of the banana Fusarium wilt fungus, *Fusarium oxysporum* f. sp. *cubense*, in Asia. PLoS One 12:e0181630

Neilands JB (1981) Iron absorption and transport in microorganisms. Annu Rev Nutr 1:27–46

Nelson EB (1990) Exudate molecules initiating fungal responses to seeds and roots. Plant Soil 129:61–73

Neupane D, Jors E, Brandt L (2014) Pesticide use, erythrocyte acetylcholinesterase level and self-reported acute intoxication symptoms among vegetable farmers in Nepal: a cross-sectional study. Environ Health 13:98

Nihorimbere V, Cawoy H, Seyer A, Brunelle A, Thonart P, Ongena M (2011) Impact of rhizosphere factors on cyclic lipopeptide signature from the plant beneficial strain *Bacillus amyloliquefaciens* S499. FEMS Microbiol Ecol 79:176–191

O'Brien M (2000) Have lessons been learned from the UK bovine spongiform encephalopathy (BSE) epidemic? Int J Epidemiol 29:730–733

O'Callaghan M, Lorenz N, Gerard EM (2006) Characterization of phylloplane and rhizosphere microbial populations using PCR and denaturing gradient gel electrophoresis (DGGE). In: Cooper JE, Rao JR (eds) Molecular approaches to soil rhizosphere and plant microorganism analysis. CABI, Oxfordshire, pp 99–115

Oclarit E, Cumagun CJR (2010) Evaluation of efficacy of *Paecilomyces lilacinus* as biological control agent of M*eloidogyne incognita* attacking tomato. J Plant Protect 49(4):337–340

Oku S, Komatsu A, Nakashimada Y, Tajima T, Kato J (2014) Identification of *Pseudomonas fluorescens* chemotaxis sensory proteins for malate, succinate, and fumarate, and their involvement in root colonization. Microbes Environ 29:413–419

Olmedo-Monfil V, Casas-Flores S (2014) Molecular mechanisms of biocontrol in *Trichoderma* spp. and their applications in agriculture. In: Gupta VK, Schmoll M, Herrera-Estrella A, Upadhyay RS, Druzhinina I, Tuohy MG (eds) Biotechnology and biology of Trichoderma. Elsevier, Waltham, pp 429–453

Ordentlich A, Elad Y, Chet I (1988) The role of chitinase of *Serratia marcescens* in biocontrol of *Sclerotium rolfsii*. Phytopathology 78:84–88

Park KS, Kloepper JW (2000) Activation of PR-1a promoter by rhizobacteria that induce systemic resistance in tobacco against *Pseudomonas syringae* pv. *tabaci*. Biol Control 18:2–9

Pearson DE, Callaway RM (2003) Indirect effects of host-specific biological control agents. Trends Ecol Evol 18:456–461

Pichersky E, Noel J, Dudareva N (2006) Biosynthesis of plant volatiles: nature's diversity and ingenuity. Science 311:808–811. https://doi.org/10.1126/science.1118510

Porter JR, Xie L, Challinor AJ et al (2014) Food security and food production systems. In: Climate change 2014: impacts, adaptation, and vulnerability. Part A: global and sectoral aspects. Contribution of Working Group II to the Fifth Assessment Report of the Intergovernmental Panel on Climate Change, pp 485–533

Pozo MJ, Azcón-Aguilar C (2007) Unraveling mycorrhiza-induced resistance. Curr Opin Plant Biol 10:393–398

Prashar P, Shah S (2016) Impact of fertilizers and pesticides on soil microflora in agriculture. In: Lichtfouse E (ed) Sustainable agriculture reviews 19. Springer, Dordrecht. https://doi.org/10.1007/978-3-319-26777-7_8

Prashar P, Kapoor N, Sachdeva S (2014) Rhizosphere: its structure, bacterial diversity and significance. Rev Environ Sci Biotechnol 13:63–77

Raaijmakers JM, Vlami M, de Souza JT (2002) Antibiotic production by bacterial biocontrol agents. Antonie Van Leeuwenhoek 81:537–547. https://doi.org/10.1023/A:1020501420831

Rahman SF, Singh E, Pieterse CMJ, Schenk PM (2017) Emerging microbial biocontrol strategies for plant pathogens. Plant Sci 267:102–111

Ramamoorthy V, Viswanathan R, Raguchander T, Prakasam V, Samiyappan R (2001) Induction of systemic resistance by plant growth promoting rhizobacteria in crop plants against pests and diseases. Crop Prot 20:1–11

Ramarathnam R, Fernando WGD, de Kievit T (2011) The role of antibiosis and induced systemic resistance, mediated by strains of *Pseudomonas chlororaphis, Bacillus cereus and B. amyloliquefaciens*, in controlling blackleg disease of canola. BioControl 56:225–235. https://doi.org/10.1007/s10526-010-9324-8

Rastogi SK, Tripathi S, Ravishanker D (2010) A study of neurologic symptoms on exposure to organophosphate pesticides in the children of agricultural workers. Indian J Occup Environ Med 14:54–57

Redman RS, Dunigan DD, Rodriguez RJ (2001) Fungal symbiosis from mutualism to parasitism: who controls the outcome, host or invader? New Phytol 151:705–716. https://doi.org/10.1046/j.0028-646x.2001.00210.x

Rice W, Olsen P, Leggett M (1995) Co-culture of Rhizobium meliloti and a phosphorus solubilizing fungus (*Penicillium bilaii*) in sterile peat. Soil Biol Biochem 27:703–705

Roberts MJ, Schimmelpfennig DE, Ashley E, Livingston MJ, Ash MS, Vasavada U (2006) The value of plant disease early-warning systems: a case study of USDA's soybean rust coordinated framework. United States Department of Agriculture, Economic Research Service

Rovira AD (1965) Interactions between plant roots and soil microorganisms. Annu Rev Microbiol 19:241–266

Ryu CM, Farag MA, Hu CH, Reddy MS, Kloepper JW, Paré PW (2004) Bacterial volatiles induce systemic resistance in Arabidopsis. Plant Physiol 134:1017–1026

Sahebani N, Hadavi N (2008) Biological control of the root-knot nematode *Meloidogyne javanica* by *Trichoderma harziannum*. Soil Biol Biochem 40(8):2016–2020

Scherm H, Ngugi H, Savelle A, Edwards J (2004) Biological control of infection of blueberry flowers caused by *Monilinia vaccinii-corymbosi*. Biol Control 29:199–206

Schwartz NA, von Glascoe CA, Torres V, Ramos L, Soria-Delgado C (2015) "Where they (live, work and) spray": pesticide exposure, childhood asthma and environmental justice among Mexican-American farmworkers. Health Place 32:83–92

Shapira R, Ordentlich A, Chet I et al (1989) Control of plant diseases by chitinase expressed from cloned DNA in *Escherichia coli*. Phytopathology 79:1246–1249

Shoda M (2000) Bacterial control of plant diseases. J Biosci Bioeng 89(6):515–521

Smith KP, Handelsman J, Goodman RM (1999) Genetic basis in plants for interactions with disease-suppressive bacteria. Proc Natl Acad Sci 96:4786–4790

Song M, Yun HY, Kim YH (2014) Antagonistic *Bacillus* species as a biological control of ginseng root rot caused by *Fusarium* cf. *incarnatum*. J Ginseng Res 38(2):136–145

Strobel GA, Dirkse E, Sears J, Markworth C (2001) Volatile antimicrobials from *Muscodor albus*, a novel endophytic fungus. Microbiology 147:2943–2950. https://doi.org/10.1099/00221287-147-11-2943

Sturz AV, Christie BR (2003) Beneficial microbial allelopathies in the root zone: the management of soil quality and plant disease with rhizobacteria. Soil Tillage Res 72(2):107–123

Sullivan NJ, O'Gara F (1992) Traits of fluorescent Pseudomonas spp. involved in suppression of plant root pathogens. Microbiol Rev 56:662–676

Takahashi H (2013) Auxin biology in roots. Plant Roots 7:49–64

Thomashow LS, Weller DM, Bonsall RF, Pierson LS (1990) Production of the antibiotic phenazine-1-carboxylic acid by fluorescent *Pseudomonas* species in the rhizosphere of wheat. Appl Environ Microbiol 56:908–912

Toyoda H, Utsumi R (1991) U.S. Patent 4:988,586

Toyoda H, Hashimoto H, Utsumi R (1988) Detoxification of fusaric acid by a fusaric acid-resistant mutant of Pseudomonas solanacearum and its application to biological control of Fusarium wilt of tomato. Phytopathology 78:1307–1311

Trivedi P, Schenk PM, Wallenstein MD, Singh BK (2017) Tiny microbes, big yields: enhancing food crop production with biological solutions. Microb Biotechnol 10:999–1003

van Dijk K, Nelson EB (2000) Fatty acid competition as a mechanism by which *Enterobacter cloacae* suppresses *Pythium ultimum* sporangium germination and damping-off. Appl Environ Microbiol 66:5340–5347

Vanitha SC, Niranjana SR, Mortensen CN, Umesha S (2009) Bacterial wilt of tomato in Karnataka and its management by *Pseudomonas fluorescens*. BioControl 54:685–695. https://doi.org/10.1007/s10526-009-9217-x

Verma M, Brar SK, Tyagi RD (2007) Antagonistic fungi, Trichoderma spp.: panoply of biological control. Biochem Eng J 37:1–20

Vitti A, La Monaca E, Sofo A et al (2015) Beneficial effects of *Trichoderma harzianum* T-22 in tomato seedlings infected by Cucumber mosaic virus (CMV). BioControl 60:135–147. https://doi.org/10.1007/s10526-014-9626-3

Vorholt JA (2012) Microbial life in the phyllosphere. Nat Rev Microbiol 10:828–840

Walker MJ, Birch RG, Pemberton JM (1988) Cloning and characterization of an albicidin resistance gene from *Klebsiella oxytoca*. Mol Microbiol 2:443–454

Wan MG, Li GQ, Zhang JB, Jiang DH, Huang HC (2008) Effect of volatile substances of *Streptomyces platensis* F-1 on control of plant fungal diseases. Biol Control 46:552–559. https://doi.org/10.1016/j.biocontrol.2008.05.015

Wang H, Chang KF, Hwang SF et al (2005) Fusarium root rot of coneflower seedlings and integrated control using *Trichoderma* and fungicides. BioControl 50:317–329. https://doi.org/10.1007/s10526-004-0457-5

Wei ZM, Beer SV (1996) Harpin from *Erwinia amylovora* induces plant resistance. Acta Hortic 411:223–225

Welbaum G, Sturz AV, Dong Z (2004) Fertilizing soil microorganisms to improve productivity of agroecosystems. Crit Rev Plant Sci 23:175–193

Weller DM, Landa BB, Mavrodi OV et al (2007) Role of 2,4-diacetylphloroglucinol-producing fluorescent *Pseudomonas* spp. in the defense of plant roots. Plant Biol 9(2007):4–20

Wheatley RE (2002) The consequences of volatile organic compound mediated bacterial and fungal interactions. A Van Leeuw J Microb 81:357–364. https://doi.org/10.1023/A:1020592802234

Whipps JM (2001) Microbial interactions and biocontrol in the rhizosphere. J Exp Bot 52:487–511

Wintermans PC, Bakker PA, Pieterse CM (2016) Natural genetic variation in *Arabidopsis* for responsiveness to plant growth-promoting rhizobacteria. Plant Mol Biol 90:623–634

Zakaria H, Kassab AS, Shamseldean M, Oraby M, El-Mourshedy MMF (2013) Controlling the root-knot nematode, *Meloidogyne incognita* in cucumber plants using some soil bioagents and some amendments under simulated field conditions. Ann Agric Sci 58:77–82. https://doi.org/10.1016/j.aoas.2013.01.011

Zhang L, Birch RG (1996) Biocontrol of sugar cane leaf scald disease by an isolate of *Pantoea dispersa* which detoxifies albicidin phytotoxins. Lett Appl Microbiol 22:132–136

Zhang L, Birch RG (1997) The gene for albicidin detoxification from *Pantoea dispersa* encodes an esterase and attenuates pathogenicity of *Xanthomonas albilineans* to sugarcane. Proc Natl Acad Sci U S A 94:9984–9989

Zhou L, Yuen G, Wang Y (2016) Evaluation of bacterial biological control agents for control of root-knot nematode disease on tomato. Crop Prot 84:8–13. https://doi.org/10.1016/j.cropro.2015.12.009

Zou CS, Mo MH, Gu YQ, Zhou JP, Zhang KQ (2007) Possible contributions of volatile-producing bacteria to soil fungistasis. Soil Biol Biochem 39:2371–2379. https://doi.org/10.1016/j.soilbio.2007.04.009

Herbicides and Plant Growth Regulators: Current Developments and Future Challenges

5

Simranjeet Singh, Vijay Kumar, Daljeet Singh Dhanjal, and Joginder Singh

Abstract

Herbicides are a class of pesticides which is used to kill nuisance plants such as weeds and grasses that compromise the yield and growth of desired crops. Most of the herbicides act on plants by affecting their biochemical processes. Certain herbicides that regulate growth in plants as growth regulators are effective on broad leaves plants but not on different species of grasses. Herbicides may either involve biological system or plant enzyme that disrupts the regular plant growth and finally death. Failure of uniform distribution of herbicide application results in crop injuries, ineffective weed control, and is directly linked with sprayer calibration. Most of the herbicides are developed after rearranging chemical groups which is not the best strategy as pests and weeds quickly develop resistance to such groups of chemicals. Current considerations are to develop herbicides which not only benefit the ecosystem by reducing dependence on soil tillage for effective weed management system but also possess better environmental and toxicological profiles. This chapter discusses the recent development and future challenges in herbicide formulation and considerations.

Simranjeet Singh, Vijay Kumar, and Daljeet Singh Dhanjal have equally contributed to this chapter.

S. Singh
Department of Biotechnology, Lovely Professional University, Jalandhar, Punjab, India

Punjab Biotechnology Incubators, Mohali, Punjab, India

Regional Advanced Water Testing Laboratory, Mohali, Punjab, India

V. Kumar
Department of Chemistry, Regional Ayurveda Research Institute for Drug Development, Gwalior, Madhya Pradesh, India

D. S. Dhanjal · J. Singh (✉)
Department of Biotechnology, Lovely Professional University, Jalandhar, Punjab, India
e-mail: joginder.15005@lpu.co.in

J. Singh, A. N. Yadav (eds.), *Natural Bioactive Products in Sustainable Agriculture*, https://doi.org/10.1007/978-981-15-3024-1_5

Keywords
Herbicides · Plant growth regulators · Mode of action · Crop tolerance · Sustainable agriculture

5.1 Introduction

Environmental and economic losses have been directly linked with nonindigenous plants (Pimentel et al. 2000). Unwanted plants such as weeds and grass damage crops worth billions of dollars worldwide. Control of these unwanted plants, weeds, and grasses in a cost-effective system is very significant to agriculture and other industries (Abbas et al. 2018). Herbicides are a class of pesticides which is used to kill nuisance plants such as weeds and grasses that compromise the yield and growth of desired crops (Gupta 2011; Singh et al. 2016a, b). Herbicides were first introduced in the late 1940s, and till then regular production of new herbicides acts as a tool in management of weed and grasses. They are also used in different industries, gardening, and landscaping (Kraehmer et al. 2014). Almost 48% of total pesticide usage in world is of herbicides. Dalapon, disodium methanearsonate, and monosodium methanearsonate are the first herbicides used to kill weeds and perennial grass such as Vasey grass and Johnson grass (Aparecida et al. 2013; Bhati et al. 2019; Kapoor et al. 2019; Kaur et al. 2017; Kumar and Singh 2018; Kumar et al. 2018a, b). Other earlier used herbicides include petroleum oils, sodium arsenate, sodium chlorate, sulfuric acid, and sodium chlorate (Gupta 2018).

Excessive or extreme use of herbicides could damage the soil system and crop system and also impart resistance to weeds which were intended for elimination (Sherwani et al. 2015). Most of the herbicides run through the soil surface directly into the water bodies (groundwater and surface water) resulting in disruption of flora and fauna ecosystem (Khatri and Tyagi 2015; Kumar et al. 2019a, b). Herbicides have deleterious effects on human health and organism by both indirect and direct approach. Herbicides are reported to cause physiological alterations, genetic damages, cancer, chromosomal damages, and structural alterations. Therefore, it is necessary to develop a balance between the optimum medium and effective strategies for maximum and best effect applications (Kubsad et al. 2019).

5.2 Types of Herbicides

On the basis of chemical name, mode of action, toxicity, and chemical composition, herbicides are broadly classified in various classes (Varshney et al. 2012). Based on apply time, they are classified as postemergence or preemergence. Two major categories based on mode of action are systemic or translocated and contact herbicides (Kumar et al. 2014; Prasad et al. 2013; Qasem 2011). Systemic or translocated herbicide changes the physiological functions of the plants by altering the changes in biochemical reactions. When systemic herbicides were applied to crops, they

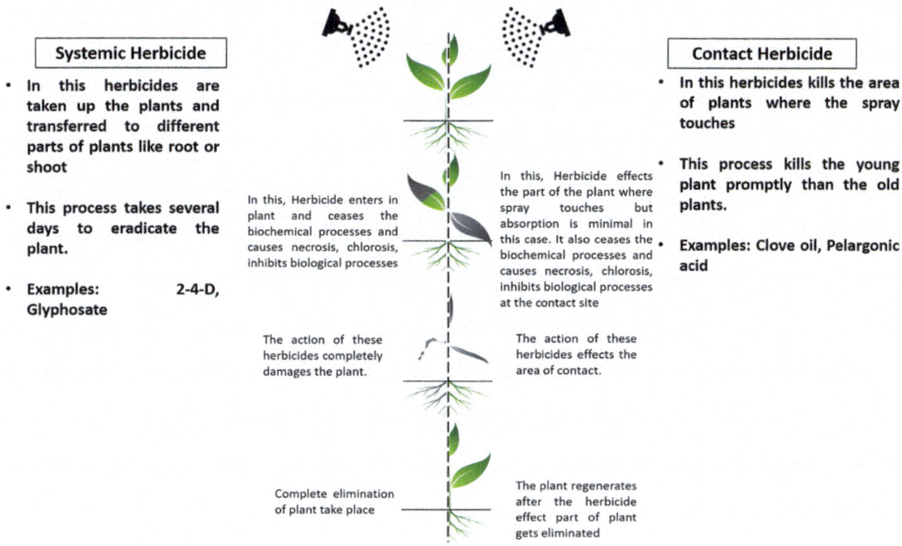

Fig. 5.1 Types of herbicides and their mechanistic action

enter the plants and move to their site of action resulting in the inhibition of various biochemical processes required for normal growth and development (Singh and Sharma 2008). They are reported to cause necrosis, chlorosis, inhibition of nucleic acid, lipid and protein synthesis, modification of protein structures, and blockage of photosynthesis. Systemic herbicides include simazine, glyphosate, atrazine, trichloroacetic acid (TCA), di-thiocarbamates, thiocarbamates, 2,4-dichlorophenoxyacetic acid (2,4-D), etc. (Au 2003; Sidhu et al. 2019; Singh et al. 2017, 2018, 2019a, b, c).

Unlike systemic herbicides, contact herbicides affect the part of the plant where they were sprayed. In contact herbicides absorption is minimal (Kumar et al. 2013; Shukla and Devine 2008). Contact herbicides are directly sprayed with water. These herbicides cause necrosis, chlorosis, generation of free radicals, inhibition of photosynthesis process, and finally death (Kumar et al. 2015a, b; Mensah et al. 2015). Examples of contact herbicides include paraquat, diquat, dinoseb and diclofop (Singh and Sharma 2008). Types of herbicides and their mechanism are well presented in Fig. 5.1.

5.3 Parameters Regulating Performance of Herbicides

The presence of aromatic amino acids in the functional groups of herbicides determines their performance. Environmental factors such as pH, temperature, moisture content, and soil texture also determine their functional ability (Barchanska et al. 2017). The selectivity of herbicides is based on their mode of action. Certain herbicides that regulate growth in plants are effective on broad leaves plants but not on different species of grasses (Wright et al. 2010).

In thiocarbamate herbicides, coarse-textured soil and high-moisture-content in soil favor quicker vaporization as compared to low-moisture clay soil. The effectiveness of thiocarbamates can be loosen by adequate rain falls and tillage (Kumar et al. 2019c, d; Scherner et al. 2018).

The efficiency of chloroacetamide pesticides depends on organic content and soil quality. Soil with higher quantity of organic matter and soil clay requires higher dose (Yamaji et al. 2016). Triazines are photosynthetic inhibitors that persist for a long time. The activity of triazine herbicides also depends on soil pH and water content. Increase in pH maintains the stability of the triazine pesticides, while low pH decreases. However, high organic matter and clay need high dose of triazine pesticides (Garcia Blanco et al. 2013).

Organophosphate herbicides are amino acid inhibitors and are usually applied after postemergence. They are used to control perennial and annual grass, broadleaf species (Gregory et al. 2013). These herbicides are nonselective and have an effect on any living tissue when they come in contact with them, although there is no effect of light on these pesticides (Nicolopoulou-Stamati et al. 2016). When they come in contact with soil, they bind with the soil particles and organic matter and become ineffective for weed control (Vats 2015).

Sulfonyl urea's herbicide starts degrading when a small portion of UV light falls on it. Other factors such as pH, temperature, organic matter, and clay content are almost similar to other herbicides (Ahmad 2017).

Benzonitriles, phenoxies, picolinic acid, and benzoic acid herbicides are applied during postemergence. pH and water have no or little effect on their persistence. However, all these four persist longer in cool and moist conditions than in warm and dry conditions (Curran 2016).

5.4 Mode of Action of Herbicides

Most of the herbicides act on plants by affecting their biochemical processes. The mechanism of action depends on the chemical structure or composition of the herbicides. The effect of herbicides is maximum in the growing stage of weeds and lessens in the maturity as it goes through flowering, maturity, and flowering, respectively (Kumar et al. 2015c, d; Sandmann et al. 2018). Herbicides have specific mode of action. They may either involve a biological system or plant enzyme that disrupts the regular plant growth and finally death (Bonny 2016).

On the basis of inhibition mechanism, they are classified into amino acid biosynthesis inhibitors, cell membrane disruptors, lipid biosynthesis inhibitors, photosynthesis inhibitors, plant growth regulators, nitrogen metabolism inhibitors, pigment inhibitors, and seedling growth inhibitors (Lushchak et al. 2018) (Fig. 5.2).

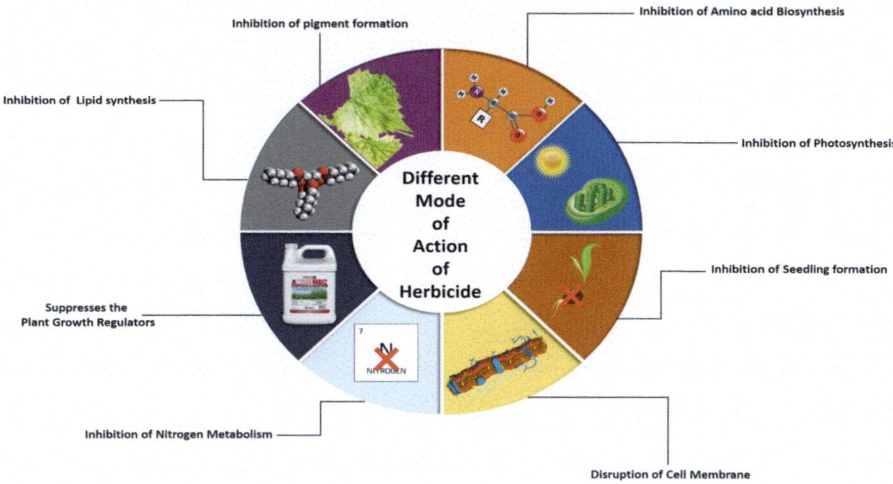

Fig. 5.2 Different modes of action of herbicides

5.4.1 Amino Acid Biosynthesis Inhibitors

They are also known as acetohydroxy acid synthase (AHAS) or acetolactate synthase (ALS) inhibitors. These are the largest classes of herbicides comprised of different chemical families of pesticides including triazolopyrimidines, sulfonylureas, sulfonylaminocarbonyltriazolinones, pyrimidinylthiobenzoates, and imidazolinones. In the presence of ALS enzyme, they minimize the production of branched amino acids causing necrosis and wilting which eventually lead to plant death (Yu and Powles 2014).

5.4.2 Cell Membrane Disruptors

Various herbicides disrupt the glutamine synthetase (GS) activity by producing ammonia, which disintegrates the proteins and lowers down the pH gradient on both sides of cell membrane. This class includes aryl triazolinone, oxadiazoles, diphenylether, thiadiazoles, oxazolidinediones, phenylpyrazoles, N-phenylphthalimides, and pyrimidindiones which are reported to destroy and penetrate the lipid bilayer of the cell resulting in the disintegration of the cell membrane (Kumar et al. 2016; Nwani et al. 2010). These herbicides are also reported to inhibit the protoporphyrinogen oxidase (PPO) activity which catalyzes heme biosynthesis and chlorophyll biosynthesis. Further proteins and lipids become oxidized resulting in the leakage of pigments (chlorophyll) and cause wilting, cell disintegration, and finally death (Dayan and Watson 2011).

5.4.3 Lipid Biosynthesis Inhibitors

They are also known as acetyl coenzyme A carboxylase (ACCase) inhibitors. They inhibit the ACCase enzyme activity which plays an eminent role in fatty acid synthesis (Kumar et al. 2015d; Yang et al. 2010). This includes phenylpyrazolin, cyclohexanedione, and aryloxyphenoxypropionate, which block production of fatty acids (phospholipids) that are useful for maintaining the cell structure (Kaundun et al. 2013).

5.4.4 Photosynthesis Inhibitors

These are also known as PSI electron divertors. Herbicides of the bipyridilium family generate herbicide radicals which interact with oxygen resulting in the generation of superoxide radicals that further generate hydroxyl and hydrogen peroxide (H_2O_2) in the presence of enzyme superoxide dismutase (Lambreva et al. 2014; Roach and Krieger-Liszkay 2014).

5.4.5 Plant Growth Regulators

Herbicides such as pyridine carboxylic acid, quinoline carboxylic acid, phenoxycarboxylic acid, and benzoic acid mimic the activity of indole acetic acid which is reported for growth enhancement and development in plants. These herbicides activate the adenosine triphosphate (ATP)ase proton pump which accelerates the enzyme function in cell wall resulting in disruption of integrity of cell membrane and nucleic acid metabolism (Grossmann 2010).

5.4.6 Nitrogen Metabolism Inhibitors

They are also known as glutamine synthesis inhibitors. These pesticides include bialophos (phosphinic acids), and glufosinate inhibits the activity of glutamine synthetase (GS), which forms glutamine from ammonia and glutamate, although GS plays an important role in re-assimilating ammonia during respiration process in nitrogen metabolism. This shuts down the PSI and PSII by uncoupling photophosphorylation (Tan et al. 2006; Dragićević et al. 2013).

5.4.7 Pigment Inhibitors

They are also called as carotenoid inhibitors. This class includes phenoxybutanamides, amides, furanones, anilidex, isoxazole, pyridines, isoxazo-lidinone, and pyridiazinones which catalyze the formation of 4-hydroxyphenyl pyruvate dioxygenase (HPPD) enzyme which inhibits the function of phytoene desaturase enzyme.

Generation of unbound lipid radicals causes lipid peroxidation which leads to decrease in the level of carotenoids (Qin et al. 2007).

5.4.8 Seedling Growth Inhibitors

They are broadly classified into seedling root and seedling shoot growth inhibitors. They include families of phosphoramidate, dinitroaniline, thiocarbamates, benzoic acid phosphorodithioates, phosphoramidate, tetrazolinone, oxyacetamide, acetamide, chloroacetamide, pyridine benzamide, and dinitroaniline chemical family (Vencill et al. 2012). They inhibit the biosynthesis of gibberellins, flavonoids, isoprenoids, proteins, and fatty acids by conjugating with sulfhydryl-containing molecules and acetyl CoA. They also inhibit the microtubules polymerization which results in the non-separation of chromosomes and nonalignment of spindle fibers (Colovic et al. 2013).

5.5 Crop Tolerance of Herbicides

Herbicide tolerance is the ability of crops to withstand with herbicides sprayed on weeds. Herbicide-tolerant plants are designed and developed to resist action of herbicides (Schütte et al. 2017). Recent advancement in the field of genetic engineering developed certain mechanisms or changes in the genetic pool to develop resistivity against different herbicides (Qasem 2011). In 1996, Monsanto (Roundup) first introduced the herbicide-resistant soya bean and later on resistant corn, and they are commonly called Roundup Ready crops. These crops are either transformed by using mutated zm-2mepsps genes from corn or cp4 epsps gene from *Agrobacterium tumefaciens* strain CP4. Later on, 19 species of different weeds were found resistant against glyphosate threatening the regular use of GR crops (Pollegioni et al. 2011). The resistance has attributed to reduced cellular transport or translocation to the plastid, gene amplification, vacuole sequestration, and altered EPSPS target site. Glyphosate-resistant weeds minimize the advantages of glyphosate-based weed management systems and also increase cost for weed control (Green and Owen 2011). Herbicides lose all their value if a weed is resistant to it. Weed species developed genetic mutations when there is repeated use of the same herbicide and develop resistance. Herbicide resistance in weeds also increases the use of herbicide in that area (Schütte et al. 2017). Herbicides usually inhibit ACCase, protoporphyrinogen oxidase (PPO), and acetolactate synthase (ALS) activity of weeds, and continuous use of the same herbicide develops resistance mechanism against such herbicides and limit their utilities. Inactivation of metabolic systems based on glutathione transferase (GST) and cytochrome P450 monooxygenases (P450) has important resistance as traits against different herbicides (Salas et al. 2016).

Herbicide-resistant crops have benefitted the ecosystem by reducing dependence on soil tillage for effective weed management system and have better environmental and toxicological profiles (Chauhan et al. 2017).

5.6 Considerations in Herbicide Application

Various factors are reported to be considered during and before herbicide spray application. A well-defined consideration for a fruitful herbicide application saves effort, money, and time (Eure et al. 2013). Factors such as weed identification, weed morphology and herbicide application, weed height and vegetative mass, crop species/varieties, crop growth and herbicide tolerance, crop morphology, herbicide status and type, time of application, herbicide formulations, labels and technical pamphlets, herbicide resistance and selection pressure, sprayers and calibration are certain considerations which should be kept in mind before and during herbicide spray application (Beckie et al. 2019). Environmental factors such as moisture level, temperature, wind speed, spraying nozzles, and boom heights also influence the effectiveness of herbicide application. Annual weeds or grasses require contact herbicides, high solution discharge, high spray volume, and spray angle (Matzenbacher et al. 2014), while perennial weeds require systemic herbicides, which is absorbed and translocated to different plant organs for functioning. Crop yield is altered by habitation of weed species and dependence on weed species (Singh et al. 2011). The first step in weed control is the identification of the weed plant in that particular area. Weed morphology and herbicide selection are the second step as it is approved that narrow-leaved weeds retain less spray quantity of herbicides as compared to broad-leaved weeds (Qasem 2011). Uniform spraying is also a difficult task to kill large and tall-growing weeds. Selectivity of crop species in response to herbicide is also a challenging task as the weed species are different in emergence, germination, duration and development, biochemical and physiological responses. Growth injury in crops is commonly observed in the early stages of germination in herbicide-treated fields (Beckie et al. 2019).

Nonselection of herbicides is termed to those who have negative influence of plant physiological functions when sprayed at high rates. However, selectivity of herbicides depend on the formulation and rate of herbicide used, method and time of application, stages of crop plants and weeds, and environmental factors (Varshney et al. 2012).

In determining the length and effectiveness of herbicide duration, time of herbicide application plays a significant role. Herbicides are applied postemergence, pre-sowing, and pre-planting which have direct influence on crop recovery (Ashraf et al. 2018). The herbicides are available in different formulation such as wetting agents (S, SL, WL), water-soluble powders (SP), water emulsions (EC, E), water-dispersed liquids (L, F, WD), granules (G), water-dispersed granules (DF, WDG), wettable powders (WP, W), and pellets. The selectivity of weeds is also influenced by formulation of herbicides. It is also reported that ester forms, granules, and powders are less stable, while emulsion herbicides are less stable than water-soluble forms (Hazra and Purkait 2019).

Reading documents like technical bulletin and herbicide labels must be read carefully before herbicide application. Labels or technical bulletin reveals common name, herbicide trade, and also information of volume of spray per unit area, time of application, persistence, formulation, method of application, active ingredient, post-application treatments, weed control spectrum, and volatility. This information provides an insight to the farmers to strictly follow the operation of weed control system which would help in regulation of weeds with no toxic or environmental problems (Lombardo et al. 2016). Herbicide resistance or tolerance in weeds is the regular use of the same herbicide or same mechanism in same agricultural fields. Regular application of the same herbicides directly imposes selection pressure within the same species of the weed population to adapt themselves against biochemical, morphological, and physiological changes which occurs during herbicide application (Busi et al. 2013).

Failure of uniform distribution of herbicide application results in crop injuries, ineffective weed control, and is directly linked with sprayer calibration. The first step before herbicide application is calibration of spraying tools. It is carried out to check the volume of solvent needed to dissolve quantity of herbicide used for a defined area (Qasem 2011). Cleaning of spraying nozzles must be checked and ensured after every spray time while applying herbicides. The nozzle determines uniformity of spray pattern and spray application, spray discharge, and drift rate. It is also recommended to use the same kind and number of nozzles while spraying. Two types of nozzles are usually employed: The first one is flat fan-type nozzle and the second one is cone-type nozzle. The orifice of flat fan-type nozzles is straight flattened, while in cone-type nozzles, the second orifice is rounded. The flat fans are used in protection of crop products, while cone type is employed for low-drift purposes (Peterson et al. 2018).

5.7 Advancement in the Herbicide Application

Advancement in the field of agricultural technologies like weed mapping, mechanical weed management system, etc. could be an alternative tool to manage the weed-related problems (Korres et al. 2019). Although chemical herbicides are directly linked to various environmental hazards, these herbicides leach out directly in the water system and affect the ecosystem. They also impose negative effect on beneficial soil microflora (Mahmood et al. 2016). Chemical synthesized herbicides cause extensive pollution in ecosystem and serious health hazards in humans also. An alternative to synthetic herbicides is the use of "bioherbicides" (Soltys et al. 2013). Bioherbicides act as microbial phytotoxins or phytopathogenic microbes which are used to control weeds as replacement of conventional herbicides (Hoagland 2000). Bioherbicides have certain advantages over conventional herbicides as they are having high specificity against targeted weed, effectiveness in maintaining herbicide-resistant weed populations, no hazardous effect on nontarget plants, and no residual formation (Harding and Raizada 2015). With advancement in recombinant DNA technology, new arrays of highly effective bioherbicides for weed control are being developed. They are designed to work against the weed defenses (Duke et al. 2015), although the outer tissue coating of weeds contains waxy layer. Using biotechnology advancement, these microbes produce certain enzymes which are able to penetrate through these defenses to fully disintegrate into the weeds (Wright et al. 2010).

5.8 Factors Influencing the Uses of Herbicides

Herbicides must overcome environmental and biological barriers to control a targeted weed plant. Various environmental conditions such as moisture, relative humidity, temperature, rain following applications, foliar canopy factors, and irradiance, influence the use of herbicides during, after, and before applying application on soil surface or plants (Powles and Yu 2010). For effective performance of soil-applied herbicides, it must be readily available for uptake by shoots or roots of seedlings of weeds. Soil properties such as organic matter level, pH, and soil texture, should also affect the activity and availability of soil-applied herbicides (Jannat et al. 2019). Systemic herbicides must be having some qualities such as high absorption rate by plants, adequate contact with plants, reach to site of action without losing its ability. Entry for each herbicide through primary site is specific. For foliar-applied herbicides, crossing of the first barrier, that is, waxy layer of weed, is an important factor that influences efficacy of herbicides (Villalobos and Fereres 2016). Only some components of leaf absorb herbicides, and it includes cell-surrounding hairs, cell-overlaying veins, and guard cells. Once they enter inside the leaves, they either move to phloem and cell of plants. In soil-applied herbicides, they are first absorbed by roots, shoots, and seeds. They are absorbed by the plants in the same manner as plants absorb water and nutrient and move through other plants via xylem (Davis and Frisvold 2017).

Some herbicide movements are mobile and immobile within the plant system. Movement is primarily accelerated by phloem and xylem vascular systems. Translocation in the xylem is the primary means of herbicide absorption to above grounded parts, while phloem is for foliar-absorbed herbicides which is regulated by transpiration (Singh et al. 2016a, b). Long exposure to low humidity and moisture reduces herbicide penetration into plant by increasing the thickness of closing stomata and cuticles. During these conditions, surfactants are added to the herbicides to enter the plant stomata and cuticle under these conditions (Dammer 2016).

Both high and low temperature slow down the metabolism of plants and minimize the effectiveness of herbicides. At high temperatures, the herbicide may crystalize or either vaporize on leaf surface where it is removed by either rain or wind or may be photo-degraded (Swinton and Van Deynze 2017).

Metabolic pathways in a plant are regulated by process of photosynthesis. Reduction in photosynthesis often leads to lower down the action and translocation of herbicides (Kniss 2017). Some of the herbicides absorb rapidly by the plants, while some require a long time to penetrate. After applying herbicides, rain is also an important factor that can influence the action of herbicides. So, we have to keep in mind to read the instruction in the label of herbicide while applying herbicides (Eure et al. 2013).

5.9 Conclusion and Future Prospects

Excessive use of chemical and fertilizers develops resistance in undesirable plants and weeds becoming a worldwide issue. They must be eliminated or contained in order to increase the productivity of the harvest. So, there is need to develop robust and proactive herbicide management strategies to minimize the agricultural loss. Various new techniques can be extrapolated to ornamental, land management, and other industries to reduce the production of herbicide-resistant pests and weeds. It is recommended to use different herbicides having different modes of action at different intervals of the year with different varieties of crops so that the pests and weeds do not develop tolerance as quick as reported in various studies. It is also recommended to decide the eco-friendliest and effective herbicide for a specific crop to reduce the risks associated with the other chemical agents. Most of the herbicides are developed after rearranging chemical groups, which is not the best strategy as pests and weeds quickly develop resistance to such groups of chemicals. Therefore, there is need to design the herbicides with maximum effect of their mode of action against weeds and pests and least capacity to acquire the resistance.

References

Abbas T, Zahir ZA, Naveed M, Kremer RJ (2018) Limitations of existing weed control practices necessitate development of alternative techniques based on biological approaches. In: Advances in agronomy. Academic, St. Paul, pp 239–280

Ahmad KS (2017) Pedospheric sorption investigation of sulfonyl urea herbicide triasulfuron via regression correlation analysis in selected soils. S Afr J Chem 70:163–170. https://doi.org/10.17159/0379-4350/2017/v70a23

Aparecida M, de Campos Ventura-Camargo B, Miyuki M (2013) Toxicity of herbicides: impact on aquatic and soil biota and human health. In: Herbicides – current research and case studies in use. InTech, Rijeka

Ashraf U, Hussain S, Sher A et al (2018) Planting geometry and herbicides for weed control in rice: implications and challenges. In: Grasses as food and feed. IntechOpen. https://doi.org/10.5772/intechopen.79579

Au AM (2003) Encyclopedia of food sciences and nutrition. Elsevier, New York

Barchanska H, Sajdak M, Szczypka K et al (2017) Atrazine, triketone herbicides, and their degradation products in sediment, soil and surface water samples in Poland. Environ Sci Pollut Res 24:644–658. https://doi.org/10.1007/s11356-016-7798-3

Beckie HJ, Ashworth MB, Flower KC (2019) Herbicide resistance management: recent developments and trends. Plan Theory 8:161. https://doi.org/10.3390/plants8060161

Bhati S, Kumar V, Singh S, Singh J (2019) Synthesis, biological activities and docking studies of piperazine incorporated 1, 3, 4-oxadiazole derivatives. J Mol Struct 1191:197–205. https://doi.org/10.1016/j.molstruc.2019.04.106

Bonny S (2016) Genetically modified herbicide-tolerant crops, weeds, and herbicides: overview and impact. Environ Manag 57:31–48. https://doi.org/10.1007/s00267-015-0589-7

Busi R, Vila-Aiub MM, Beckie HJ et al (2013) Herbicide-resistant weeds: from research and knowledge to future needs. Evol Appl 6:1218–1221. https://doi.org/10.1111/eva.12098

Chauhan BS, Matloob A, Mahajan G et al (2017) Emerging challenges and opportunities for education and research in weed science. Front Plant Sci 8. https://doi.org/10.3389/fpls.2017.01537

Colovic MB, Krstic DZ, Lazarevic-Pasti TD et al (2013) Acetylcholinesterase inhibitors: pharmacology and toxicology. Curr Neuropharmacol 11:315–335. https://doi.org/10.2174/1570159x11311030006

Curran WS (2016) Persistence of herbicides in soil. Crop Soils 49:16. https://doi.org/10.2134/cs2016-49-0504

Dammer K-H (2016) Real-time variable-rate herbicide application for weed control in carrots. Weed Res 56:237–246. https://doi.org/10.1111/wre.12205

Davis AS, Frisvold GB (2017) Are herbicides a once in a century method of weed control? Pest Manag Sci 73:2209–2220. https://doi.org/10.1002/ps.4643

Dayan FE, Watson SB (2011) Plant cell membrane as a marker for light-dependent and light-independent herbicide mechanisms of action. Pestic Biochem Physiol 101:182–190. https://doi.org/10.1016/j.pestbp.2011.09.004

Dragićević M, Platiša J, Nikolić R et al (2013) Herbicide phosphinothricin causes direct stimulation hormesis. Dose-Response 11:344–360. https://doi.org/10.2203/dose-response.12-039.Simonovic

Duke SO, Scheffler BE, Boyette CD, Dayan FE (2015) Biotechnology in weed control. In: Kirk-Othmer encyclopedia of chemical technology. Wiley, New York, pp 1–25

Eure PM, Jordan DL, Fisher LR, York AC (2013) Efficacy of herbicides when spray solution application is delayed. Int J Agron 2013:1–7. https://doi.org/10.1155/2013/782486

Garcia Blanco FM, de Almeida SDB, Barifouse M (2013) Herbicide – soil interactions, applied to maize crop under Brazilian conditions. In: Herbicides – current research and case studies in use. InTech, Croatia

Green JM, Owen MDK (2011) Herbicide-resistant crops: utilities and limitations for herbicide-resistant weed management. J Agric Food Chem 59:5819–5829. https://doi.org/10.1021/jf101286h

Gregory E, Lyn A, Jason A, Sellers BA (2013) Herbicides for natural area weed management. In: Herbicides – current research and case studies in use. InTech, Croatia

Grossmann K (2010) Auxin herbicides: current status of mechanism and mode of action. Pest Manag Sci 66:113–120

Gupta PK (2011) Chapter 39 – herbicides and fungicides. Elsevier Inc., Boca Raton

Gupta PK (2018) Toxicity of herbicides. In: Veterinary toxicology: basic and clinical principles, 3rd edn. Elsevier, Boca Raton, pp 553–567

Harding DP, Raizada MN (2015) Controlling weeds with fungi, bacteria and viruses: a review. Front Plant Sci 6:659

Hazra DK, Purkait A (2019) Role of pesticide formulations for sustainable crop protection and environment management: a review. J Pharmacogn Phytochem 8:689–693

Hoagland RE (2000) Bioherbicides: phytotoxic natural products. ACS Symp Ser 774:72–90

Jannat N, Begum M, Salam A, Monira S (2019) Weed seed distribution in different depths of soil in a crop field. Fundam Appl Agric 1. https://doi.org/10.5455/faa.19969

Kapoor D, Singh S, Kumar V, Romero R, Singh J (2019) Antioxidant enzymes regulation in plants in reference to reactive oxygen species (ROS) and reactive nitrogen species (RNS). Plant Gene 19:100182. https://doi.org/10.1016/j.plgene.2019.100182

Kaundun SS, Hutchings S-J, Dale RP, McIndoe E (2013) Role of a novel I1781T mutation and other mechanisms in conferring resistance to acetyl-CoA carboxylase inhibiting herbicides in a black-grass population. PLoS One 8:e69568. https://doi.org/10.1371/journal.pone.0069568

Kaur S, Kumar V, Chawla M, Cavallo L, Poater A, Upadhyay N (2017) Pesticides curbing soil fertility: effect of complexation of free metal ions. Front Chem 5:1–9. https://doi.org/10.3389/fchem.2017.00043

Khatri N, Tyagi S (2015) Influences of natural and anthropogenic factors on surface and groundwater quality in rural and urban areas. Front Life Sci 8:23–39. https://doi.org/10.1080/21553769.2014.933716

Kniss AR (2017) Long-term trends in the intensity and relative toxicity of herbicide use. Nat Commun 8. https://doi.org/10.1038/ncomms14865

Korres NE, Burgos NR, Travlos I et al (2019) New directions for integrated weed management: modern technologies, tools and knowledge discovery. In: Advances in agronomy. Academic, New York, pp 243–319

Kraehmer H, Laber B, Rosinger C, Schulz A (2014) Herbicides as weed control agents: state of the art: I. weed control research and safener technology: the path to modern agriculture. Plant Physiol 166:1119–1131. https://doi.org/10.1104/pp.114.241901

Kubsad D, Nilsson EE, King SE et al (2019) Assessment of glyphosate induced epigenetic transgenerational inheritance of pathologies and sperm epimutations: generational toxicology. Sci Rep 9. https://doi.org/10.1038/s41598-019-42860-0

Kumar V, Singh S (2018) Interactions of acephate, glyphosate, monocrotophos and phorate with bovine serum albumin. Indian J Pharm Sci 80:1151–1155. https://doi.org/10.4172/pharmaceutical-sciences.1000467

Kumar V, Upadhyay N, Wasit AB, Singh S, Kaur P (2013) Spectroscopic methods for the detection of organophosphate pesticides – a preview. Curr World Environ 8:313–319. https://doi.org/10.12944/CWE.8.2.19

Kumar V, Upadhyay N, Kumar V, Kaur S, Singh J, Singh S, Datta S (2014) Environmental exposure and health risks of the insecticide monocrotophos-a review. J Biodivers Environ Sci 5:111–120

Kumar V, Upadhyay N, Kumar V, Sharma S (2015a) A review on sample preparation and chromatographic determination of acephate and methamidophos in different samples. Arab J Chem 8:624–631. https://doi.org/10.1016/j.arabjc.2014.12.007

Kumar V, Singh S, Singh J, Upadhyay N (2015b) Potential of plant growth promoting traits by bacteria isolated from heavy metal contaminated soils. Bull Environ Contam Toxicol 94:807–815. https://doi.org/10.1007/s00128-015-1523-7

Kumar V, Upadhyay N, Manhas A (2015c) Designing syntheses characterization, computational study and biological activities of silver-phenothiazine metal complex. J Mol Struct 1099:135–140. https://doi.org/10.1016/j.molstruc.2015.06.055

Kumar V, Kumar V, Upadhyay N, Sharma S (2015d) Interactions of atrazine with transition metal ions in aqueous media: experimental and computational approach. 3Biotech 5:791–798. https://doi.org/10.1007/s13205-015-0281-x

Kumar V, Kumar V, Kaur S, Singh S, Upadhyay N (2016) Unexpected formation of N-phenyl-thiophosphorohydrazidic acid O,S-dimethyl ester from acephate: chemical biotechnical and computational study. 3 Biotech 6:1–11. https://doi.org/10.1007/s13205-015-0313-6

Kumar V, Singh S, Singh R, Upadhyay N, Singh J, Pant P, Singh R, Shrivastava B, Singh A, Subhose V (2018a) Spectral, structural and energetic study of acephate, glyphosate, monocrotophos and phorate: an experimental and computational approach. J Taibah Univ Sci 12:69–78. https://doi.org/10.1080/16583655.2018.1451109

Kumar V, Chawla M, Cavallo L, Wani AB, Manhas A, Kaur S, Poater A, Chadar H, Upadhyay N (2018b) Complexation of trichlorosalicylic acid with alkaline and first row transition metals as a switch for their antibacterial activity. Inorg Chim Acta 469:379–386

Kumar V, Singh S, Upadhyay N (2019a) Effects of organophosphate pesticides on siderophore producing soils microorganisms. Biocatal Agric Biotechnol 21:101359. https://doi.org/10.1016/j.bcab.2019.101359

Kumar V, Singh S, Singh A, Subhose V, Prakash O (2019b) Assessment of heavy metal ions, essential metal ions, and antioxidant properties of the most common herbal drugs in Indian Ayurvedic hospital: for ensuring quality assurance of certain Ayurvedic drugs. Biocatal Agric Biotechnol 18:101018. https://doi.org/10.1016/j.bcab.2019.01.056

Kumar V, Singh S, Srivastava B, Bhadouria R, Singh R (2019c) Green synthesis of silver nanoparticles using leaf extract of holopteleaintegrifolia and preliminary investigation of its antioxidant, anti-inflammatory, antidiabetic and antibacterial activities. J Environ Chem Eng 7:103094. https://doi.org/10.1016/j.jece.2019.103094

Kumar V, Singh S, Bhadouria R, Singh R, Prakash O (2019d) Phytochemical, analytical and medicinal studies of holoptelea integrifolia roxb. Planch – a review. Curr Trad Med 5:270–277. https://doi.org/10.2174/2215083805666190521103308

Lambreva M, Russo D, Polticelli F et al (2014) Structure/function/dynamics of photosystem II plastoquinone binding sites. Curr Protein Pept Sci 15:285–295. https://doi.org/10.2174/1389203715666140327104802

Lombardo L, Coppola G, Zelasco S (2016) New technologies for insect-resistant and herbicide-tolerant plants. Trends Biotechnol 34:49–57
Lushchak VI, Matviishyn TM, Husak VV et al (2018) Pesticide toxicity: a mechanistic approach. EXCLI J 17:1101–1136
Mahmood I, Imadi SR, Shazadi K et al (2016) Effects of pesticides on environment. In: Plant, soil and microbes, Volume 1: implications in crop science. Springer, Cham, pp 253–269
Matzenbacher FO, Vidal RA, Merotto A, Trezzi MM (2014) Environmental and physiological factors that affect the efficacy of herbicides that inhibit the enzyme protoporphyrinogen oxidase: a literature review. Planta Daninha 32:457–463. https://doi.org/10.1590/S0100-83582014000200024
Mensah PK, Palmer CG, Odume ON (2015) Ecotoxicology of glyphosate and glyphosate-based herbicides – toxicity to wildlife and humans. In: Toxicity and hazard of agrochemicals. InTech, Rijeka
Nicolopoulou-Stamati P, Maipas S, Kotampasi C et al (2016) Chemical pesticides and human health: the urgent need for a new concept in agriculture. Front Public Heal 4. https://doi.org/10.3389/fpubh.2016.00148
Nwani CD, Lakra WS, Nagpure NS et al (2010) Toxicity of the herbicide atrazine: effects on lipid peroxidation and activities of antioxidant enzymes in the freshwater fish *Channa punctatus* (Bloch). Int J Environ Res Public Health 7:3298–3312. https://doi.org/10.3390/ijerph7083298
Peterson MA, Collavo A, Ovejero R et al (2018) The challenge of herbicide resistance around the world: a current summary. Pest Manag Sci 74:2246–2259. https://doi.org/10.1002/ps.4821
Pimentel D, Lach L, Zuniga R, Morrison D (2000) Environmental and economic costs of nonindigenous species in the United States. Bioscience 50:53. https://doi.org/10.1641/0006-3568(2000)050[0053:eaecon]2.3.co;2
Pollegioni L, Schonbrunn E, Siehl D (2011) Molecular basis of glyphosate resistance – different approaches through protein engineering. FEBS J 278:2753–2766
Powles SB, Yu Q (2010) Evolution in action: plants resistant to herbicides. Annu Rev Plant Biol 61:317–347. https://doi.org/10.1146/annurev-arplant-042809-112119
Prasad R, Upadhyay N, Kumar V (2013) Simultaneous determination of seven carbamate pesticide residues in gram wheat lentil soybean fenugreek leaves and apple matrices. Microchem J 111:91–97. https://doi.org/10.1016/j.microc.2012.12.014
Qasem JR (2011) Herbicides applications: problems and considerations. In: Kortekamp A (ed) Herbicides and environment. InTech, Rijeka, pp 643–664
Qin G, Gu H, Ma L et al (2007) Disruption of phytoene desaturase gene results in albino and dwarf phenotypes in Arabidopsis by impairing chlorophyll, carotenoid, and gibberellin biosynthesis. Cell Res 17:471–482. https://doi.org/10.1038/cr.2007.40
Roach T, Krieger-Liszkay A (2014) Regulation of photosynthetic electron transport and photoinhibition. Curr Protein Pept Sci 15:351–362. https://doi.org/10.2174/138920371566614032710 5143
Salas RA, Burgos NR, Tranel PJ et al (2016) Resistance to PPO-inhibiting herbicide in Palmer amaranth from Arkansas. Pest Manag Sci 72:864–869. https://doi.org/10.1002/ps.4241
Sandmann G, Misawa N, Böger P (2018) Steps towards genetic engineering of crops resistant against bleaching herbicides. In: Herbicide- resistant crops. CRC Press, Boca Raton, pp 189–200
Scherner A, Schreiber F, Fomsgaard IS et al (2018) Effect of tillage systems on the dissipation of prosulfocarb herbicide. Weed Technol 32:195–204. https://doi.org/10.1017/wet.2017.103
Schütte G, Eckerstorfer M, Rastelli V et al (2017) Herbicide resistance and biodiversity: agronomic and environmental aspects of genetically modified herbicide-resistant plants. Environ Sci Eur 29:5
Sherwani SI, Arif IA, Khan HA (2015) Modes of action of different classes of herbicides. In: Herbicides, physiology of action, and safety. InTech, Rijeka
Shukla A, Devine MD (2008) Basis of crop selectivity and weed resistance to Triazine herbicides. In: The Triazine herbicides. Elsevier, Amsterdam, pp 111–118

Sidhu GK, Singh S, Kumar V, Datta S, Singh J (2019) Environmental toxicity, monitoring and biodegradation of organophosphate pesticides: a review. Crit Rev Environ Sci Technol 49:1135–1187. https://doi.org/10.1080/10643389.2019.1565554

Singh M, Sharma SD (2008) Benefits of Triazine herbicides and other weed control technology in citrus management. In: The Triazine herbicides. Elsevier, Amsterdam, pp 199–209

Singh RP, Singh RK, Singh MK (2011) Impact of climate and carbon dioxide change on weeds and their management – a review. Indian J Weed Sci 43:1–11

Singh S, Singh N, Kumar V, Datta S, Wani AB, Singh D, Singh K, Singh J (2016a) Toxicity monitoring and biodegradation of the fungicide carbendazim. Environ Chem Lett 1:41–13. https://doi.org/10.1007/s10311-016-0566-2

Singh VP, Dhyani VC, Singh SP et al (2016b) Effect of herbicides on weed management in dry-seeded rice sown under different tillage systems. Crop Prot 80:118–126. https://doi.org/10.1016/j.cropro.2015.09.018

Singh S, Kumar V, Upadhyay N, Singh J, Singla S, Datta S (2017) Efficient biodegradation of acephate by Pseudomonas pseudoalcaligenes PS-5 in the presence and absence of heavy metal ions [Cu(II) and Fe(III)], and humic acid. 3 Biotech 7(2017):262

Singh S, Kumar V, Chauhan A, Datta S, Wani AB, Singh N, Singh J (2018) Toxicity, degradation and analysis of the herbicide atrazine. Environ Chem Lett 16:211–237

Singh S, Kumar V, Singh J (2019a) Kinetic study of the biodegradation of glyphosate by indigenous soil bacterial isolates in presence of humic acid, Fe(III) and Cu(II) ions. J Environ Chem Eng 7:103098. https://doi.org/10.1016/j.jece.2019.103098

Singh S, Kumar V, Singh S, Singh J (2019b) Influence of humic acid, iron and copper on microbial degradation of fungicide Carbendazim. Biocatal Agric Biotechnol 20(2019):101196. https://doi.org/10.1016/j.bcab.2019.101196

Singh S, Kumar V, Sidhu GK, Datta S, Singh J (2019c) Plant growth promoting rhizobacteria from heavy metal contaminated soil and their plant growth promoting attributes for Pisumsativum L. Biocatal Agric Biotechnol 17:665–671. https://doi.org/10.1016/j.bcab.2019.01.035

Soltys D, Krasuska U, Bogatek R, Gniazdowsk A (2013) Allelochemicals as bioherbicides — present and perspectives. In: Herbicides – current research and case studies in use. InTech, Rijeka

Swinton SM, Van Deynze B (2017) Hoes to herbicides: economics of evolving weed management in the United States. Eur J Dev Res 29:560–574. https://doi.org/10.1057/s41287-017-0077-4

Tan S, Evans R, Singh B (2006) Herbicidal inhibitors of amino acid biosynthesis and herbicide-tolerant crops. Amino Acids 30:195–204

Varshney S, Hayat S, Alyemeni MN, Ahmad A (2012) Effects of herbicide applications in wheat fields: is phytohormones application a remedy? Plant Signal Behav 7:570

Vats S (2015) Herbicides: history, classification and genetic manipulation of plants for herbicide resistance. Springer, Cham, pp 153–192

Vencill WK, Nichols RL, Webster TM et al (2012) Herbicide resistance: toward an understanding of resistance development and the impact of herbicide-resistant crops. Weed Sci 60:2–30. https://doi.org/10.1614/ws-d-11-00206.1

Villalobos FJ, Fereres E (2016) Application of herbicides and other biotic control agents. In: Principles of agronomy for sustainable agriculture. Springer, Cham, pp 471–485

Wright TR, Shan G, Walsha TA et al (2010) Robust crop resistance to broadleaf and grass herbicides provided by aryloxyalkanoate dioxygenase transgenes. Proc Natl Acad Sci U S A 107:20240–20245. https://doi.org/10.1073/pnas.1013154107

Yamaji Y, Honda H, Hanai R, Inoue J (2016) Soil and environmental factors affecting the efficacy of pyroxasulfone for weed control. J Pestic Sci 41:1–5. https://doi.org/10.1584/jpestics.D15-047

Yang X, Guschina IA, Hurst S et al (2010) The action of herbicides on fatty acid biosynthesis and elongation in barley and cucumber. Pest Manag Sci 66:794–800. https://doi.org/10.1002/ps.1944

Yu Q, Powles SB (2014) Resistance to AHAS inhibitor herbicides: current understanding. Pest Manag Sci 70:1340–1350. https://doi.org/10.1002/ps.3710

Insecticides Derived from Natural Products: Diversity and Potential Applications

6

Preeti Mishra, Arpita Tripathi, Anupam Dikshit, and Anand Pandey

Abstract

Termites are one among the pests that pose serious threat to live plants as well as cause harm to plant-based products. The urban environment and agriculture are most threatened by them as they are the most challenging pests. The annual and perennial crops face significant losses, and in the semiarid and subhumid tropics around the globe, wooden components are also destroyed by them. Among the successful methods, chemical control is of great importance, but the side effects of such chemicals lead to the production of serious health issues, especially respiratory problems, if present in the environment. Plant-based natural products are promising replacements for these chemical pesticides. Botanicals used for pest remediation especially the one derived from essential oils are target-specific. A safe environment and food free from residues are attained by their use as they are required in little amount and also possess a quick decomposition. The mode of action of these herbal pesticides may vary. Some may act as direct toxicant, while others might act upon as antifeedant, repellent or behavior modifiers, morphogenetic agents, or phagostimulants. The current chapter aims to discuss the different aspects of various natural herbal plants parts and their potential application in the management of pest particularly termites. A sneak peek to their mechanism of action will also be taken into consideration. Essential oil-bearing plants being abundantly grown for various purposes become a possible herbal alternate to chemical pesticides.

P. Mishra · A. Tripathi · A. Dikshit (✉)
Biological Product Laboratory, Department of Botany, University of Allahabad, Prayagraj, Uttar Pradesh, India

A. Pandey (✉)
Department of Plant Sciences, Avvaiyar Government College for Women, University of Puducherry, Karaikal, Puducherry, India

J. Singh, A. N. Yadav (eds.), *Natural Bioactive Products in Sustainable Agriculture*, https://doi.org/10.1007/978-981-15-3024-1_6

Keywords
Termite · Agriculture · Tropics · Pesticides · Antifeedant · Repellent · Phagostimulant

6.1 Introduction

The substances that function as a pest-controlling agent are commonly known as insecticides or pesticides. They are very beneficial as they provide the protection to crops, food material preservation, and the restriction of diseases caused by the vectors. They are also used in different fields like agriculture, forestry, medicine, aquaculture, food industry, processing, transportation, and storage of wood and other biological products. Gross use of insecticide causes damage to public health and ecosystem. Indiscriminate use of pesticides induced resistance, resurgence, and residues in harvested produce and adverse impact on other organisms in the ecosystem including human beings. This resulted in the need for the development of ecologically safer chemicals with target specificity. The use of chemical pesticides is manifold in developing countries as in developed countries. In recent years, the disadvantages of the chemicals use in agriculture have been noticed. Therefore, a search for various efforts to find out the alternatives for chemical pesticides has been started throughout the world. The use of eco-friendly technologies for the production as well as protection of crops is necessary for sustainable agriculture. The renewed interest in environment and eco-friendly technologies has widened the scope for the use of biopesticides or natural insecticide, including plant derivatives and biocontrol agents. The ultimate aim is to get pesticide residue-free food for human consumption within the country and for export purposes.

In the agriculture and urban environment, termites are one of the most troublesome pests. Globally, approximately 2800 termite species are identified. In India, out of 300 species, 35 are known for their damaging effects (Lewis 1997). The highly susceptible crops of various regions of India have been recorded for the severe losses (Rajagopal 2002). The preventive measure that is commonly used to minimize the termite infestation is achieved by the means of chemical control. Across the world, various termiticides are known under different brand names such as spinosad, disodium octaborate tetrahydrate (DOT), calcium arsenate, chlorpyriphos, etc. Though termite protection by chemical means is a proven method, their extensive application is hazardous. For the termite control, researchers are seeking for new ways. Plant-derived natural products/botanicals possess low mammalian toxicity, no risk of developing pest resistance, less hazard to non-target organisms, and are less expensive and easily available. Biomass of certain plants possesses the insecticidal activity that can be exploited for the controlling termites.

6.2 Termite Biology

Termites belong to the group *Isoptera* (Meyer 2005). The large colonies of termites depend either on wood entirely or on the entire or somewhat decomposed woody tissues. Termite colony basically comprises reproductively active termites, soldiers, sterile workers, and the young ones. The active reproductive termites are categorized into two main types: primary and supplementary. The king and queen (primary reproductive) are adults with pigmented and fully developed wings. They are responsible for production of egg and distribution by the means of colonizing flights. About 3000 eggs per day are laid down by the queen by the aid of an enlarged abdomen (Thompson 2000). The life span of the queen is approximately 25 years. The eggs are whitish-yellow colored, and after the incubation period of 50–60 days, they usually hatch. There is only one pair of primary reproductive in most of the colonies, but after their death replacement is usually carried out by the supplementary reproductive which are more pigmented and possess than workers. The absence of eyes and wings is generally found in sterile castes; the soldiers and workers are wingless and generally deficient in eyes (Myles 2005). Worker and soldier termites are 6 mm long and pale cream colored; the head is nearly half their body length with visible black jaws in the case of soldiers. The distinctive shelter tubes responsible mainly for damage are made by the workers and food collection for the members of the colony. In order to obtain certain secretions, they usually groom each other. The numbers of individuals in different castes are regulated by these secretions (Philip 2004). The maturation of workers takes around a year with a life span of 3–5 years. The maturation of soldiers occurs within a year and has 5 years of a life span (Myles 2005). In the month of April and May, winged reproductive (alates) appears in a mass nuptial flight. The first indication of termite infestation is shown by these flights (Philip 2004). The wings are shed by alates after a short period of flight. Nesting sites are searched by the females along with the males closely behind. The moist crevice with wooden materials is often selected by the pair for the formation of royal chamber and egg laying (Su and Scheffrahn 2000).

6.3 Termite Management

Termites are the most dominant social insects forming termitarium for large colonies. They mainly feed on wood or other cellulose-containing material as food source. It is a devastating pest of agricultural crops and forest. They are soft-bodied creatures, having pale cream complexion, with mouth parts used for biting and chewing to take cellulose as a food source; they belong to group of *Isopterans*. The termite colony consists of winged (king and queen) and wingless (soldiers, sterile workers, and immature individuals) reproductive forms. Pesticides are natural or synthetic agents which synthesize to kill or repel insects, weeds, rodents, fungi, and other organisms. United Nations Organization for Food and Agriculture (FAO) defines pesticides as a mixture of substances used for growing, destroying, and preventing any kind of pests, including vectors of human disease or animal disease. They are also used for controlling unwanted species of plants or animals that cause damage or are otherwise inquisitive in the production, processing, storage, or hampering transport and marketing facility. Agricultural products, wood and wood products, or animal feedstuffs or which may be administered to animals can be prevented from insects, arachnids, or other pests in or on their bodies (FAO Report 2018).

Although, termites have shown considerable contribution to most of the world's ecosystem diversity. Termites are of the most important species in recycling woody and other plant material. Their tunneling helps to aerate soils. Because of this activity, they improve soil composition and aeration fertility. Instead they turn to be a serious pest in urban ecosystem; they attack wooden dwellings and crops. Therefore, efficient methods are adopted to control them. Different control methods have been adopted and fundamentally accepted worldwide, and some of the novel methods are discussed further.

6.4 Physical Control Methods

There are several physical methods that can be used around the building's foundation to prevent the entry of termites; these can be classified into two classes: toxic and nontoxic. Toxic physical barrier: Chlorfenapyr is potent toxic chemical which shows delayed toxicity and is non-repellent and is frequently used for treatment of termite which is used in soil treatment before construction (Rust and Saran 2006).

Nontoxic physical barrier: It mainly includes a mechanical barrier which prevents termites to penetrate through them. Commonly used mechanical barriers are metal mesh or sheeting, sand or gravel aggregates, etc. Basaltic termite barrier (BTB) is made up of gravel, widely used in Hawaii before construction for preventing termite penetrance (Grace 1996). Many different forms of mechanical barriers such as nongraded stone products have become popular in some continents like Australia, Hawaii, Texas, and other parts of the USA in recent years. It includes solid steel sheets, stainless steel mesh, polymer sheets, and copper shields (Potter 2004; Baker 2005).

Australia developed and patented Termimesh®, a stainless steel wire mesh sized 0.66 × 0.45 mm which is a noncorrosive bendable steel mesh. It is planted within the walls of the foundations and in various field trials for effective prevention of termites (Lenz and Runko 1994; Grace 1996; Kard 1999). Various physical treatments too come under physical methods such as heating, freezing, electrical method, electromagnetic waves, and microwaves.

Heat treatment: In this method, termites are killed by using nylon tarps which are used to cover buildings, and then the temperature is raised to 45 °C (120 F) for a time of 35 min and at 50 °C (130 F) for 1 h (Myles 2005). *Cryptotermes brevis* has been controlled using this method, experiment performed by Woodrow and Grace (1998). Freezing treatment termites can be killed by low temperature by pumping liquid nitrogen (−20 °F) into the infected area.

Electrical treatment: Electro-Gun is used to give electric shock to the termite infected area of wooden block or board, having low current (~0.5 amps), high voltage (90,000 volts), and high frequency (60,000 cycles) (Myles 2005).

6.5 Chemical Methods

Control of termites by chemical methods is the most efficient and commonly used solution so far. Bifenthrin, chlorfenapyr, cypermethrin, fipronil, imidacloprid, permethrin spinosad, disodium octaborate tetrahydrate, calcium arsenate, and chlorpyriphos are the variety of termiticides with active constituents which are registered for termite control all over the world.

Bifenthrin, chlorpyrifos, endosulfan, imidacloprid, and lindane are some common insecticides which are at present being used for controlling termites in stored wood and also for crops (Su et al. 1999). Smith and Rust (1990) reported that cypermethrin, bifenthrin and permethrin, chlorpyrifos (≥50 ppm), fenvalerate (<100 ppm) and isofenphos, chlordane, bifenthrin, cypermethrin, and permethrin formulations are used to treat soil and are the most effective chemical insecticides. High concentrations of cypermethrin (100 ppm), imidacloprid (1000 ppm), and fenvalerate (500 ppm) were found to be very helpful against subterranean termites (Kuriachan and Gold 1998). *Trinervitermes trinervius*, *Odontotermes meathmani*, and *Amitermes lucifer* were observed to cause lesser damage in cotton, rice, maize, sorghum, and sugarcane crop with the application of insecticides, namely, chlorpyriphos, lindane, or thiamethoxam (Ahmed et al. 2006; Bhanot and Singal 2007). Chlorpyriphos was most effective among all the insecticides studied against subterranean termite in sugarcane crop. Wheat, barley, and gram have been standardized by seed treatment with chlorpyrifos, endosulfan, formothion, and monocrotophos (Bhanot et al. 1991a, b, 1995). Chromated copper arsenate is used to treat southern yellow pine and radiata pine for wood preservation from attack of Formosan subterranean termites, *Coptotermes formosanus* (Grace 1998).

Subterranean termites are controlled by applying soil termiticide injection. It is done by drilling the groundwork wall/slab and injecting the termiticide under wall and also into the soil which is in contact with the base. Chemical fumigation is one of the best methods for controlling dry-wood termite infestation. Chemical fumigants used in this method are sulfuryl fluoride, methyl bromide, and carbon dioxide. Firstly, tent is used to cover the whole building, and after that fumigant is pumped in the construction; then the tent is separated, and pumping fumigants is also stopped. Those materials are also removed which absorb chemicals and allow to be fumigated (Myles 2005).

Carbon dioxide (asphyxiant), methyl bromide, phosphine, and sulfuryl fluoride (metabolic poison) are active ingredients present in a variety of fumigants; among all most common fumigant used is methyl bromide. Methyl bromide is a highly toxic gas, marking the concerns which involve atmospheric ozone layer, having aroma in the domestic materials treated with this, and also the long ventilation time for fumigants makes it inadequate for use (UNEP Report 2000). Photo-immobilizing the bifenthrin-embedded chitosan on the wooden exterior is the eco-friendly method for preventing termite growth which has been given by Guan et al. (2011). They united the efficacy of bifenthrin against termites and used chitosan as a carrier to insert bifenthrin in photo-immobilization technique and gave ultraviolet treatment on the wooden surface for immobilization. Immobilized bifenthrin shows very high competence in opposition to termite at a dose of 2.5 mg/cm^2 which gives long-term stability and protection.

Biological resistant method has been studied against *Anacanthotermes vagans* termite. Copper- and zinc-salicylate were used to treat wood particles, and then it is exposed to termites. Minimum lightness was recorded in the specimen treated with copper-salicylate (4.7%), whereas maximum lightness was recorded in control (16.3%). The result established considerable defense of copper salicylate against termite. It showed that copper-salicylate can be used for prevention against *A. vagans* termite in industries (Bayatkashkoli et al. 2016).

The most effective method for controlling termite growth is chemical methods. However, their extreme use results in side effects which are toxic in nature, pesticide remaining in soil and water, and pest resistance having adverse effect on human and environment. Alternative methods can be used to substitute and lessen the use of chemical insecticides being practiced worldwide for termite management. Eco-friendly and new methods are being developed by the researchers to prevent termite growth. Some unusual methods for termite control are produced from natural products derived from plants, bacteria, nematodes, and some entomopathogenic fungi. Botanicals which are used as termiticides are discussed with other biological methods.

6.6 Diversity and Applications of Natural Insecticides for Termite Control

A number of efforts have been conceded toward the attainment of controlling insect growth with the help of biopesticides and green pesticides which means naturally occurring pesticides having partial or no harmful effect on the surroundings and non-targeted organisms including humans (Stevenson et al. 2017; Benelli 2018; Fig. 6.1).

6.6.1 Biological Control Strategies

Plants are used as latent alternatives to control insects as they comprise an immense supply of bioactive chemical components. More than 2000 plants belonging to 60 plant families are recognized to have insecticidal actions (Dev and Koul 1997). Insecticides, ecdysones, insect growth regulators (IGRs), juvenile hormones, anti-feedants, attractants, repellents, arrestants, etc., are present as bioactive chemicals, showing that plants can be an important substitute for chemical pesticides (Kannaiyan 1999). Natural products and biocontrol methods derived from plants show potential replacement to chemical control. Initiative for optional strategies to control termites was started in 1935, where citrollenellic acid was used to block cellulose blocks to check beside termites (Trikojus 1935). Different parts of plants as leaves, fruit, bark, root, flower, stem, resin etc., as well as their extracts, were tested

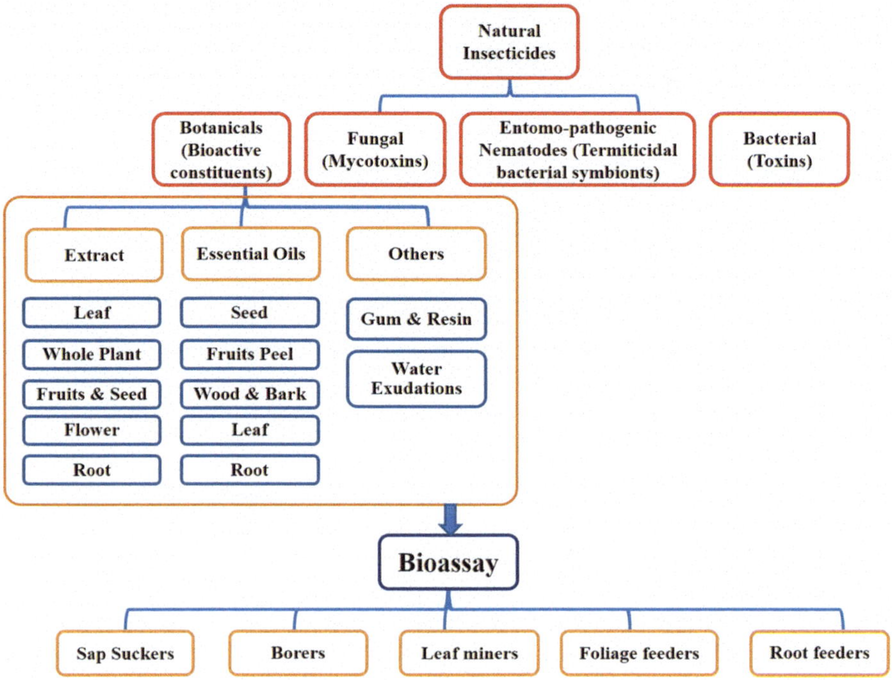

Fig. 6.1 Development of different natural insecticides for termite control and their bioassay

by many scientists worldwide to take advantage of their pest control competence. The botanicals used in this purpose have low cost and are locally present; they are highly effective in termite control. Laboratory and field studies have renowned studies showing broad range of plants with toxic and repulsive and antifeedant properties in opposition to termites; numerous have been well studied in order to use as insecticides (Stoll 1986; Gerrits and Van Latum 1988; Harborne 1988).

In fact botanical pesticides are being used in Indian agriculture for more than a century in order to reduce fatalities that are caused by different pests and diseases (Prakash et al. 1990; Parmar and Devkumar 1993). Botanical pesticidal formulations are much advantageous than synthetic pesticides, which include the following:

1. Pesticides obtained from botanicals add up to least or no health issue and ecological pollution as they have low mammalian toxicity.
2. If the products are used in its innate form, then there is no use to develop pest resistance.
3. They are least hazardous toward nontargeting organisms, and only synthetic pyrethroids have been reported as effective against pests.
4. It has undesirable effects on plant growth along with viability of seeds and therefore on cooking of grains.
5. Botanical pesticidal formulations are cheap and accessible because they are available naturally in oriental countries. Such are neem, bel, senwar, pyrethrum,

tobacco, karanj, mahua, sweet flag etc., which have previously attained the position of potential pesticides to be used as IPM in crop fields against insects as well as in storage ecosystems (Prakash and Rao 1997).

6.6.2 Mechanism of Botanical Action on Termites

The mechanism of botanical action of any termiticidal component depends on a variety of factors, which mainly include chemical composition, mode of entry of insecticide into the body of the insect, mode of action, toxicity, and stage specificity.

6.6.3 Essential Oils

Essential oils being volatile in nature possess powerful aromatic attribute, characteristic odor, and taste. Plants usually have generally lower density than that of water. Plants produce secondary metabolites that are by-products of plant metabolism which contains volatile oils; those are lipophilic (fat soluble) in nature found in glandular hairs of plant cell walls, and they are present in different parts such as droplets of fluid in the bark, leaves, stems, flowers, roots, fruits, etc. They are found in different compositions such as monoterpenes, phenols, and sesquiterpenes which are complex mixtures of plant secondary metabolites.

Essential oils have many characteristic properties like insecticidal in nature, oviposition deterrents, antifeedants, repellents, growth regulation, and antivector against many disastrous pests and plant pathogenic fungi (Koul et al. 2008). Table 6.1 enlists the termiticidal properties of various plant essential oils derived from secondary metabolites. Nakashima and Shimizu (1972) evaluated insecticidal activity of essential oil. Isman (2000) concluded that there are some plants containing essential oils not only used as repellants toward insects but also have fumigating insecticidal activities against specific pest.

Antitermite activity was evaluated in fruit essential oil of *Myristica fragrans* against *Microcerotermes beesoni*, in which fruit essential oil contains LC50 value 28.6 mg/g and the compound studied *myristicin* caused 100% mortality against termites at a dosage of 5 mg/g after 14-day experiment (Pal et al. 2011).

6.6.4 Plant Resin

Tree species of *dipterocarp* (two-winged fruit tree) timbers are familiar to be immune from attack by pests. High resistance was shown by *Shorea robusta* (sal tree) against the termite species *Microcerotermes beesoni* and *Heterotermes indicola* (Sen-Sarma 1963; Sen-Sarma and Chatterjee 1968). Particle boards and other furniture items which are made from sal wood were also protected from *Cryptotermes cynocephalus* (Moi 1980). Due to this resistance property of

Table 6.1 Plant essential oil from various plant parts used in termite control

S. no.	Plant name	Plant part	Active compound	Termite	Effect	References
1.	*Azadirachta indica*	Seed	Limonoid	*Reticulitermes speratus*	Antifeedant	Serit et al. (1992)
2.	*Taiwaniacryptomeriodes*	Wood	Cedroland α-cadinol	*Coptotermus formosanus Shiraki*	Toxic	Chang et al. (2001)
3.	*Tagetes erecta*	Leaf	(z)-Ocimene	*Odontotermes obesus*	Mortality	Singh et al. (2002)
4.	*Citrus*	Peel	d-Limonene	*C. formosanus*	Toxic	Raina et al. (2007)
5.	*Moneses uniflora*	Aerial part	Naphthoquinones, 2,7-dimethyl-1,4-napthoquinone & 3 hydroxy2,7 dimethyl-1,4 naphthoquinone	*C. formosanus*	Toxic	Kobaisy et al. (2001)
6.	*Piper nigrum*	Seed	Guineensine	*C. formosanus*	Toxic	Meepagala et al. (2006)
7.	*Kalopanax septemlobus*	Wood	Saponins	*C. formosanus*	Toxic	Saeki et al. (1970)
8.	*Cryptomeria japonica*	Wood	Wood vinegar	*R. speratus*	Toxic	Yatagai et al. (2002)
9.	*Ephedra distachya*	Areal part	Ethyl benzoate benzaldehyde	*C. formasanus*	Toxic	Chang and Cheng (2002)
10.	*Taxodium distichum*	Wood	Ferruginol, manool, nezukol	*C. formosanus*	Toxic	Scheffrahn et al. (1997)
11.	*Thujopsis*	Wood	B-Thujaplicin and carvacrol		Toxic	Nakashima and Shimizu (1972)
12.	*Chamaecyparis pisifera*	Wood	Chamaecyone and isochamaecyone	*Coptotermus formosanus*	Toxic	Saeki et al. (1973)
13.	*Cryptomeria japonica*	Wood	β- Eudesmol and cedrol	–	Toxic	Yatagai et al. (1991)
14.	*Vetiveria zizanioides*	Root	Nootkatone (a sesquiterpene, alcohol, and cedrene)	*C. formosanus*	Arrestants, antifeedant repellent, and toxic	Maistrello et al. (2001), Zhu et al. (2001) Maistrello et al. (2003), Nix et al. (2006)
15.	*Tagetes erecta*	Leaf	(Z)-Ocimene	*Odentermes obesus*	Mortality	
16.	*Lepidium meyenii*	Leaf	Benzylthiocynate 3-methoxyphenylace-tonitrile and β ionone	*C. formosanus*	Feeding deterrent	Tellez et al. (2002)

(continued)

Table 6.1 (continued)

S. no.	Plant name	Plant part	Active compound	Termite	Effect	References
17.	*Lantana camara var. aculeate*	Leaves	Triterpenoid, 22 β-acetoxylantic acid	*O. obesus*	Toxic	Verma and Verma (2006)
18.	*Jatropha curcas*	Seed oil	Anacardic acid, cardanol, methyl anacardate	*Microcerotermes beesoni*	Toxicity	Singh and Kumar (2008)
19.	*Ocimum canum*	Whole plant	Alkaloids, matrine, and oxymatrine	*C. formosanus Shiraki*	Antifeedant, repellent	Owusu et al. (2008)
20.	*Ocimum gratissimum*	Whole plant	Alkaloids, matrine, and oxymatrine	*C. formosanus Shiraki*	Antifeedant, repellent	Owusu et al. (2008)
21.	*Protium javonica*	Leaves	Scopoletin	*C. formosanus*	Toxic	Adfa et al. (2010)
22.	*Xylopia aethiopica*	Fruits and seeds	Diterpenes and amides	*R. speratus*	Antifeedant	Lajide et al. (1995)
23.	*Calotropis procera*	Leaves	Monoterpenes and sesquiterpenes	*Reticulitermes* sp.	Toxic	Tellez et al. (2001)
24.	*Sophora flavescens Aiton*	Seed	Alkaloid, matrine, and oxymatrine	*C. formosauns Shiraki*	Antifeedant and acute residual toxicity	Mao and Henderson (2007)
25.	*Ganophyllum falcatum*	Wood	o- Methoxycinnamaldehyde and torreyal	*C. formosauns*	Toxic	Yazaki (1982)
26.	*Pometia pinnata*	Wood	Saponins	*R. flavipes*, *R. virginicus*, and *C. formosauns*	Toxic	McDaniel (1989)
27.	*Ternstroemia japonica*	Wood extract	Barrigenol glycoside (saponin)	*C. formosauns*	Toxic	Saeki et al. (1968)
28.	*Myracrodruon urundeuva*	Wood extract	Lectin	*Nasutitermes corniger*	Repellent and termiticidal	Sa et al. (2008)
29.	*Sextonia rubra*	Wood	Rubrynolide	*R. flavipes*	Toxic	Rodrigues et al. (2011)

dipterocarp woods, it causes significant increase in mortality rates of insects feeding on them.

Moi (1980) observed in a 3-month test period on sal species that 86% to 99% mortality occurred when termites fed on this plant, while with non-dipterocarp *Dyera costulata*, only a 13% death rate was recorded. Wood resins contain an array of terpenoids which have been found in the Palaeotropical plant family Dipterocarpaceae (Bisset et al. 1966, 1971; Diaz et al. 1966). Bee nests have been protected from termites due to the presence of fresh resins of *Anisoptera thurifera* Purified fractions of *Dipterocarpus kerrii* crude resin containing composition of four sesquiterpenoids, which are chemically similar to α-gurjunene, are responsible for termiticidal action against *Zootermopsis angusticollis.*

6.6.5 Termite Control by Fungi

Pathogenic fungal species were the first cosmopolitan organisms to be used as biocontrol agents for insect pests. They are the most satisfying alternative to chemically control pests. The entomopathogenic fungal species play a considerable role in integrated pest management (IPM) (Carrunthus et al. 1991). The actual meaning of "entomogenous" means arising in insects. Fungus that attacks houseflies was first explained by De Geer (1776) and De Geer (1782). More than 750 species (56 genera) of fungus are known to be pathogenic toward insects, many of which possess great probability toward pest management. It is the pathogenic capability of a fungus to perforate the cuticle lining of insect and compromise the insect immune system. In order to prevent such fungal attack, the insect shows complex relationship with fungal growth. Huxham et al. (1989) suggested that specific pathogenic strains of fungal species attacks particular host species; it will not show the same growth pattern and pathogenicity in another insect species.

Stranes et al. (1993) reported that fungal species of both tropical and temperate regions *Beauveria bassiana* (Balsamo) Vuillemin reveal to be highly pathogenic toward many insect species. There are some fungal species that are very sensitive and require special care and handling technique, such as *Metarhizium aniospliae* (BioBlast), which is a biological control agent. *Streptomyces avermitilis* fungus produces ivermectin, a metabolite, which is lethal toward termites, and it also decreases their food consumption and tunneling ability of workers of species *C. formosanus* (Mo et al. 2006).

6.6.6 Termite Control by Nematode

Entomopathogenic nematodal species have excellent potential for controlling insects; these are being merchandized and used to control many pests. Nematode families such as Steinernematidae and Heterorhabditidae are obligate insect parasites (Poinar Jr 1979), which are associated by symbiotic relationship with bacterial species *Xenorhabdus* and *Photorhabdus* (Boemere et al. 1993; Forst et al. 1997).

These bacterial species are nonmotile, gram-negative, and anaerobic; they are broadly classified as biological control agents (Gaugler and Kaya 1990). The infantile stage of such nematodal species (free living in soil) infects the insect host and the symbiotic relationship of bacteria, and nematode are used and released into the insect hemocoel, causing septicemia and death of pests (Kaya and Gaugler 1993).

Trudeau (1989) found high mortality in termite *Reticulitermes flavipes* (Kollar) by using nematode, although experiments with termite species *R. tibialis* (Epsky and Capinera 1988) and *Coptotermes formosanus* (Pemberton 1928) were not successfully recorded. Weeks and Baker (2004) worked on two entomopathogenic nematodal species *Heterorhabdidtis bacteriophora* (Poinar) and *Steinernema carpocapsae* (Weiser), respectively; based on their survivability, detectability, and mortality of *Heterotermes aureus* (subterranean termite) nematode, *S. carpocapsae* proved to be more effective in causing termite mortality in *H. Aureus* as compared to *H. bacteriophora*.

6.6.7 Termite Control by Bacteria

There are many rhizospheric bacteria which produce hydrogen cyanide (HCN) into the rhizosphere (region of soil in the vicinity of plant roots). Release of HCN by rhizospheric bacteria into the soil is reported poisonous to other subterranean animals. It was found that *Pseudomonas aeruginosa* produces HCN which have toxic effects on nematodes (Darby et al. 1999; Gallagher and Manoil 2001). Termite pests are also controlled by rhizospheric bacteria which produce HCN; these are selectively introduced into termite mounds due to localized use of cyanide production; it minimizes injurious effects on other soil fauna.

To facilitate biocontrol of termites, there are various nonparasitic rhizobacterial species which produce harmful metabolites. Devi et al. (2006) experimentally concluded that three different species of HCN-producing rhizobacteria, *Alcaligenes latus*, *Rhizobium radiobacter*, and *Aeromonas caviae*, were experimented to check the mortality rate of *O. obesus* which was found to be more effective toward termite control under in vitro conditions.

6.7 Conclusion and Future Prospects

Agrochemicals such as chemical insecticides and fertilizers have been used for a long time although they are considered as a major approach to control pests in agriculture field, and it also increases the productivity. But, due to increased use of agrochemicals and its application on a large scale, it resulted to a pest-resistant population, reduction in fertility of soil, and negative impact on human as well as environment. After years of the profound use of agrochemicals, it has been realized now that they are causing critical damage to human health, ecosystems, and groundwater. If the trend continues, this will be a threat to present and future generations. There is an imperative need to move toward environment-affable methods to enrich fertility of soil and disease caused by pests in crop field also controlled. This can be accomplished with the transition to sustainable agriculture. Products of the new generation have been created by using biopesticides as an influential tool. They are the most prone alternatives to some of the most difficult chemical pesticides used recently. Even though synthetic chemicals are more effective than bioinsecticides, it is harmful for other life forms, and it also increases resistant populations. On the other hand, biopesticides are not as harmful; it does not emerge a resistant population and also minimizes the elimination of beneficial species and also lessens human health hazards. Biopesticides are being used in IPM programs, and nowadays it has become an effective and eco-friendly substitute. Termites are the major pests of wooden structures in buildings, human dwellings, and agricultural and forestry crops. So far the only method of their control is to permanently utilize chemical insecticides; also these chemicals are reported to cause injurious effects to ecosystem; the search for alternative means is ongoing with the use of biocontrol agents (fungi, bacteria) and botanicals. Though much research has been done on botanicals in controlling a number of pests and plant diseases, not much success has been achieved in controlling termites. Hence, it is imperative and worthwhile to work on botanicals separately as well as in combination to get success in this case. Insecticidal activities also have been found in around more than 2000 plants belonging to some 60 families. This vast pool of natural pesticides should be explored and tested alone and in combination to obtain a potent termiticidal product.

References

Adfa M, Yoshimura T, Komura K, Koketsu M (2010) Antitermite activity of Coumarin Derivatives and scopoletin from *Protiumjavanicum* (Burm). F. J Chem Ecol 36:720–726

Ahmed S, Mustafa T, Riaz MA, Hussain A (2006) Efficacy of insecticides against subterranean termites in sugarcane. Int J Agric Biol 8:508–510

Baker P (2005) Termite management for homeowners – Cooperative Extension. The University of Arizona Cooperative Extension. http://extension.arizona.edu/sites/extension.arizona.edu/files/pubs/az1356.pdf

Bayatkashkoli A, Taghiyari HR, Kameshki B, Ravan S, Shamsian M (2016) Effects of zinc and copper salicylate on biological resistance of particleboard against *Anacanthotermesvagans*termite. Int Biodeterior Biodegrad 115:26–30

Benelli G (2018) Plant borne compounds and nanoparticles; challenges for medicine, parasitology and entomology. Environ Sci Pollut Res 25:10149–10150

Bhanot JP, Singal SK (2007) Management of termites in field crops. In: Bhanot JP, Rohilla HR, Kalra VK (eds) Recent trends in biology and management of polyphagous pests of agricultural importance. CCS Haryana Agricultural University, Hisar, pp 166–172

Bhanot JP, Batra GR, Verma AN (1991a) Effect of seed treatment with different insecticides on germination, termite damage and grain yield of barley. J Insect Sci 4:61–63

Bhanot JP, Verma AN, Batra GR (1991b) Effect of seed treatment with different insecticides on germination, termite (*Microtermesobesi*) damage and grain yield of wheat (*Triticumaestivum*). Indian J Agric Sci 61:689–691

Bhanot JP, Sharma AK, Batra GR, Verma AN (1995) Influence of different levels of irrigation and fertilizer on termite damage and yield of gram crop raised from aldrin treated and untreated seed. J Insect Sci 7:115–116

Bisset NG, Diaz MA, Ehret C, Ourisson G, Palmad M, Patil F, Pesnelle P, Streith J (1966) Etudes chimio-taxonomiquesdans la famille des diptgrocarpacge- II. Phytochemistry 5:865–880
Bisset NG, Chavanel V, Lantz JP, Wolff RE (1971) Constituants sesquiterpéniques et triterpéniques des résines du genre *Shorea*. Phytochemistry 10:2451–2463
Boemere NE, Akhurst RJ, Mourant RG (1993) DNA relatedness between *Xenorhabdus* spp. (Enterobacteriacae), symbiotic bacteria of entomopathogenic nematodes, and a proposal to transfer Xenorhabdusluminescens to a new genus, *Photorhabdus* gen. Nov. Int J Syst Bacteriol 43:249–255
Carrunthus RI, Sawer AJ, Hural K (1991) Use of fungal pathogens for biological control of insect pests. In: Sustainable agriculture research and education in the field. National Academy Press, Washington, DC, pp 336–339
Chang ST, Cheng SS (2002) Antitermitic activity of leaf essential oils and components from *Cinnamomumosmophleum*. J Agric Food Chem 50:1389–1392
Chang ST, Cheng SS, Wang SY (2001) Antitermitic activity of essential oils and components from Taiwania (*Taiwania cryptomerioides*). J Chem Ecol 27:717–724
Darby C, Cosma CL, Thomas JH (1999) Lethal paralysis of Caenorhabditiselegans by *Pseudomonas aeruginosa*. Proc Natl Acad Sci U S A 96(15):202–215
De Geer C (1776) Memories Pour Servita =L 'this toriededs insect, vol 5. Pierre Hasselberg, Stolkholm
De Geer C (1782) Aabhandlungen, Zen Geschit der Insekten (Translated from German by Goeze JAE)
Dev S, Koul O (1997) Higher plants. In: Insecticides of natural origin. Harwood Academic Press, Amsterdam, p 365
Devi KK, Seth N, Kothamasi S, Kothamasi D (2006) Hydrogen cyanide-producing rhizobacteria kill subterranean termite Odontotermesobesus (Rambur) by cyanide poisoning under in vitro conditions. Curr Microbiol 54:74–78
Diaz MA, Ourisson G, Bisset NG (1966) Etudes chimio-taxonomiquesdans la famill des diptero-carpacaes–I. Phytochemistry 5:855–863
Epsky N, Capinera JL (1988) Efficacy of the entomogenous nematode *Steinernema feltiae* against a subterranean termite, *Reticulitermes tibialis* (Isoptera: Rhinotermitidae). J Econ Entomol 81:1313–1317
FAO-Food and Agricultural Organization of United Nations. Internal code of conduct on the distribution and use of pesticides, Rome (2002), p 36. [Cited on 2018 April 7] available from http://www.fao.org/docrep/005/y5544e/y4544e00html
Forst S, Dowds B, Boemare N, Stackebrandt E (1997) *Xenorhabdus* and *Protorhabdus* spp.: bugs that kill bugs. Annu Rev Microbiolol 51:47–72
Gallagher LA, Manoil C (2001) *Pseudomonas aeruginosa* PAO1 kills *Caenorhabditis elegans* by cyanide poisoning. J Bacteriol 183:6207–6214
Gaugler R, Kaya HK (eds) (1990) Entomopathogenic nematodes in biological control. CRC Press, Boca Raton, p 365
Gerrits R, Van Latum EBJ (1988) Plant-derived pesticides in developing countries: possibilities and research needs. Netherlands' Ministry of Housing. Physical Planning and Environment, The Hague, p 104
Grace JK (1996) Biological control strategies for suppression of termites. J Agric Urban Entomol 14:281–289
Grace JK (1998) Resistance of pine treated with chromated copper arsenate to the Formosan subterranean termite. For Prod J 48:79–82
Guan YQ, Chen J, Tang J, Yang L, Liu JM (2011) Immobilizing bifenthrin on wood for termite control. Int Biodeterior Biodegrad 65:389–395
Harborne JB (1988) Introduction to ecological biochemistry, 3rd edn. Academic, London, p 356
Huxham LM, Samueli KDZ, Hele JB, McCorkindale NJ (1989) *In vivo* and *in vitro* assays for pathogenicity of wild type and mutant strains of *Metarhizium anisopliae* for three insect species. J Invertebr Pathol 53:143–151
Isman MB (2000) Plant essential oils for pest and disease management. Crop Prot 19:10603–10608

Kannaiyan S (1999) Botanicals in pest control. Inaugural Address in the Training Programme on Botanicals in Pest Management, Tamilnadu Agriculture University, Coimbatore, India, December 1–10, 1999

Kard BM (1999) Mesh may fit in as a termite barrier. Pest Control 67:50–53

Kaya HK, Gaugler R (1993) Entomopathogenic nematodes. Annu Rev Entomol 38:181–191

Kobaisy M, Tellez MR, Osbrink WL (2001) Anti termite naphthoquinones from *Moneses uniflora*. In: Abstracts of papers, 222nd ACS national meeting, Chicago, IL, United States, August 26–30, 2001

Koul O, Walia S, Dhaliwal GS (2008) Essential oils as green pesticides: potential and constraints. Biopestic Int 4(1):63–84

Kuriachan I, Gold RE (1998) Evaluation of the ability of *Reticulitermes flavipus* Kollar, a subterranean termite (Isoptera: Rhinotrmitideae) to differentiate between termiticide treated and untreated soils in laboratory tests. Sociobiology 32:151–166

Lajide L, Escoubas P, Mizutani J (1995) Termite antifeedant activity in *Detarium microcarpum*. Phytochemistry 40:1101–1104

Lenz M, Runko S (1994) Protection of buildings, other structures and materials in ground contact from attack by subterranean termites (Isoptera) with a physical barrier-a fine mesh of high grade stainless steel. Sociobiology 24:1–16

Lewis VR (1997) Alternative control strategies for termites. J Agric Entomol 14:291–307

Maistrello L, Henderson G, Laine RA (2001) Efficacy of vetiver oil and nootkatone as soil barriers against formosan subterranean termite (Isoptera: Rhinotermitidae). J Econ Entomol 94:1532–1153

Maistrello L, Henderson G, Laine RA (2003) Comparative effects of vetiver oil, nootkatone and disodium octaborate tetrahydrate on *Coptotermes formosanus* and its symbiotic fauna. Pest Manag Sci 59:58–68

Mao L, Henderson G (2007) Antifeedant activity and acute and residual toxicity of alkaloids from *Sophora flavescens* (Leguminosae) against formosan subterranean termites (Isoptera: Rhinotermitidae). J Chem Ecol 100:866–870

McDaniel CA (1989) Major termiticidal components of heartwood of Port-Orfordcedar, *Chaemecyparis lawsoniana* (A. Murr.) Parl. Mater Org 24:1–15

Meepagala KM, Osbrink W, Ballew A, Lax AR, Duke SO (2006) Natural product based amides against Formosan subterranean termites (*Coptotermes formosanus*). In: Abstracts of papers, 232nd ACS national meeting, San Francisco, CA, United States, September 10–14, 2006

Meyer JR (2005) Isoptera. Department of Entomology, NC State University, Online at: http://www.cals.ncsu.edu/course/ent425/compendium/termites.html

Mo J, Wang Z, Song X, Guo J, Cao X, Cheng J (2006) Effects of sublethal concentration of ivermectin on behaviour of *Coptotermes formasanus*. Sociobiology 47:687–696

Moi LC (1980) A new laboratory method for testing the resistance of particle to the drywood termite *Cryptotermes cynocephalus*. Malays For 43:350–355

Myles TG (2005) Termite biology. Urban Entomology Programme. http://www.utoronto.ca/forest/termite/termite.htm

Nakashima Y, Shimizu K (1972) Antitermitic activity of *Thujopsis dolabrata* var *Hondai*. III. Components with a termiticidal activity. Miyazaki Daigaku Nogakubu KenkyuHokoku 19:251–259

Nix KE, Henderson G, Zhu BCR, Laine RA (2006) Evaluation of vetiver grass root growth, oil distribution, and repellency against Formosan subterranean termites. Hortic Sci 41:167–171

Owusu EO, Akutse KS, Afreh-Nuamah K (2008) Effect of some traditional plant components on the control of termites, *Macrotermus* spp (Isoptera: Termitidae). Afr J Sci Technol 9(2):82–89

Pal M, Verma RK, Krishan S, Tewari SK (2011) Anti-termite activity of essential oil and its components from *Myristica fragrans* against *Microcerotermes beesoni*. J Appl Sci Environ Manage 15(3):559–561

Parmar BS, Devkumar C (1993) Botanicals and biopesticides. Westvill Publishing House, New Delhi, p 199

Pemberton CE (1928) Nematodes associated with termites in Hawaii, Borneo and Celebes. Proc Hawaiian Entomol Soc 7:147
Philip H (2004) Biology and control of the subterranean termite. Pest Management Factsheet 98-01. http://www.agf.gov.bc.ca/cropprot/termite.htm
Poinar GO Jr (1979) The natural history of nematodes. Prentice Hall, Englewood Cliffs
Potter MF (2004) Termites. In: Moreland D (ed) Handbook of pest control: the behavior, life history, and control of household pests. GIE Media, Cleveland, pp 216–361
Prakash A, Rao J (1997) Botanical pesticides in agriculture. CRC Lewis Publishers, Boca Raton
Prakash A, Tiwari SN, Rao J (1990) Exploitation of natural plant Products for management of pests and diseases in rice ecosystems. In: Proceedings of symposium of growth development resource conservation, pp 23–36
Raina A, Bland J, Doolittle M, Lax A, Boopathy R, Folkins M (2007) Effect of orange oil extract on the formosan subterranean termite (Isoptera: Rhinotermitidae). J Econ Entomol 100:880–885
Rajagopal D (2002) Economically important termite species in India. Sociobiology 40(1):33–46
Rodrigues AM, Amusant N, Beauchene J, Eparvier V, Lemenager N, Baudasse C, Espindola LS, Stien D (2011) The termiticidal activity of Sextonia rubra (Mez) vander Werff (Lauraceae) extract and its active constituent rubrynolide. Pest Manag Sci 67:1420–1423
Rust MK, Saran RK (2006) Toxicity, repellency, and transfer of chlorfenapyr against Western Subterranean termites (Isoptera: Rhinotermitidae). J Econ Entomol 99:864–872
Sa RA, Napoleao TH, Santos NDL, Gomes FS, Albuquerque AC, Xavier HS, Coelho LCBB, Bieber LW, Paiva PMG (2008) Induction of mortality on *Nasutitermescorniger* (Isoptera, Termitidae) by *Myracrodruon urundeuva* heartwood lectin. Int Biodeterior Biodegrad 62:460–464
Saeki I, Sumimoto M, Kondo T (1968) Antitermitic substance of *Ternstroemia japonica* wood. III. Biological tests of the antitermitic substance. Mokuzai Gakkaishi 14:110–114
Saeki I, Sumimoto M, Kondo T (1970) Termiticidal substances from the wood of *Podocarpus macrophyllus*. Holzforschung 24:83–86
Saeki I, Sumimoto M, Kondo T (1973) Termiticidal substances from the wood of *Chamaecyparis pisifera*. Holzforschung 27:93–96
Scheffrahn RH, Su NY, Busey P (1997) Laboratory and field evaluation of selected chemical treatment and field evaluation of selected chemical treatment for control of drywood termites (Isoptera: Kalotermitidae). J Econ Entomol 90:492–502
Sen-Sarma PK (1963) Studies on the natural resistance of timbers to termites. I. Observations on the longevity of the test termite *Heterotermes indicola* Wasm. in the saw dust from forty common Indian timbers. Indian For Bull (N.S.) Entomolol 220:1–3
Sen-Sarma PK, Chatterjee PN (1968) Studies on the natural resistance of timbers to termite attack. V. Laboratory evaluation of the resistance of Indian wood *Microcerotermes beesoni* Snyder (Termitidae, Amitermitinae). Indian For 94:694–704
Serit M, Ishida M, Nakata K, Kim M, Takahashi S (1992) Antifeedency potency of neem (*Azadirachta indica*) extractives and limonoids against termite (Reticulitermessperatus). J Pestic Sci 17:267–227
Singh N, Kumar S (2008) Antitermite activity of *Jatropha curcas* Linn. biochemicals. J App Sci Env Manage 12(3):67–69
Singh G, Singh OP, Lampasona MP, Cesar AN (2002) Studies on essential oils. Part 35: chemical and biocidal investigations on *Tageteserecta* leaf volatile oil. Flavour Fragr 18:62–65
Smith JL, Rust MK (1990) Tunneling response and mortality of the western subterranean termite (Isoptera: Rhinotermitidae) to soil treated with termiticides. J Econ Entomol 83:1395–1401
Stevenson PC, Isman MB, Belmain SR (2017) Pesticidal plants in Africa: a global vision of new biological control products from local uses. Ind Crop Prod 110:2–9
Stoll G (1986) Natural crop protection. Josef Margraf, Langen, p 188
Stranes RL, Liu L, Marrane PG (1993) History, use and future of microbial insecticides. Am Entomol 39:83–91
Su NY, Scheffrahn RH (2000) Formosan subterranean termite. University of Florida. http://creatures.ifas.ufl.edu/urban/termites/fst.10htm

Su NY, Ban PM, Scheffrahn RH (1999) Longevity and efficacy of pyrethroid and organophosphate termiticides in field degradation studies using miniature slabs. J Econ Entomol 92:890–898

Tellez M, Estell R, Fredrickson E, Powell J, Wedge D, Schrader K, Kobaisy M (2001) Extracts of *Flourensiacernua* (l): volatile constituents and antifungal, antialgal, and antitermite bioactivities. J Chem Ecol 27:11

Tellez MR, Khan IA, Kobaisy M, Schrader KK, Dayan FE, Osbrink W (2002) Composition of the essential oil of *Lepidiummeyenii* (Walp.). Phytochemistry 61:149–155

Thompson G (2000) Termites. Tropical Topics News Letter No. 64, Tropical Savanna, Australia

Trikojus VM (1935) Some synthetic and natural antitermitic substances. Aust Chem Inst J Proc 2:171–176

Trudeau D (1989) Selection of entomophilic nematodes for control of the eastern subterranean termite, *Reticultermes flavipes* (Kollar) (Isoptera: Rhinotermitidae). Master's thesis, University of Toronto, Toronto, Ontario, Canada, p 93

UNEP (2000) Finding alternatives to Persistent Organic Pollutants (POPs) for termite management. Prepared by members of the UNEP/FAO/Global IPM Facility Expert Group on Termite Biology and Management. http://www.chem.unep.ch/pops/termites/termite

Verma RK, Verma SK (2006) Phytochemical and termiticidal study of *Lantana camara*var. *aculeata* leaves. Fitotrapia 77:466–468

Weeks B, Baker P (2004) Subterranean termite (Isoptera: Rhinotermitidae) mortality due to entomopathogenic nematodes (Nematoda: *Steinernematidae, Heterorhabditidae*). University of Arizona College of Agriculture Turfgrass and Ornamental Research Report, index at: http://cals.arizona.edu/pubs/crops/az1359/

Yatagai M, Miyazaki Y, Morita S (1991) Extractives from Yakusugi bogwood andtheirtermiticidal activity and growth regulation effects on plant seeds. For. Prod. Res. Inst., Tsukuba, Japan. Mokuzai Gakkaishi 37:345–351

Yatagai M, Nishimoto M, Ohira KHT, Shibata A (2002) Termiticidal activity of wood vinegar, its components and their homologues. J Wood Sci 48:338–342

Yazaki Y (1982) Termiticidal extracts from the wood of *Ganophyllumfalcatum* B1. Holzforschung 36:249–253

Zhu BCR, Henderson G, Chen F, Maistrello L, Laine RA (2001) Nootkatone is a repellent for Formosan subterranean termite (*Coptotermes formosanus*). J Chem Ecol 27:523–531

Juvenoids and Its Application in Crop Management

7

Simranjeet Singh, Vijay Kumar, Daljeet Singh Dhanjal, and Joginder Singh

Abstract

Pest management is one of the major growing concerns worldwide. Moreover, these insects are important for natural ecosystem as they perform various functions such as organic matter decomposition and facilitating food for birds, fishes, and reptiles. Juvenoids, the chemical compound which mimics the juvenile hormones and inhibits the metamorphosis process, have gained significant attention among researchers. From the past few decades, intensive research has been done on biochemical and physiological effects of juvenile hormones and their chemical analogs in which they regulate reproduction and metamorphosis of pests. Juvenile hormones are the derivatives of fatty acid which are produced by neurosecretory cells. These juvenoid hormones conserve natural fauna and flora and minimize the chemical pesticide usage. Currently, numerous artificial juvenoids are commercially available and more effective than traditional juvenoids. These artificial juvenoids possess less toxicity and show no teratogenic or mutagenic effects. The juvenoids have inhibition effect on insect morphogenesis as individual specific cells may show an inflexible response and only few cells show

Authors Simranjeet Singh and Vijay Kumar have been equally contributed to this chapter.

S. Singh
Department of Biotechnology, Lovely Professional University, Jalandhar, Punjab, India

Punjab Biotechnology Incubators, Mohali, Punjab, India

Regional Advanced Water Testing laboratory, Mohali, Punjab, India

V. Kumar
Department of Chemistry, Regional Ayurveda Research Institute for Drug Development, Gwalior, Madhya Pradesh, India

D. S. Dhanjal · J. Singh (✉)
Department of Biotechnology, Lovely Professional University, Jalandhar, Punjab, India
e-mail: joginder.15005@lpu.co.in

J. Singh, A. N. Yadav (eds.), *Natural Bioactive Products in Sustainable Agriculture*, https://doi.org/10.1007/978-981-15-3024-1_7

sensitivity to juvenoids at a particular time. This makes juvenoid usage advantageous over traditional insecticides and a valuable chemical in crop management. In this chapter, we have discussed its various classes, chemistry, mode of action, and application in crop management systems.

Keywords
Juvenoids · Analogues · Mode of action · Pest management

7.1 Introduction

Insects are one of the most primitive and diverse and living animals on this earth. These insects are found in all habitats like deserts, forest, swamps, and harsh environment except open oceans (Daglish et al. 2018). Undoubtedly, insects are a highly adaptable form of life as there are about five million species of insects, which is very high in comparison to animals. Nearly about 60–70% of stored crops are damaged by pests every year. Pest management is one of the growing concerns worldwide (Clapham et al. 2016). Moreover, these insects are important for natural ecosystem as they perform various functions like organic matter decomposition and facilitate food for birds, fishes, and reptiles. In contrary to that, they also act as herbivores, hosts, parasites, predators, preys, pollinators and saprophages, which shows the ecological importance of these insects in the environment (Losey and Vaughan 2006). Most of the various insect species show negative effect on both the environment and humans. Thus, due to detrimental effect of the insects on the environment, agricultural researchers have started exploring the compatible, eco-friendly, and safer alternative to regulate the growth of these pests (Kenis et al. 2009). Regrettably, various insect species also show negative effect on both the environment and humans (Deroy et al. 2015). Juvenoids, the chemical compound which mimics the juvenile hormones and inhibits the metamorphosis process, have gained significant attention among researchers. Juvenile hormones are the derivatives of fatty acid which are produced by neurosecretory cells. These hormones have methyl ester of epoxy farnesoic acid and have a terpenoid-related structure. This hormone exists in different forms, out of which JH III is the most common and present in species of lepidoptera (Sláma 2014). Some commonly synthesized juvenoids are represented in Fig. 7.1 and natural in Fig. 7.2.

The JH hormones were discovered undoubtedly from ingenious and thoughtful experiments, and the first study was reported by Williams in 1956 in which he synthesized the lipid extracts having juvenoid hormones from the abdomen of *Cecropia moths* and suggested that these extracts could be used as insecticides being nontoxic in nature (Dhadialla et al. 1998). Further experimental findings concluded that these extracts of juvenoid hormones are also present in other invertebrates, plants, and microorganisms (Takatsuka et al. 2017). In the late 1950s, scientists started synthesizing terpenoids and other related compounds to estimate their JH activity (Cherney and Baran 2011). Efforts of Maag Agrochemicals, HLR Sciences, Inc., and Dr.

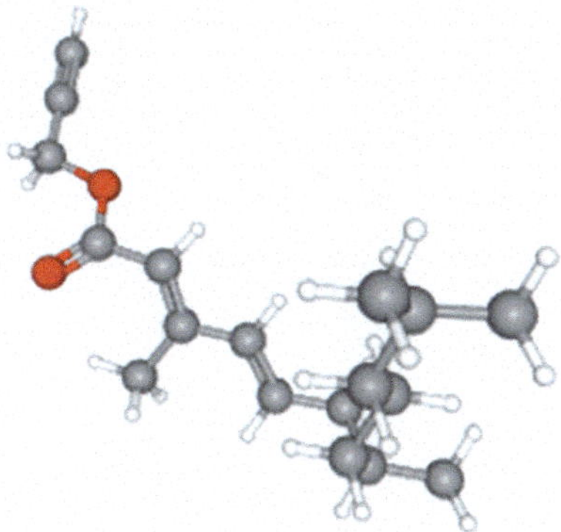

1) Kinoprene

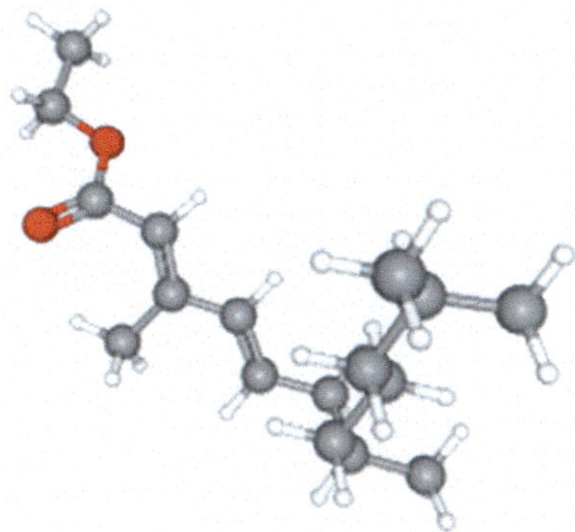

2) Hydroprene

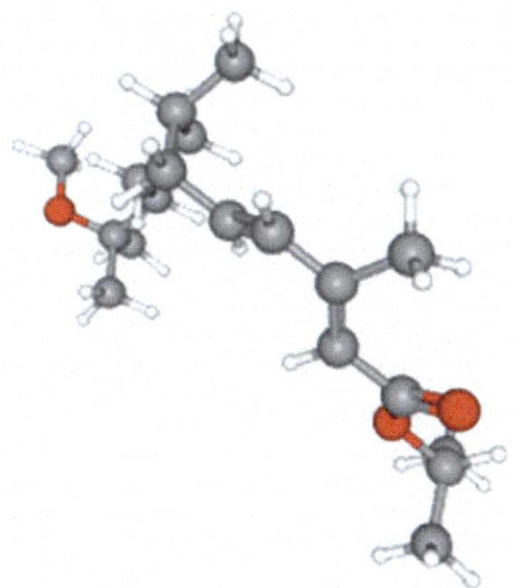

3) Methoprene

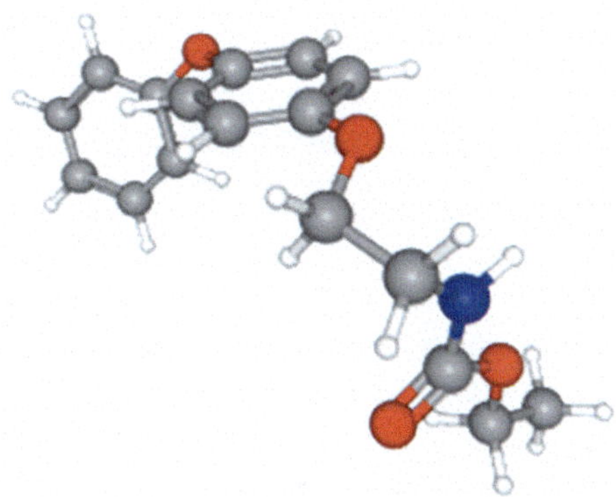

4) Fenoxycarb

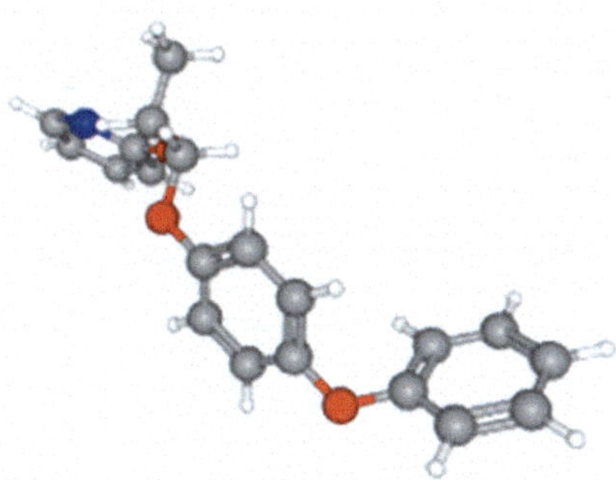

5) Pyriproxyfen

Fig. 7.1 3D conformer of few synthesized juvenoids

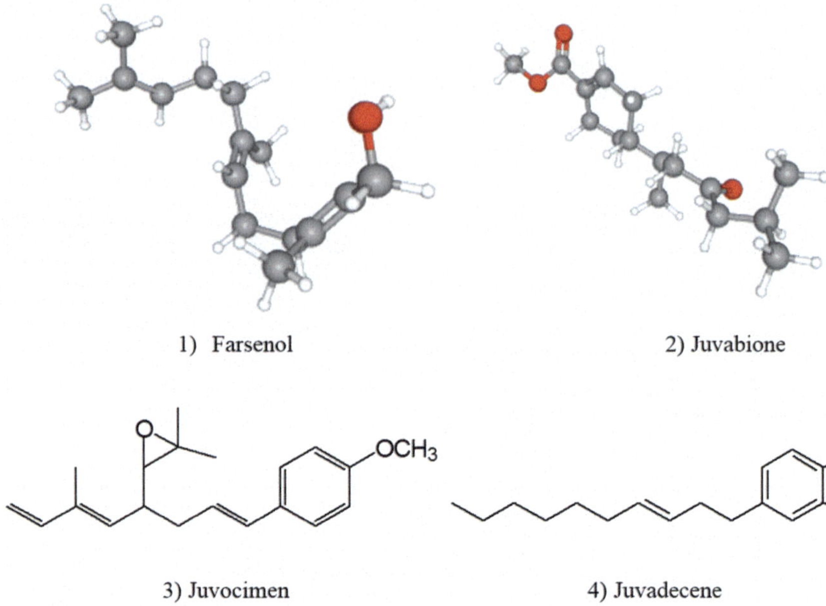

Fig. 7.2 2D and 3D conformer of few natural juvenoids

Maag Ltd., Dielsdorf (Switzerland) developed and discovered non-isoprenoid juvenoid which is found to be highly effective, which was further registered for insect control (Mian 2014). The most effective molecule in non-isoprenoid series was ethyl [2-(p-phenoxyphenoxy)ethyl]carbamate (fenoxycarb), which is used to control ants, moths, cockroaches, scale insects, and moths and insects attacking fruit, cotton, olives, etc. (Ramaseshadri et al. 2012). However, though ethyl [2-(p-phenoxyphenoxy)ethyl]carbamate has a carbamate moiety, it has no effect on cholinesterase activity. These juvenoids hormones conserve natural fauna and flora and minimize the chemical pesticide usage (Goncu and Parlak 2012). In this chapter, we will discuss the various classes of juvenoids, chemistry, mode of action, and their application in crop management systems.

7.2 Classes and Type of Juvenoids

Depending upon the mode of action, juvenile hormones are of four types: JH 0, JH I, JH II, and JH III. Out of these JH III is the most widely accepted and in use (Noriega 2014). Juvenoid hormones are isolated from various other sources such as dehydro-juvabione and juvabione from *Abies pseudotsuga*; farnesyl-methyl ether from insect excreta, yeast, and plants; sesame from the sesame oil; echinolone from *Echinacea angustifolia*; and juvocimen from *Ocimum basilicium* (Nëmec 1993). Exogenous use of juvenoid compounds can disrupt and retard the insect growth in various ways (Lee et al. 2015). Juvenoid hormones are sesquiterpenoid compounds which regulate metamorphosis and growth of insects when interacted with ecdysteroids. Juvenoids have well-defined molecular size and certain structural features (Qu et al. 2015). They also have certain properties like lipophilicity, low polarity, less or more pronounced volatility, etc. The common action of juvenoids is inhibition of metamorphosis by preventing adult and pupal emergence (Wheeler and Nijhout 1982). More than 4000 man-made juvenoid analogs have been discovered so far. Juvenoids cause malformation, prolongation of larval stage, disturbance in organ formation, gametogenesis, etc. (Srivastava and Srivastava 1983). Various commercially available juvenoids are well represented in Table 7.1.

Table 7.1 List of commercially available juvenoids (Wheeler and Nijhout 1982)

Juvenoids	Dosage	Active substance	Application
Altosid	280 g/ha or more	Methoprene	Regulate the growth of horn fly and mosquitoes
Apex	1 g/m^3	Methoprene	Regulate the growth of sciarid flies during mushroom culturing
Diacon	1.12 g a.i./100kg (spray)	Methoprene	Regulates the growth of pest affecting the stored products like peanuts
Dianex	0.0207 g a.i./m^3 (aerosol)	Methoprene	Regulates the growth of pest affecting the stored products like peanuts
Enstar 5E	0.5% or 0.1 g /m^2 (spray)	Kinoprene	Regulates the growth of homopterans in greenhouse
Gencor	0.03 g a.i./m^2	Hydroprene	Regulates the growth of cockroaches
Gencor Plus	0.03 g a.i./m^2	Hydroprene (0.6%) + permethrin (0.25%)	Regulates the growth of cockroaches
Kabat	100 ppm a.i. (spray)	Methoprene	Regulates the growth of *Ephestia elutella* and *Lasioderma serricorne*
Manta	2.5 ppm a.i. (spray)	Methoprene	Enhances the production of silk
Precor	–	Methoprene	Regulates the growth of fleas
Raid Flea Killer II Plus	–	Methoprene + pyrethroid	Regulates the growth of fleas
Ro 13–5223	250 g/ha per 2–3 applications; 4–110 ppm; 0.03–0.1 ppm and 5–10 mg/colony	Fenoxycarb	Regulates the growth of fire ants, Homoptera, mosquitos, and Tortricidae
S-71639	100 g/ha	Pyriproxyfen	Regulates the growth of mosquitos

7.3 Methods of Juvenoid Application

7.3.1 Direct Application

Considering the chemical nature of the juvenoids (i.e., lipophilic), they can be dissolved in wax, acetone, or oil. Then, this can be either directly applied on the surface of insects or injected in the body cavity (Baker et al. 1986). Since a large number of insects have smaller size, both processes require precision to transferring 0.1–1.0 μl solution of juvenoids with <5% (standard error). These approaches are usually used in in vitro analysis for assessing the juvenoid activity on the insects (Žďárek and Denlinger 1975).

7.3.2 Indirect Application

This approach is designed in such a way to check out the responses of growing insects at given dose of juvenoids. While applying juvenoids using indirect method, special equipment is used for spraying juvenoid solution per square unit area (Kamya et al. 2017). This method includes dipping and spraying of plants in solution of juvenoids, etc. To increase the stability of juvenoids, they are linked to certain substrates (sugar, fatty acids). The resultant compounds are called juvenogens which when enter the target body, are broken down by enzymes, releasing the juvenoid hormones and triggering its action (Walker 1976).

7.3.3 Pest Management by Juvenoids

Various reports have documented the potent role of juvenoids in pest management of lepidopterous pests of forestry and field crops. From the past few decades, intensive research has been done on biochemical and physiological effects of juvenile hormones and their chemical analogs in which they regulate reproduction and metamorphosis of pests (Deb and Chakravorty 1981).

Juvenoid hormones are divided into two main groups: first one is the phenoxy juvenoids like pyriproxyfen and fenoxycarb, and the second one is terpenoid compounds which include kinoprene, hydroprene, and methoprene (Palma et al. 1993). All these hormones are highly effective against the homopteran and dipteran insects but ineffective against lepidopteran pests (Horowitz and Ishaaya 2004). Terpenoid juvenoid hormones such as kinoprene and methoprene have been used against mealybugs, scales, and mosquito larvae. Phenoxy juvenoid hormones cause ovicidal action, infertility, and morphogenic effects (Boina et al. 2010). Fenoxycarb, the first commercially registered juvenoid, is used against sucking insects, fleas, fire ants, apple wooly aphid, etc. Pyriproxyfen offers to be effective against whiteflies, psyllids, and aphids which cause damage to crops (El-Kareim et al. 1988).

All the three juvenile hormones are found in most of the insects; JH I and JH II are the two hormones which regulate metamorphosis, whereas JH III plays gonadotropic role. In adults, juvenile hormones act to induce ovarian development, spermatogenesis, accessory gland, and vitellinogenesis development in males. Insects which produce cyclic batches are more affected by action of juvenile hormones (Hartfelder 2000).

Juvenile hormones also play an important role in color polymorphism, caste and phase determination, pigmentation, diapause, pheromone production, etc. They are also reported to influence good range of physiological process both on mature and developing insects (Riddiford 2012).

Juvenile hormones and their chemically synthesized structures have various effects when sprayed on insects during their developmental stage. Applications of juvenoids result in disruption or developmental retardation and inhibition of function or formation of reproductive system (Elvira et al. 2010). Figure 7.3 describe the effects of juvenoid hormones on pests.

Juvenile hormones are also involved in the regulation or programming of post-embryonic development. Lower doses of juvenile hormones result in successful pupation leading to deformities in pupal adult intermediates (Daimon et al. 2012). Most of the insects are highly sensitive to juvenoids during early pupal and final larval stages. Morphogenetic effects are also seen in penultimate larval instar in hemipteran groups which are found to be sensitive (Truman and Riddiford 2019). Juvenoid biosynthesis hormones are controlled by various neuronal and neuroendocrine in species and complex stage-specific ways. Juvenoid hormones show action via two receptors, that is, nuclear and membrane receptors. SRC and Met, two bHLH PAS transcription factors which form nuclear receptor, interact with juvenoid hormone response element in the promoter of Kruppel homolog 1 and trypsin that activate the transcription of these genes (Miyakawa et al. 2014; Li et al. 2019).

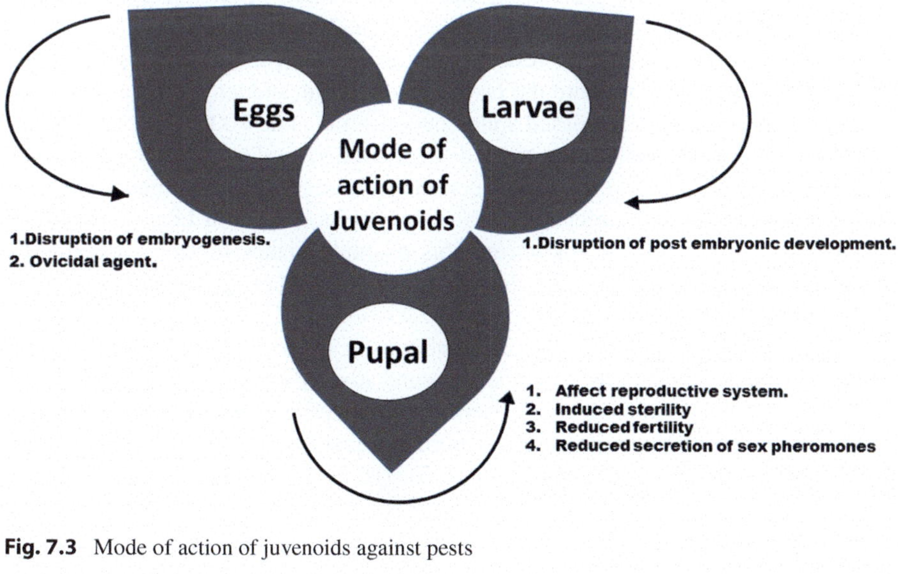

Fig. 7.3 Mode of action of juvenoids against pests

Mostly juvenoid hormones are reported to affect all the three stages of insect life which include eggs, larval, and adults or pupal. In eggs, usually disruption in embryonic stages of developments was reported along with the ovicidal effects (Oouchi 2005). Juvenile hormones also have a negative effect on disruption of the embryogenesis process which results in ovicidal effects. Juvenoid treatment also leads to disruption of the female reproductive system (Eliahu et al. 2007). Excess of juvenoids on female insects has direct impact on fertility and their physiological processes. Prolonged treatment of juvenoids causes diapause or hemolymph in larvae of insects. However, in adult female of silkworms, juvenoids change the voltinism of larval offsprings. Juvenoids also disrupt the reproduction, general physiology, behavior, and development of treated insects (Mojaver and Bandani 2010).

In larval stages, juvenoid hormones lead to formation with adult features which finally lead to death. Sometimes, it disrupts postembryonic development of larvae. In adults or pupal of insects, the juvenile hormones act to disrupt the diapause. They also reported to affect the reproduction system, induce sterility, affect color polymorphism, reduce fertility, etc. (Ishimaru et al. 2016). Various authors have also reported the applications of juvenoids as pest managements against large number of lepidopterous pests of forestry and field crops. Juvenile hormones are also used to control manure-breeding and water-breeding flies (Singh and Kumar 2015).

Juvenoids were also reported to act on *Lepidopterous* and *Coleopterous* species. Juvenile hormones are also reported to kill the pest in store products. The effect of juvenoid hormones on three pest species *Callosobruchus chinensis*, *Stegobium paniceum* (drugstore beetle), and *Tribolium castaneum* (rust-red flour beetle) were also studied by Thomas and Bhatnagar in which juvenoid hormones have morphogenetic effects at very low doses (Thomas and Bhatnagar-Thomas 1968). These effects include reduction of ovipositions in species of *S. paniceum* and *C. chinensis*. Juvenoid hormones also prevent imaginal differentiation in the *Tenebrio molitor* species of yellow mealworm (Ochoa-Sanabria et al. 2019).

Juvenoid hormones were also reported to enhance the production of silk. When JH I and JH II were injected into the larvae of silkworm, sudden increase in size of larvae was observed with increased spun cocoons (Tan et al. 2005).

When silkworm females are treated with methylenedioxyphenyl-substituted juvenoids, increase in egg size and production was observed. The synthetic juvenoid methoprene is the first commercially registered juvenoid used in Japan to increase the yield and production of silkworm. This example makes juvenoid specials to increase the yield of insects rather than to minimize the damage caused from their overproductivity (Nair et al. 2012).

Recent advancement in the field of biochemistry led to increase the stability of juvenoid hormones both in the insect species and adverse environmental conditions. For this purpose, uses of catabolic enzymes are used to improve the hormone's stability. Usually, endogenous enzymes are used to release hormones from long chains of fatty acid esters which acts as pro-juvenoids (Hui et al. 2010).

7.4 Factors Influencing the Uses of Juvenoids

Various factors are responsible for influencing the use of juvenoid hormones. It includes choice of target species, resistance, synergism, metabolism, and some biological factors. Biological factors mean the developmental stage of an organism influences the sensitivity of pests when treated with juvenoids (Jindra and Bittova 2019). Timing of application is also a crucial factor which limits the stability of juvenile hormones when used against various pest species. Several other factors such as identification of pests and techniques for spraying the hormones against the pests also play an important role in influencing the effectivity of juvenoid hormones (Ishmuratov et al. 2015). These things should be carefully planned in order to get the desired effect on population of pests which are highly susceptible to both reproductive and morphogenetic effects. Selective management of pest populations also influences the applications of juvenile hormones (Fathpour et al. 2007).

The half-life of most of the juvenile hormones especially JH I is less than 1 h which makes them highly unstable. These juvenoids are highly sensitive to UV radiation and undergo degradation when poured in aqueous media. Juvenoid hormones are usually metabolized in two simple pathways using enzymes JH-esterase and JH-epoxide hydrase in which they form 10-diol derivative, 11-diol derivatives, and 1 molecule of acid derivatives (Bassal et al. 2018).

Sometimes, juvenoid hormones are also reported to bind with certain proteins termed as carrier proteins which protect juvenoid hormones from being metabolized by nonspecific enzymes. These carrier proteins also facilitate the movement of molecules of lipophilic juvenoid hormones to deliver the product into the site of action in stable conditions (Tanaka et al. 2019).

7.5 Future Prospects

The juvenile hormones (JH) of insects are involved in various physiological processes like development, morphogenesis, and reproduction in insect pests. Hence, molecules interact with JH or mimic as JH, which either inhibit the biosynthesis of JH or interfere in their catabolism process (Noriega et al. 2006). Since our knowledge related to JH is now significantly expanding, there are chances that some new more target-specific juvenoids will come into existence. The chemical effect in inhibiting the JH biosynthesis can act as persuasive insecticide (El-Ibrashy 1987). Arylpyridyl-thiosemicarbazones and 1,5-dissubtitutedimidazoles are the juvenoid compounds which have been discovered, and they exhibit inhibitory action during JH biosynthesis (Němec 1993). Therefore, intense insight about these juvenoids and their action mechanism at gene level as endocrinological level will pave the advance way to develop novel strategies to hinder the life cycle of insect pests. Later, when we will be able to clone the DNA stretches which encode for the receptors that are involved in JH biosynthesis process, they can be used for in vitro assay for JH as it will alleviate the juvenoid exploration process (Ahmad et al. 2018).

7.6 Conclusion

In this chapter, we have attempted to compile data on the applications of juvenoids in pest control system. The uniqueness of juvenoid hormones has offered various advancements in the development of effective pesticides which are not having negative effect on animals, plants, and animals. The juvenoids hold the potential by serving as analogs of insect hormones to regulate the insect pests. These juvenoids are analogs of true juvenile hormones, as they have similar distribution of polar functional groups and molecular size as that of juvenile hormones and nowadays are practically used in agricultural fields. Currently, thousands of artificial juvenoids are commercially available and more effective than traditional juvenoids. Moreover, these artificial juvenoids have less toxicity and no teratogenic or mutagenic effect. Primarily, these juvenoids have inhibition effect on insect morphogenesis as individual specific cells may show an inflexible response and only few cells show sensitivity to juvenoids at a particular time. Hence, these advantages of juvenoids over traditional insecticides make it a valuable approach in crop management.

References

Ahmad T, Nabi S, Humera Q (2018) Biotechnology: a tool in termite management. In: Termites and sustainable management. Springer, Cham, pp 289–315

Baker FC, Miller CA, Tsai LW et al (1986) The effect of juvenoids, anti juvenile hormone agents, and several intermediates of juvenile hormone biosynthesis on the in vivo juvenile hormone levels in *Manduca sexta* larvae. Insect Biochem 16:741–747. https://doi.org/10.1016/0020-1790(86)90109-5

Bassal TTM, Shaaban MR, Mohamed AA, Almadiy AA (2018) Insect Juvenoids: structural qualifications and structure-activity relationships of Alkylphenoxy-substituted terpenes as insect growth disruptors against *Spodoptera littoralis* (Boisduval) (Lepidoptera: Noctuidae) – revisited. African Entomol 26:522–535. https://doi.org/10.4001/003.026.0522

Boina DR, Rogers ME, Wang N, Stelinski LL (2010) Effect of pyriproxyfen, a juvenile hormone mimic, on egg hatch, nymph development, adult emergence and reproduction of the Asian citrus psyllid, *Diaphorina citri* Kuwayama. Pest Manag Sci 66:349–357. https://doi.org/10.1002/ps.1880

Cherney EC, Baran PS (2011) Terpenoid-alkaloids: their biosynthetic twist of fate and total synthesis. Isr J Chem 51:391–405

Clapham ME, Karr JA, Nicholson DB et al (2016) Ancient origin of high taxonomic richness among insects. Proc R Soc B Biol Sci 283. https://doi.org/10.1098/rspb.2015.2476

Daglish GJ, Nayak MK, Arthur FH, Athanassiou CG (2018) Insect pest management in stored grain. In: Recent advances in stored product protection. Springer, Berlin Heidelberg, pp 45–63

Daimon T, Kozaki T, Niwa R et al (2012) Precocious metamorphosis in the juvenile hormone-deficient mutant of the silkworm, *Bombyx mori*. PLoS Genet 8. https://doi.org/10.1371/journal.pgen.1002486

Deb DC, Chakravorty S (1981) Effect of a juvenoid on the growth and differentiation of the ovary of *Corcyra cephalonica* (Lepidoptera). J Insect Physiol 27:103–111. https://doi.org/10.1016/0022-1910(81)90116-5

Deroy O, Reade B, Spence C (2015) The insectivore's dilemma, and how to take the West out of it. Food Qual Prefer 44:44–55

Dhadialla TS, Carlson GR, Le DP (1998) New Insecticides with Ecdysteroidal and Juvenile Hormone Activity. Ann Rev Entomol 43:545–569

Eliahu M, Blumberg D, Horowitz AR, Ishaaya I (2007) Effect of pyriproxyfen on developing stages and embryogenesis of California red scale (CRS), *Aonidiella aurantii*. Pest Manag Sci 63:743–746. https://doi.org/10.1002/ps.1369

El-Ibrashy MT (1987) Juvenoids and related compounds in tropical pest management: a review. Int J Trop Insect Sci 8:743–753. https://doi.org/10.1017/s1742758400022864

El-Kareim AIA, Darvas B, Kozár F (1988) Effects of the juvenoids fenoxycarb, hydroprene, kinoprene and methoprene on first instar larvae of *Epidiaspis leperii* Sign. (Hom., Diaspididae) and on its ectoparasitoid, *Aphytis mytilaspidis* (Le Baron) (Hym., Aphelinidae). J Appl Entomol 106:270–275. https://doi.org/10.1111/j.1439-0418.1988.tb00593.x

Elvira S, Williams T, Caballero P (2010) Juvenile hormone Analog technology: effects on larval cannibalism and the production of *Spodoptera exigua* (Lepidoptera: Noctuidae) Nucleopolyhedrovirus. J Econ Entomol 103:577–582. https://doi.org/10.1603/ec09325

Fathpour H, Noori A, Zeinali B (2007) Effects of a juvenoid pyriproxyfen on reproductive organ development and reproduction in German cockroach (Dictyoptera: Blattellidae). Iran J Sci Technol Trans A Sci 31:89–98. https://doi.org/10.22099/ijsts.2007.2319

Goncu E, Parlak O (2012) The effect of the juvenile hormone analog, fenoxycarb, on ecdysone receptor B1 expression in the midgut of Bombyx mori during larval-pupal metamorphosis. Folia Histochem Cytobiol 50:52–57. https://doi.org/10.2478/18696

Hartfelder K (2000) Insect juvenile hormone: from "status quo" to high society. Brazilian J Med Biol Res 33:157–177. https://doi.org/10.1590/S0100-879X2000000200003

Horowitz AR, Ishaaya I (2004) Biorational insecticides — mechanisms, selectivity and importance in pest management. In: Insect Pest Management. Springer, Berlin Heidelberg, pp 1–28

Hui JHL, Hayward A, Bendena WG et al (2010) Evolution and functional divergence of enzymes involved in sesquiterpenoid hormone biosynthesis in crustaceans and insects. Peptides 31:451–455. https://doi.org/10.1016/j.peptides.2009.10.003

Ishimaru Y, Tomonari S, Matsuoka Y et al (2016) TGF-β signaling in insects regulates metamorphosis via juvenile hormone biosynthesis. Proc Natl Acad Sci U S A 113:5634–5639. https://doi.org/10.1073/pnas.1600612113

Ishmuratov GY, Legostaeva YV, Garifullina LR et al (2015) Ozonolysis of unsaturated compounds in the synthesis of insect pheromones and Juvenoids. Chem Nat Compd 51:199–219. https://doi.org/10.1007/s10600-015-1246-3

Jindra M, Bittova L (2019) The juvenile hormone receptor as a target of juvenoid "insect growth regulators.". Arch Insect Biochem Physiol. https://doi.org/10.1002/arch.21615

Kamya PZ, Byrne M, Mos B et al (2017) Indirect effects of ocean acidification drive feeding and growth of juvenile crown-of-thorns starfish, *Acanthaster planci*. Proc R Soc B Biol Sci 284. https://doi.org/10.1098/rspb.2017.0778

Kenis M, Auger-Rozenberg MA, Roques A et al (2009) Ecological effects of invasive alien insects. In: Ecological Impacts of Non-Native Invertebrates and Fungi on Terrestrial Ecosystems. Springer, Dordrecht, pp 21–45

Lee SH, Oh HW, Fang Y et al (2015) Identification of plant compounds that disrupt the insect juvenile hormone receptor complex. Proc Natl Acad Sci U S A 112:1733–1738. https://doi.org/10.1073/pnas.1424386112

Li K, Jia QQ, Li S (2019) Juvenile hormone signaling – a mini review. Insect Sci 26:600–606

Losey JE, Vaughan M (2006) The economic value of ecological services provided by insects. Bioscience 56:311. https://doi.org/10.1641/0006-3568(2006)56[311:tevoes]2.0.co;2

Mian LS (2014) Juvenile hormones and Juvenoids: Modeling biological effects and environmental fate. J Am Mosq Control Assoc 30:71–72. https://doi.org/10.2987/14-6399.1

Miyakawa H, Toyota K, Sumiya E, Iguchi T (2014) Comparison of JH signaling in insects and crustaceans. Curr Opin Insect Sci 1:81–87

Mojaver M, Bandani AR (2010) Effects of the insect growth regulator Pyriproxyfen on immature stages of Sunn Pest, Eurygaster Integriceps Puton. Scutelleridae, Heteroptera

Nair KS, Gopal N, Kumar SN (2012) Improvement in cocoon yield induced by a juvenile hormone analogue, SB-515 in the bivoltine silkworm (*Bombyx mori* L.) Hybrid, CSR2 × CSR4 and its reciprocal combination. J Entomol 9:50–56. https://doi.org/10.3923/je.2012.50.56

Němec V (1993) Juvenoids: from basic research to practical use: a short review. Bull Insectol 48:67–74

Noriega FG (2014) Juvenile hormone biosynthesis in insects: what is new, what do we know, and what questions remain? Int Sch Res Not 2014:1–16. https://doi.org/10.1155/2014/967361

Noriega FG, Ribeiro JMC, Koener JF et al (2006) Comparative genomics of insect juvenile hormone biosynthesis. Insect Biochem Mol Biol 36:366–374. https://doi.org/10.1016/j.ibmb.2006.01.013

Ochoa-Sanabria C, Hogan N, Madder K et al (2019) Yellow mealworm larvae (*Tenebrio molitor*) fed sustainable protein source for animal feed ? Toxins (Basel) 11. https://doi.org/10.3390/toxins11050282

Oouchi H (2005) Insecticidal properties of a juvenoid, pyriproxyfen, on all life stages of the diamondback moth, *Plutella xylostella* (Lepidoptera: Yponomeutidae). Appl Entomol Zool 40:145–149. https://doi.org/10.1303/aez.2005.145

Palma KG, Meola SM, Meola RW (1993) Mode of action of pyriproxyfen and methoprene on eggs of *Ctenocephalides felis* (Siphonaptera: Pulicidae). J Med Entomol 30:421–426. https://doi.org/10.1093/jmedent/30.2.421

Qu Z, Kenny NJ, Lam HM et al (2015) How did arthropod Sesquiterpenoids and Ecdysteroids Arise? Comparison of hormonal pathway genes in noninsect arthropod genomes. Genome Biol Evol 7:1951–1959. https://doi.org/10.1093/gbe/evv120

Ramaseshadri P, Farkaš R, Palli SR (2012) Recent Progress in Juvenile Hormone Analogs (JHA) research. In: Advances in Insect Physiology. Academic Press Inc., Amsterdam, pp 353–436

Riddiford LM (2012) How does juvenile hormone control insect metamorphosis and reproduction? Gen Comp Endocrinol 179:477–484

Singh S, Kumar K (2015) Comparative efficacy of phenoxy derivative JHAs Pyriproxyfen and Diofenolan against polyphagous pest *Spodoptera litura* (Fabricius) (Noctuidae: Lepidoptera). Phytoparasitica 43:553–563. https://doi.org/10.1007/s12600-015-0473-2

Sláma K (2014) Insect hormones: more than 50-years after the discovery of insect juvenile hormone analogues (JHA, juvenoids). Terr Arthropod Rev 6:257–333. https://doi.org/10.1163/18749836-06041073

Srivastava US, Srivastava RC (1983) Juvenoid-induced supernumerary larval instars in certain stored grain insects. Proc Anim Sci 92(3):263–276

Takatsuka J, Nakai M, Shinoda T (2017) A virus carries a gene encoding juvenile hormone acid methyltransferase, a key regulatory enzyme in insect metamorphosis. Sci Rep 7. https://doi.org/10.1038/s41598-017-14059-8

Tan A, Tanaka H, Tamura T, Shiotsuki T (2005) Precocious metamorphosis in transgenic silkworms overexpressing juvenile hormone esterase. Proc Natl Acad Sci U S A 102:11751–11756. https://doi.org/10.1073/pnas.0500954102

Tanaka T, Iguchi T, Miyakawa H (2019) Establishment of a high-sensitivity reporter system in mammalian cells for detecting juvenoids using juvenile hormone receptors of *Daphnia pulex*. J Appl Toxicol 39:241–246. https://doi.org/10.1002/jat.3713

Thomas PJ, Bhatnagar-Thomas PL (1968) Use of a juvenile hormone analogue as insecticide for pests of stored grain. Nature 219:949

Truman JW, Riddiford LM (2019) The evolution of insect metamorphosis: a developmental and endocrine view. Philos Trans R Soc B Biol Sci 374:20190070. https://doi.org/10.1098/rstb.2019.0070

Walker WF (1976) Juvenoid stimulation of egg production in *Oncopeltus fasciatus* on non-host diets. Environ Entomol 5:599–603. https://doi.org/10.1093/ee/5.3.599

Wheeler DE, Nijhout HF (1982) Juvenile hormone and the physiological basis of insect polymorphisms. Q Rev Biol 57

Žďárek J, Denlinger DL (1975) Action of ecdysoids, juvenoids, and non-hormonal agents on termination of pupal diapause in the flesh fly. J Insect Physiol 21. https://doi.org/10.1016/0022-1910(75)90087-6

Pyrethroids: A Natural Product for Crop Protection

8

Kanchan Bhardwaj, Ruchi Sharma, Jayanthi Abraham, and Pankaj Sharma

Abstract

Pyrethroids are the synthetic compounds derived from the *Chrysanthemum cinerariaefolium* plant. The first synthetic pyrethroids developed in the United States are allethrin and bioallethrin. According to the World Health Organization, the classification pyrethroids has a place in the fourth group of insecticides and includes 42 substances. More than 30% of pyrethroid insecticides are used worldwide. In the year 2015, the global market of pyrethroid insecticides has been estimated at USD 4.67 billion and is expected to touch USD 6.45 billion by the year 2021. Pyrethroid insecticides are potent against an extensive variety of pests belonging to the orders *Coleoptera*, *Diptera*, *Hemiptera*, *Hymenoptera*, *Lepidoptera*, *Orthoptera*, and *Thysanoptera*. Pyrethroid insecticides interrupt the functioning of the peripheral nervous system by reacting with the voltage-gated sodium channels and cause a series of bursts and paralyses. The low tendency to accumulate in organisms, short biodegradation period, and economic value have led to the overuse of pyrethroids with unavoidable consequences. The increase in the production of mites in cotton, in tea, and in vegetables was reported by the constant use of synthetic pyrethroids. Even at a very low

K. Bhardwaj (✉)
Department of Botany, Shoolini University of Biotechnology and Management Sciences, Solan, Himachal Pradesh, India

R. Sharma
School of Bioengineering & Food Technology, Shoolini University of Biotechnology and Management Sciences, Solan, Himachal Pradesh, India

J. Abraham
School of Bio Sciences and Technology, Vellore Institute of Technology VIT, Vellore, Tamil Nadu, India

P. Sharma
Himachal Pradesh State Biodiversity Board, Vigyan Bhawan, Shimla, Himachal Pradesh, India

J. Singh, A. N. Yadav (eds.), *Natural Bioactive Products in Sustainable Agriculture*, https://doi.org/10.1007/978-981-15-3024-1_8

concentration in water, pyrethroids are strongly absorbed by the gills of fish due to their lipophilic nature and lead to their toxicity and even altered homing ability in honeybees. Pyrethroid metabolites were detected in the breast milk of women in various parts of the world. Long-term exposure of pyrethroids leads to aggressive behavior in humans due to the leftover traces of pyrethroid metabolites in urine. Further research should be done to prove the toxicity of pesticides in the ecosystem, the effects of pesticide residues, and their interaction with nutrients.

Keywords
Pyrethroids · Insecticides · Sodium channel · Toxicity · Pyrethroid degradation, World Health Organization

8.1 Introduction

Pyrethroids are synthetic insecticides and are the structural modifications of pyrethrins which are extracted from *Chrysanthemum cinerariaefolium* flowers of the genus *Chrysanthemum.* Pyrethrins are the esters of cyclopentenolone alcohol and cyclopropane carboxylic acid, and in the existence of sunlight, moisture, and water, they increase the insecticidal potency and longevity (Elliott 1995). Pyrethroids keep the acid/alcohol configuration of pyrethrins and have similar chemical structures across the class. As a result, concern of health effects can be made on the full class of pyrethroids. Over dozens of pyrethroid molecules are registered in various regions of the world which find application in many products of agriculture, household, veterinary, and in the field of medicine. Pyrethroid class includes allethrin, bioallethrin, bifenthrin, cyfluthrin, cypermethrin, deltamethrin, d-phenothrin, esfenvalerate, fenvalerate, fenpropathrin, flumethrin, fluvalinate-tau, lambda-cyhalothrin, permethrin, prallethrin, resmethrin, tefluthrin, and tetramethrin.

From the last 20 years, pyrethroid use has risen with their extensive exposure to environment, humans, and aquatic animals. Pyrethroids have a half-life of less than 8 h (Kim et al. 2008; Godin et al. 2010). In population sampling programs, urinary metabolites of pyrethroid have been confirmed (Health Canada 2013; Dewailly et al. 2014; Lewis et al. 2014; CDC 2015). Only recognition does not determine that an antagonistic health consequence will arise. So, there is an unending interest in the possible relations of pyrethroid exposure and health effects, especially on environment levels.

8.2 History

The pyrethrum flowers were firstly used by Caucasian tribes in the early 1800s to control body lice and later on produced on a commercial level in Armenia in 1828. In Dalmatia (Yugoslavia) the production started in about 1840 and was centered there until the First World War, in Japan before the Second World War, and after that in East Africa. Insect powder was first imported into the United States in about 1860, and several attempts were made for the next 90 years to produce the flowers commercially in this country but remained unsuccessful (Casida 1980). In the year 1940, Schechter and his colleagues developed the first synthetic pyrethroids, allethrin and bioallethrin, in the United States (Sanders and Taff 1954; BBSRC 2014). Pyrethroids are 20 times more successful in killing insects than dichloro-diphenyl-trichloroethane (DDT) without affecting environmental and human health (BBSRC 2014). Elliott's team at Rothamsted (United Kingdom) in the year 1962 developed a synthetic pyrethroid resmethrin by changing the molecular arrangement of naturally arising pyrethrin, and later on it was developed by the researchers at Sumitomo, a chemical company in Japan.

Later on, in the year 1967, the Elliott team isolated an active compound from synthetic pyrethroid resmethrin and again produced a first-generation pyrethroid, bioresmethrin, which is a mixture of four diverse isomers. Permethrin, the first pyrethroid to be used for agricultural purposes which does not collapse quickly in sunlight, was developed in 1972 by Michael Elliott. With the growing concern of the bioaccumulation of pesticides, for incidence, the breakdown of DDT in sunlight and its persistence in the environment lead to its ban by the United States in the same year. Two new extremely potent insecticides cypermethrin and deltamethrin were developed by Michael Elliott along with his colleague Izuru Yamamoto in Japan in 1976. Sumitomo, a chemical company, developed fenvalerate pyrethroid in 1976. Pyrethroids generate 25.1% of the worldwide insecticide market in the year 1983, and around 33 million hectares of crops were treated with pyrethroids (Wirtz et al. 2009). Owing to the low toxicity of pyrethroids, deltamethrin, and permethrin to humans and other mammals, the World Health Organization (WHO) in the 1980s recommend their use in insecticide-treated nets (BBSRC 2014). The annual sales of synthetic pyrethroids reached US $1.2 billion in the early 1990s (Housset and Dickmann 2009). A study was conducted in the rural region of Gambia in 1991 in which children under the age of 5 are treated by using permethrin-treated mosquito nets, and a reduction in the number of deaths by around two-thirds was observed (Alonso et al. 1991). In the year 2002, deltamethrin turned out to be the world's major-selling pyrethroid with yearly sale of US $208 million (BBSRC 2014). A product is developed by the researchers at Rothamsted (UK) in 2004, which releases an enzyme inhibitor to disable the insect's resistance mechanism to overcome resistance to pyrethroid insecticides that is emerging in a number of insect crop pests (BBSRC 2014). In the year 2007, global sales of insecticides reached US $8 billion with 17% of global insecticides as pyrethroids (Davies et al. 2007; BBSRC 2014). To tackle the cases of malaria, WHO in the year 2011 recommended the utilization of long-lasting insecticidal mosquito nets (LLINs) which were developed at Rothamsted (UK) (Lengeler 2004).

8.3 Classes of Pyrethroids

Pyrethroids are categorized into two classes, namely, class I and class II, on the basis of physical and toxicological properties (Gajendiran and Abraham 2018). Pyrethroids of class I contain cyclopropane carboxylic ester in their structure and include resmethrin, phenothrin, allethrin, tefluthrin, bifenthrin, permethrin, and tetramethrin. Class II pyrethroids contain a cyano group and include fenpropathrin, flumethrin, tralomethrin, deltamethrin, cyfluthrin, cyhalothrin, cypermethrin, fenvalerate, flucythrinate, and fluvalinate. These pyrethroids cause choreoathetosis and salivation. Pyrethroids are proficient in contrast to an extensive variety of pests which belong to the order Coleoptera, Hemiptera (Homoptera and Heteroptera), Diptera, Hymenoptera, Lepidoptera, Orthoptera, and Thysanoptera. Mostly, they are used for domestic purposes, for example, as a grain protectant, active in animal houses, fields, greenhouses, and in veterinary medicines (ATSDR 2003) (Table 8.1).

8.4 Mode of Action of Pyrethroids

The molecular targets of pyrethroid class of insecticides are the same in case of mammals and insects. Mode of action includes voltage-gated sodium, nicotinic receptors, chloride and calcium channels, intercellular gap junctions, gamma-aminobutyric acid (GABA)-gated chlorine channels, and membrane depolarization (Forshaw and Ray 1990; Song and Narahashi, 1996a, b). Mammals are vulnerable to pyrethroid toxicosis in small amount as compared to insects, the primary reason being higher body temperatures, rapid metabolic clearance, and a lower sympathy for pyrethroids (Song and Narahashi 1996b; Gammon et al. 2012). This particular insecticidal class slows the opening and closing of the sodium channels, causing the subsequent excitation of the cell (Marban et al. 1989). The action potential for type II pyrethroids is more durable than for type I. The direct exposure of pyrethroids causes paresthesia of the sensory nerve endings. This leads to the repetitive firing of the fibers. Sodium channels must be reformed by the insecticide to produce definite neurological signs and symptoms. In higher concentrations, pyrethroids of class II may act on GABA-gated chloride channels (Bloomquist et al. 1986) and control the cell excitability when it comes in contact with the voltage-dependent chloride channels existing in the brain, nerve, muscle, and salivary gland. Different forms of functional chloride channels are present when related to the sodium channels. Most of the insecticide-sensitive channels have been found to be linked with the Maxi chloride channel class, which gets triggered by various modes of excitation such as depolarization and protein kinase C phosphorylation. This particular channel has high conductivity and is calcium-independent.

There are a number of ways by which pyrethroids can penetrate into the body of an organism. One way is non-stereospecific in which pyrethroids permeate quickly from the epidermis, followed by uptake by the blood or hemolymph carrier proteins and continuously delivered all over the body. The main route of pyrethroid delivery to the central nervous system is along the epidermis cells. They directly enter into

Table 8.1 Classes of the pyrethroids used for crop protection, household, animal house, and veterinary

Pyrethroid	Class	Trade name	Molecular structure	Types of crop	Insecticidal role	Insecticidal role other than crop
Permethrin	I	Ambush	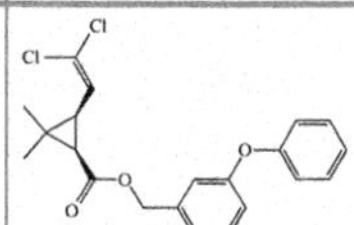	Cotton, wheat, maize, alfalfa, potato, spinach, green pepper, mushroom	Beetle, bollworm, budworm, termites, and weevils, moths	Ants, fleas, flies, lice, mosquitoes
Phenothrin	I	Sumithrin	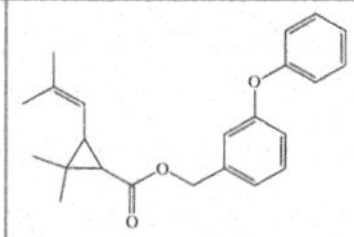	NA	NA	Flies, gnats, mosquitoes, cockroaches, and lice
Resmethrin	I	Chrysron	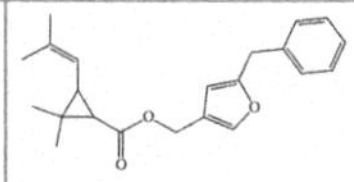	NA	NA	Flies, gnats, mosquitoes, fleas, ticks, and black flies
Tefluthrin	I	Force	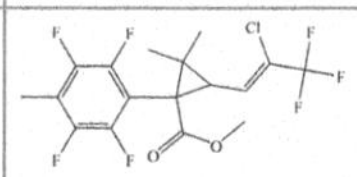	Sugar beet, cabbage, maize, carrot	White grub, southern corn leaf beetle, flea beetle, and chinch bug	NA
Bifenthrin	I	Brigade	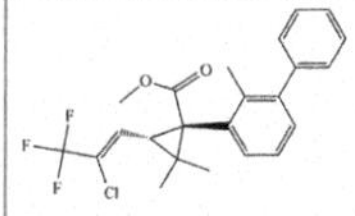	Corn, hops, raspberries	Beetles, weevil, aphids, moths, locust	Mosquitoes, lice, bedbugs, cockroaches
Allethrin	I	Pynamin	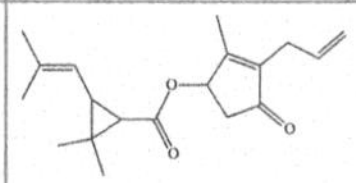	NA	NA	Flies, mosquitoes, and ants
Tetramethrin	I	Neo-Pynamin	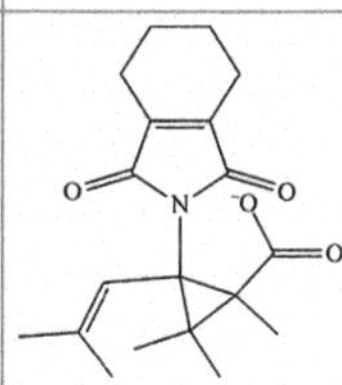	NA	NA	Wasps, hornets, roaches, ants, fleas, and mosquitoes

(continued)

Table 8.1 (continued)

Pyrethroid	Class	Trade name	Molecular structure	Types of crop	Insecticidal role	Insecticidal role other than crop
Cyfluthrin	II	Baythroid	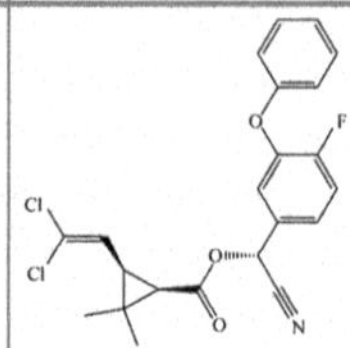	Artichokes, brassica crops, caneberries, canola, *Crambe*, rapeseed, cilantro, coriander, citrus, com, cotton, cucurbits, beans, peas, grapes, hops, lettuce (head), mayhaw, okra, peanut, pears, roots crops, soybean, spinach, tobacco, tomato	Aphids, cabbage stem flea beetle, rape winter stem weevil	Houseflies, cockroaches, mosquitoes
Cyhalothrin	II	Commodore	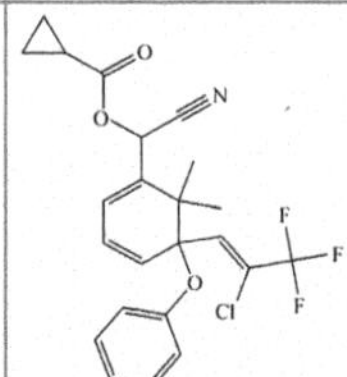	Tomato, Bengal gram, chili, grapes, onion, soya bean	Bedbugs, beetles, moths, weevils	Houseflies, ked, lice, mosquitoes
Cypermethrin	II	Ammo, Cymbush	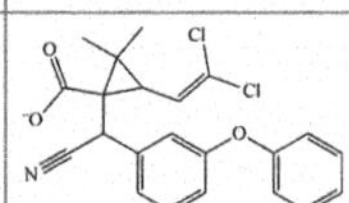	Wheat, okra, sunflower, cabbage	Moths	Cockroaches, mosquitoes, flies
Fenpropathrin	II	Danitol, Meothrin	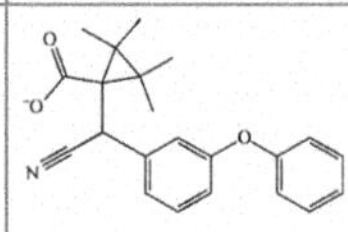	Tea, chili	Aphids, beet armyworm, mealybug, potato leafhopper, moths, leafrollers, and lace bugs	Mites

(continued)

Table 8.1 (continued)

Pyrethroid	Class	Trade name	Molecular structure	Types of crop	Insecticidal role	Insecticidal role other than crop
Fenvalerate	II	Pydrin, Ectrin	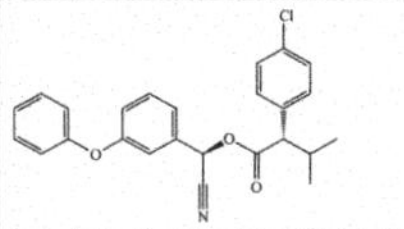	Cauliflower	Beetles, locusts, moths	Cockroaches, flies, mosquitoes
Flucythrinate	II	Cybolt	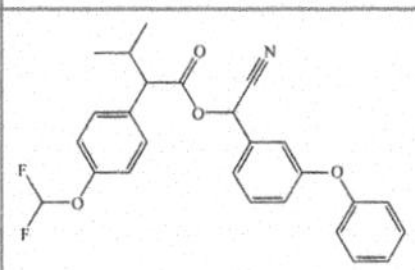	Lettuce, apples, cabbage, maize, cotton seed	Bollworms, leafworms, sucking insects, whiteflies, and beetles	NA
Fluvalinate	II	Klartan, Mavrik	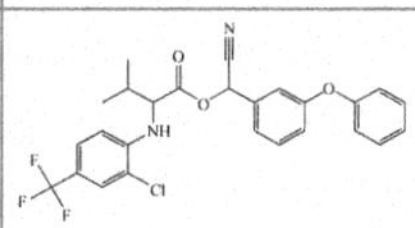	Cotton	Aphids, leafhoppers, moths, spider mites, thrips, and whiteflies	NA
Deltamethrin	II	Butoflin, Butoss	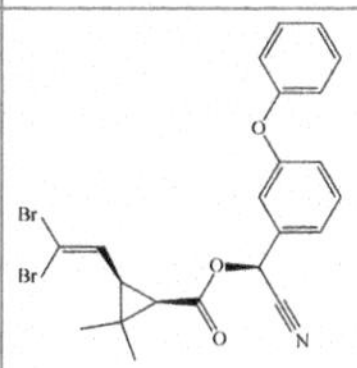	Wheat, rice	Aphids, beetles, bollworm, budworm, caterpillars, cicadas, moths, tortrix moths, weevils, whitefly, and winter moths	NA
Tralomethrin	II	Scout X-TRA	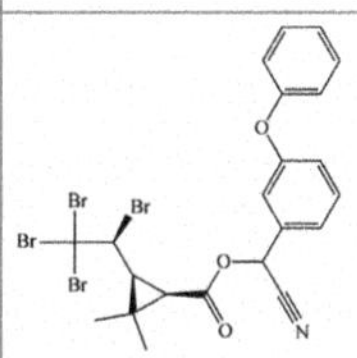	NA	NA	Ants and cockroaches

NA Not Applicable
Adapted from Gajendiran and Abraham (2018)

the central nervous system (CNS) via acting together with sensory organs of the peripheral nervous system. Also, they enter the body through the air in the vapor phase. Invertebrates and vertebrate insects are delicate to pyrethroids (Soderlund and Bloomquist 1989).

The peripheral and central nervous system of insects both are affected with the pyrethroids. Initially, they stimulate the nerve cells for the production of repetitive discharges which eventually cause paralysis. For the production of repetitive discharges, only a minor section of the sodium channel inhabitants is reformed by pyrethroids as with DDT. After alteration by pyrethroids, the sodium channels retain their capability to conduct Na^+, and the channels will remain open as the insecticide interferes with it and get closed either by inactivation or deactivation. The membrane potential is moved for the functioning of nerve cells in a comparatively stable form of abnormal hyperexcitability. In insects a sublethal effect known as "knockdown" is produced. Due to greater lipophilicity, the pyrethroids enters to the target more quickly and delivers better knockdown levels. Type I pyrethroids (e.g., permethrin) are capable of influencing repetitive firing in axons, restlessness, un-coordination, and hyperactivity followed by prostration and paralysis and are usually good knockdown agents as shown in Fig. 8.1. Pyrethroids of the class II (e.g., deltamethrin) with cyano group at the α-benzylic position (the α-carbon of the 3-phenoxybenzyl alcohol) caused a noticeable uncontrollable stage bringing about better kill because depolarization of the nerve axons and terminals is unalterable as shown in Fig. 8.1 (Bloomquist 1996).

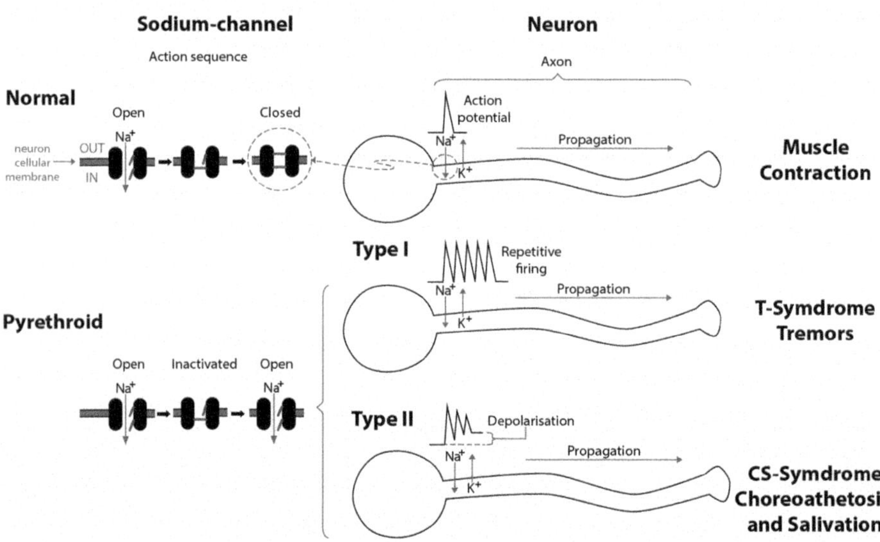

Fig. 8.1 Mode of action of pyrethroids on neurons. The top diagram shows the normal functioning of sodium channels which open, allowing sodium to pass, but then close after the action potential. This single action potential propagates through the nerve tail (axon) and triggers muscle contraction. Upon exposure to pyrethroids, the sodium channels malfunction and may remain open instead of returning to a closed state after initiation of the action potential. This will lead to repetitive firing (in type I pyrethroids) or depolarization (in type II pyrethroids) leading to tremors or involuntary movements (choreoathetosis) depending on the type of pyrethroid. Note that the T (fine tremors) and CS (choreoathetosis and salivation) syndromes are not as clearly differentiated as initially characterized in the pyrethroid literature and mixed symptoms may occur. (Adopted from Hénault-Ethier et al. 2016)

8.5 Current Uses of Pyrethroids

From the past 20 years, synthetic pyrethroids have been used in various crops to control pests (Maund et al. 2001), but they are becoming more and more popular even after the ban on the usage of cholinesterase-retarding insecticides (Feo et al. 2010; Luo and Zhang 2011). The ban on the use of two commonly used organophosphate (OP) pesticides, chlorpyrifos and diazinon, by the Environmental Protection Agency (EPA) in the year 2000–2001 resulted in the substantial rise in the marketplace diffusion of the pyrethroid products (EPA 2000, 2001). Due to the wide spectrum, high efficiency, low toxicity to mammals and avian, and biodegradability, the pyrethroids have a large share in the insecticidal market (Pap et al. 1996).

Nowadays, more than 30% of insecticides are used worldwide mostly in the field of horticulture, agriculture, forestry, public health and household purposes (Barr et al. 2010; Feo et al. 2010). The usage of synthetic pyrethroids and pyrethrins to control vector has been accepted by WHO and recommended the use of pyrethroids (lambda-cyhalothrin, bifenthrin, deltamethrin, cyfluthrin) for spraying indoor against malarial vectors (Walker 2000; Raghavendra et al. 2011). Pyrethroids are also applied on bed nets to control malarial vector (WHOPES 2005; Raghavendra et al. 2011).

According to the Environmental Protection Agency (EPA) data, about 1 million kg permethrin are used every year in agricultural, in household, and in public health fields (Feo et al. 2010). In 2015, the global market of pyrethroid insecticides has been evaluated at USD 4.67 billion and is expected to touch USD 6.45 billion by the year 2021 (Business Wire 2016).

8.6 Toxicity

Skin exposure is the most common route of entry for the insecticide pyrethroids (Gammon et al. 2012; Anadon et al. 2013). Its bioavailability usually accounts to 1% when exposed dermally. Absorption usually occurs via the stomach after an oral exposure in humans and mostly accounts to 36%. Soon after absorption, the insecticide gets quickly dispersed due to their lipophilicity and produce uncontrollable effects such as increased salivation and hyperexcitability. Majority of the pyrethroid formulations which are marketed contain solvents which are also the main cause of toxicity (Malik et al. 2010; Ensley 2018).

The half-life of this particular class of insecticide is usually hours (in blood plasma), while oral exposure is relatively shorter than the dermal exposure. Cyfluthrin has a half-life of 19–86 min. Acute toxicity is the major neurotoxicity caused due to pyrethroid exposure. Fishes are highly sensitive to pyrethroid (Ansari and Kumar 1988; Ensley 2018). Household exposure of fish to the insecticide can arise when the premises are sprayed with it. Birds are considered to be tolerant toward pyrethroid but they tend to be carriers. It has been reported that the LD_{50} value is greater than 1000 mg/Kg (Mueller-Beilschmidt 1990). Half-life values of the different pyrethroid compounds have been enlisted in Table 8.2. Clinical signs

Table 8.2 Half-life of pyrethroid compounds in environment

	Photolysis		Soil degradation	
Pyrethroid	Half-life in water	Half-life in soil	Aerobic soil	Anaerobic soil
Bifenthrin	408	96.9	96.3	425
Cyfluthrin	0.673	5.02	11.5	33.6
Cypermethrin	30.1	165	27.6	55
Deltamethrin	55.5	34.7	24.2	28.9
Esfenvalerate	17.2	10	38.6	90.4
Fenpropathrin	603	4.47	22.3	276
γ-Cyhalothrin	24.5	53.7	42.6	–
Pennethrin	110	104	39.5	197
Tralomethrin	2.47	3.87	3.25	5

Adopted from Laskowski (2002)

and symptoms after exposure have been observed to be almost similar when it comes to mammals such as cats and dogs. Some of which are as follows: salivation, vomiting, seizures, dyspnea, prostration, weakness, and eventually death (Ensley 2018). Apart from neurotoxicity, pyrethroids can also cause dermal, hepatic, renal, cardiac, endocrine disruption, reproduction, and developmental effects in mammals (Drago et al. 2014; Atmaca and Aksoy 2015; Hossain et al. 2015; Botnariu et al. 2016; Ben Slima et al. 2016; Malik et al. 2017; Ensley 2018).

8.7 Effect on Human Health

Usage of permethrin in household causes allergies and asthma, chiefly in children. A research conducted on 300 children residing in the Baltimore region presented a decline in the anti-inflammatory level IL-10 (Interleukin) in plasma as when related to people who are not in touch with pyrethroids (Skolarczyk et al. 2017). Similarly, when 5% permethrin was applied on the skin of 20-month-old child travailing from scabies showed symptoms of nausea, metabolic acidosis, respiratory distress, vomiting, and tachycardia (Goksugur et al. 2015). Metabolites of permethrin in concentrations 1.45–24.2 ng/g were recognized in the breast milk of women in Spain, Brazil, and Columbia (Corcellas et al. 2014). Long-term exposures of permethrin in children were described to cause an increase in the level of urine, behavioral changes, and an increase in aggressive behaviors shown in Fig. 8.2 (Outhlote and Bouchard 2013).

Similarly, exposure of deltamethrin at a dose level of 0.25–1% to humans for a long time through insecticidal mosquito nets caused lacrimation, limb spasms, abdominal pain, weakness, nausea, headaches, diarrhea, vomiting, apathy, ataxia, convulsions, and allergic reactions (Kumar et al. 2011). Permethrin metabolites were examined in the urine of 6-year-old children residing in Brittany (France) (Glorennec et al. 2017). The recommended dosage level of pyrethroids on a daily basis is 0.01 mg/kg, and poisoning symptoms occur after dosage of 2–250 mg/kg body weight. Aggregation of deltamethrin takes place in brain neurons when

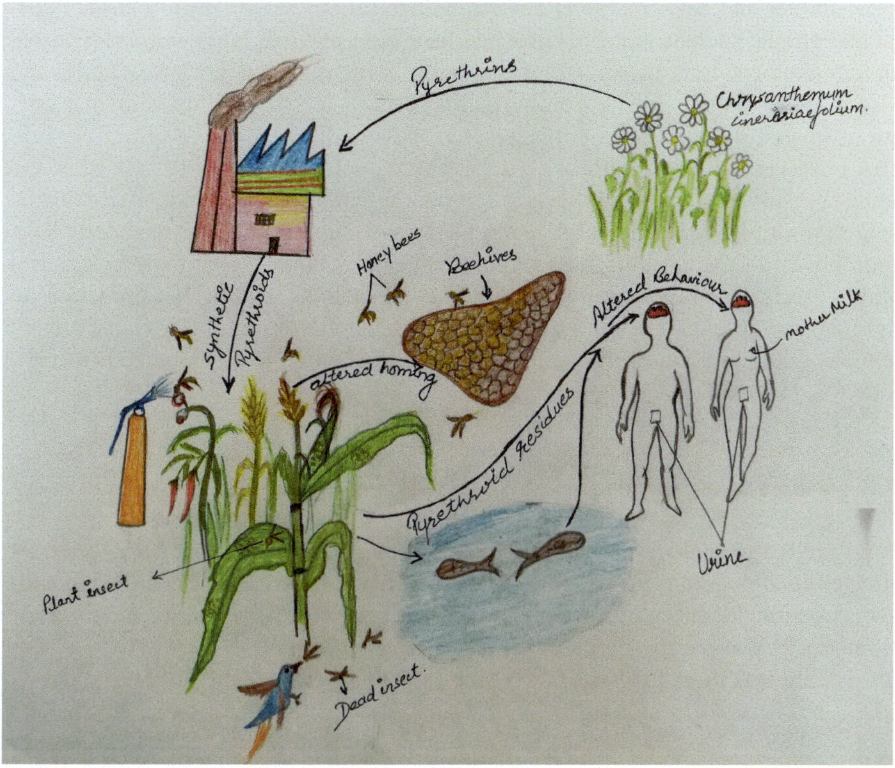

Fig. 8.2 Biomagnification of the pyrethroids in the ecosystem

administered orally or through the skin (Husain et al. 1996; Kim et al. 2008; Viel et al. 2015). Even exposure of deltamethrin in the course of pregnancy results in detrimental health effects such as fetal central nervous system (Husain et al. 1996; Viel et al. 2015). Children undergo sleep disorders, memory impairment, poor verbal abilities, and decrease in intelligence (Elwan et al. 2006; Viel et al. 2015). Deltamethrin contributes to Parkinson's disease by acting on the neuronal dopamine carrier (Elwan et al. 2006).

Alpha-cypermethrin metabolites were observed in the urine of people working in the cotton fields which further caused skin abrasions on the face and neck (Singleton et al. 2014). In contrast to permethrin, prolonged exposure of alpha-cypermethrin affects the central nervous system and induces complications with motor coordination and learning, but aggressive behaviors have not been observed (Manna et al. 2005). Through free radical formation, cypermethrin induces neurotoxicity, reduces the antioxidant defense mechanism, and inhibits the acetylcholinesterase (AChE) activity by acting together with the anionic substrate binding site (Sharma et al. 2014). Resveratrol improved the brain damage caused by cypermethrin by reducing oxidative stress and enhancing AChE activity in Wistar rats (Sharma et al. 2014).

Mcdaniel and Moser (1993) have concluded that Cypermethrin causes detrimental effects such as neurobehavioral changes in pawing, burrowing, salivation, choreoathetosis, hypothermia, and reduction in the motor activity (Mcdaniel and Moser 1993). Noticeable neuromuscular weakness, lateral head movements, variations in stimuli, equilibrium changes, retropulsion, and increased urination were also observed in cypermethrin toxicity (Mcdaniel and Moser 1993). Both the acute and toxic reactions of cypermethrin on the seminal gland, a rise in the height, multiplying of the cells, and a progressive appearance of mast cells have been observed (Mun et al. 2005; Rodriguez et al. 2009). Cypermethrin stands a highly used pesticide in agricultural practices as well as in household practices to fight against insects, but their consistent use may cause chronic toxicity among humans that may disturb the male fertility in upcoming years and also affect the food (Manna et al. 2005).

8.8 Effect on Animal Health

Pyrethroids are highly lethal to fish as they affect them indirectly through insecticide-affected food materials (WHO 2014; Hossain et al. 2017). Deltamethrin is the most toxic insecticide and allethrin as the least toxic followed by intermediately toxic pyrethroids, fenvalerate, permethrin, and cypermethrin (WHO 2014). LC_{50} values for fish are less than 1.0 parts per billion (ppb) in 40% cases. Fenvalerate mainly affects the nervous system of the teleost fish. There is an alteration in the calcium uptake, abnormal excretion rates of sodium and potassium, and increase in level of urine osmolality due to the production of osmoregulatory imbalance from fenvalerate (Shafer et al. 2008; Omotoso et al. 2014; Dohlman et al. 2016). This insecticide histologically damages the gill surface of fish by accumulating in the gills and causes mucus secretion, increases the aeration capacity, and decreases oxygen uptake efficiency in gills. Fenvalerate poisoning in fish causes reduction in the schooling behavior, inability to swim close to the surface of water, hyperactivity, buoyancy loss, raised cough level, increase in the secretion of gill mucus, head shaking, and lethargy prior to death (Kotila and Yön 2015).

Alteration in the behavior of honeybees to maintenance, feeding, and communication were observed when they are exposed to permethrin. Bees which receive surface exposure in the concentration of 0.001 μg permethrin were involved in trembling dances, self-cleaning, rotation, leg rubbing, and abdomen tucking than the nonexposed bees (Cox and Wilson 1984). About 90% of bees arrive to their hive within 30 s of journey, while among the deltamethrin-treated bees, only 9% were capable to return within this time. A change in the flight patterns and homing abilities was observed when forager honeybees were exposed to 2.5 ng deltamethrin per bee in relation to nonexposed bees (Vandame et al. 1995). Permethrin-exposed bees spend less time in walking, giving food, and antennae touching. Dietary exposures of pyrethroid concentrations (i.e., as in nectar or syrup) cause irregularities in behavior and fall in the fertility. The bees which feed on syrup comprising 940 μg/L deltamethrin were reported to exhibit learned alignment toward an odor stimulus by nearly 11–24% (Decourtye et al. 2005). Bifenthrin or deltamethrin fed diet at concentration level of 4.0, 7.9, 15.5, 30.6, and 60.2 mg/L or 20.0, 36.0, 64.8, 116.6, and 210.0 mg/L caused adverse impact in honeybees. Similarly, ingestion of bifenthrin and deltamethrin reduced the production of egg and the period in the egg stage. Exposure of deltamethrin lowered the capping frequency and prolongs the extent of the undeveloped stage (Dai et al. 2010).

The pyrethroids also influence birds because of the threat to their food supply. Small insectivorous and waterfowl are more prone to pyrethroids (Peter et al. 1996). They are mostly unaffected by pyrethroids as compared to mammals (Addy-Orduna et al. 2011). Quail ejected fenvalerate more quickly and showed poorer absorption and fast metabolism. The LD_{50} value of 4000 mg/kg body weight and 450 mg/kg body weight in quail and rat was observed when fenvalerate was administered orally which is nearly an order of 10 magnitudes higher (Dayal et al. 2003).

8.9 Degradation of Pyrethroid Residues

On the basis of clinical information and laboratory work, the pyrethroids hold estrogenic and antiprogestagenic actions and are categorized as endocrine disruptors (Garey and Wolff 1998). As a result, it is vital to create quick and proficient degradation methods to eradicate or decrease their amount in the environment. Biotic and abiotic methods comprising of photooxidation, chemical oxidation, and biodegradation degrade pyrethroids in the natural environment (Abraham and Silambarasan 2014; Abraham and Silambarasan 2016). Mainly, they are degraded by chemicals and native microorganisms present in the soil. Microorganisms play a substantial role in degradation of pyrethroids in the soil and sediments. Degradation frequency lies mainly on the type of pyrethroids, soil, climate, and the kind of microorganism and the size of their population. *Pseudomonas aeruginosa* CMG 154 make use of cypermethrin as the source of carbon (Thatheyus and Selvam 2013). The effectiveness of *Enterobacter asburiae* and *Pseudomonas stutzeri* for degradation of cypermethrin at concentration of 500 ppm was predicted (Thatheyus and Selvam 2013).

Lee et al. (2003) studied the capability of six bacterial strains and transformed bifenthrin and permethrin by isolating these bacteria from contaminated sediments. A degradation of permethrin and bifenthrin in the aqueous phase and reduction in their half-life from 700 h to 30–131 h were observed by using *Stenotrophomonas acidaminiphila*. Permethrin isomers can be degraded by using *Aeromonas sobria*, *Erwinia carotovora*, and *Yersinia*, and reduction by tenfold in the half-life of *cis*- and *trans*-permethrin was observed. Permethrin, deltamethrin, Fastac, fenvalerate, and fluvalinate were also degraded by using *Bacillus cereus*, *Achromobacter* spp., and *Pseudomonas fluorescens*. Of all the pyrethroids and deltamethrins, permethrin has a half-life of 21–28 days and can be degraded quickly (Maloney et al. 1988). The isolation of *Serratia plymuthica* and *Pseudomonas fluorescens* from synthetic pyrethroids-contaminated (SPs) farmland was noticed to degrade SPs by at least 50%. Biodegradation is a practical and suitable way for purifying SPs before disposing them either into soil, dip trough, or into the river (Grant et al. 2002).

8.10 Conclusion and Future Prospects

The resistance to change under the influence of radiant energy property of pyrethroids leads to discovery of the first pyrethroid, permethrin, and consequently this increases their use for management of pests. Pyrethroids are the broad-spectrum insecticides, that is, represent various compounds that are very toxic to nontarget land-dwelling insects and many aquatic organisms. The environmental providence and physical properties of pyrethrins and pyrethroids are clearly understood. Pyrethroids are sustained in the soil and sediments with a half-life greater than 30 days, but in contrast to legacy pesticide DDT, their half-lives are considerably lower. The sediment-residing invertebrates are mostly influenced by the pyrethroids because of their extensive half-lives mainly in urban areas where these insecticides are mostly used. Pyrethroids can be immediately biodegraded and are not biomagnified through different levels of the food chain. The research and expansion in the discovery of pyrethroids on commercial basis have mostly come to an end since the late 1990s, but in spite of this, efforts are going on to bring together isomer combinations of compounds like cypermethrin and cyhalothrin. With the ban on the use of fenvalerate and esfenvalerate, there is progress in the development of pyrethroids by many manufacturers in Japan which developed metofluthrin for commercial use; pyrethroid development appears to be well past its maximum (Matsuo et al. 2005). The pseudo-pyrethroids like etofenprox are the key for the continued commercialization of pyrethroids in Europe and the United States that are widely used and present lower acute toxicity to aquatic organisms. After 1984, pyrethrins, pyrethroids, and their synergists that were registered are presently experiencing process reviews in the United States to evaluate the efficacy of recent regulatory decisions and to consider new data. The registration review is concentrated on the progressive neurotoxicity. Pyrethroids are commonly being used for the past 40 years even though they are not pest-specific. However, they are target specific to an extensive range of pests and have low application amount, low mammalian toxicity, and a favorable environmental providence outline. The pyrethrins and pyrethroids will keep on being utilized in the future provided their utilization in a suitable way, and rules for them should be based on scientific indications.

References

Abraham J, Silambarasan S (2014) Biomineralization and formulation of endosulfan degrading bacterial and fungal consortiums. Pestic Biochem Physiol 116:24–31

Abraham J, Silambarasan S (2016) Biodegradation of chlorpyrifos and its hydrolysis product 3,5,6-trichloro-2-pyridinol using a novel bacterium *Ochrobactrum* sp. JAS2: a proposal of its metabolic pathway. Pestic Biochem Physiol 126:13–21

Addy-Orduna LM, Zaccagnini ME, Canavelli SB et al (2011) Formulated beta-cyfluthrin shows wide divergence in toxicity among bird species. J Toxicol 2011:803451

Agency for Toxic Substances and Disease Registry (2003) Toxicological profile for pyrethrins and pyrethroids. https://www.atsdr.cdc.gov/toxprofiles/tp155.pdf. Accessed 1 April 2019

Alonso PL, Lindsay SW, Armstrong JR et al (1991) The effect of insecticide-treated bed nets on mortality of Gambian children. Lancet 337:1499–1502

Anadón A, Arés I, Martínez MA et al (2013) Pyrethrins and synthetic pyrethroids:use in veterinary medicine. In: Ramawat KG, Mérillon J-M (eds) Natural products. Springer, Berlin, pp 4061–4086

Ansari BA, Kumar K (1988) Cypermethrin toxicity: effect on the carbohydrate metabolism of the Indian catfish, *Heteropneustes fossilis*. Sci Total Environ 72:161–166

Atmaca E, Aksoy A (2015) D-phenothrin-induced oxidative DNA damage in rat liver and kidney determined by HPLC-ECD/DAD. Environ Toxicol 30:607–613

Barr DB, Olsson AO, Wong LY et al (2010) Urinary concentrations of metabolites of pyrethroid insecticides in the general U.S. population: National Health and nutrition examination survey 1999–2002. Environ Health Perspect 118:742–748

BBRSC (2014) The history of the pyrethroid insecticides. https://bbsrc.ukri.org/documents/pyrethroid-timeline-pdf/. Accessed 2 May 2019

Ben Slima A, Chtourou Y, Barkallah M et al (2016) Endocrine disrupting potential and reproductive dysfunction in male mice exposed to deltamethrin. Hum Exp Toxicol 36:218–226

Bloomquist JR (1996) Ion channels as targets for insecticides. Annu Rev Entomol 41:163–190

Bloomquist JR, Adams PM, Soderlund DM (1986) Inhibition of gamma-aminobutyric acid-stimulated chloride flux in mouse brain vesicles by polychloroalkane and pyrethroid insecticides. Neurotoxicology 7:11–20

Botnariu G, Birsan C, Podoleanu C et al (2016) Skin necrosis caused by prallethrin- a worldwide used insecticide. Environ Toxicol Pharmacol 43:103–104

Business Wire (2016) Global pyrethroid insecticide market-growth, trends and forecasts (2016–2021). https://www.businesswire.com/news/home/20161206006254/en/Global-Pyrethroid-Insecticide-Market-Reach-6-Billion. Accessed 29 April 2019

Casida JE (1980) Pyrethrum flowers and pyrethroid insecticides. Environ Health Perspect 34:189–202

CDC (2015) Fourth report on human exposure to environmental chemicals, updated tables. https://www.cdc.gov/biomonitoring/pdf/FourthReport_UpdatedTables_Feb2015.pdf. Accessed 30 April 2019

Corcellas C, Feo ML, Torres JP et al (2014) Pyrethroids in human breastmilk: occurrence and nursing daily intake estimation. Environ Int 47:17–22

Cox RL, Wilson WT (1984) Effects of permethrin on the behavior of individually tagged honey bees, *Apis mellifera* L. (Hymenoptera: Apidae). Environ Entomol 13:375–378

Dai PL, Wang Q, Sun JH et al (2010) Effects of sublethal concentrations of bifenthrin and deltamethrin on fecundity, growth, and development of the honeybee *Apis mellifera ligustica*. Environ Toxicol Chem 29:644–649

Davies TG, Field LM, Usherwood PN et al (2007) DDT, pyrethrins, pyrethroids and insect sodium channels. IUBMB Life 59:151–162

Dayal M, Parmar D, Dhawan A et al (2003) Effect of pretreatment of cytochrome P450 (P450) modifiers on neurobehavioral toxicity induced by deltamethrin. Food Chem Toxicol 41:431–437

Decourtye A, Devillers J, Genecque E et al (2005) Comparative sublethal toxicity of nine pesticides on olfactory learning performances of the honeybee *Apis mellifera*. Arch Environ Contam Toxicol 48:242–250

Dewailly E, Forde M, Robertson L et al (2014) Evaluation of pyrethroid exposures in pregnant women from 10 Caribbean countries. Environ Int 63:201–206

Dohlman TM, Phillips PE, Madson DM et al (2016) Effects of label dose permethrin administration in yearling beef cattle: I. Bull reproductive function and testicular histopathology. Theriogenology 85:1534–1539

Drago B, Shah NS, Shah SH (2014) Acute permethrin neurotoxicity: variable presentations, high index of suspicion. Toxicol Rep 1:1026–1028

Elliott M (1995) Chemicals in insect control. In: Casida JE, Quistad GB (eds) Pyrethrum flowers: production, chemistry, toxicology, and uses. Oxford University Press, New York, pp 3–31

Elwan MA, Richardson JR, Guillot TS et al (2006) Pyrethroid pesticide-induced alterations in dopamine transporter function. Toxicol Appl Pharmacol 211:188–197
Ensley SM (2018) Pyrethrins and Pyrethroids. In: Gupta RC (ed) Veterinary toxicology basic and clinical principles. Academic Press, New York, pp 515–520
EPA (2000) Chlorpyrifos revised risk assessment and agreement with registrants. http://www.ibiblio.org/london/NAFEX/message-archives/old/pdf00000.pdf. Accessed 8 May 2019
EPA (2001) Diazinon revised risk assessment and agreement with registrants. https://www.ag.ok.gov/cps/epaagree.pdf. Accessed 8 May 2019
Feo ML, Eljarrat E, Barceló D (2010) Determination of pyrethroid insecticides in environmental samples. Trends Anal Chem 29:692–705
Forshaw PJ, Ray DE (1990) A novel action of deltamethrin on membrane resistance in mammalian skeletal muscle and non-myelinated nerve-fibers. Neuropharmacology 29:71–81
Gajendiran A, Abraham J (2018) An overview of pyrethroid insecticides. Front Biol 13:79–90
Gammon DW, Chandrasekaran A, Elnaggar SF (2012) Comparative metabolism and toxicology of pyrethroids in mammals. In: Marrs TC (ed) Mammalian toxicology of insecticides. Royal Society of Chemistry, Cambridge, UK, pp 137–183
Garey J, Wolff MS (1998) Estrogenic and antiprogestagenic activities of pyrethroid insecticides. Biochem Biophys Res Commun 251:855–859
Glorennec P, Serrano T, Fravallo M et al (2017) Determinants of children's exposure to pyrethroid insecticides in western France. Environ Int 104:76–82
Godin SJ, DeVito MJ, Hughes MF et al (2010) Physiologically based pharmacokinetic modeling of deltamethrin: development of a rat and human diffusion-limited model. Toxicol Sci 115:330–343
Goksugur SB, Karatas Z, Goksugur N et al (2015) Metabolic acidosis in an infant associated with permethrin toxicity. Pediatr Dermatol 32:15–17
Grant RJ, Daniell TJ, Betts WB (2002) Isolation and identification of synthetic pyrethorid-degrading bacteria. J Appl Microbiol 92:534–540
Health Canada (2013) Second report on human biomonitoring of environmental chemicals in Canada. https://www.canada.ca/content/dam/hc-sc/migration/hc-sc/ewh-semt/alt_formats/pdf/pubs/contaminants/chms-ecms-cycle2/chms-ecms-cycle2-eng.pdf. Accessed 20 May 2019
Hénault-Ethier L, Soumis N, Bouchard M (2016) Health and environmental impacts of pyrethroid insecticides: What we know, what we don't know and what we should do about it. https://equiterre.org/sites/fichiers/health_and_environmental_impacts_of_pyrethroid_insecticides_full_report_en.pdf. Accessed 18 May 2019
Hossain MM, DiCicco-Bloom E, Richardson JR (2015) Hippocampal ER stress and learning deficits following repeated pyrethroid exposure. Toxicol Sci 143:220–228
Hossain MM, Liu J, Richardson JR (2017) Pyrethroid insecticides directly activate microglia through interaction with voltage-gated sodium channels. Toxicol Sci 155:112–123
Housset P, Dickmann R (2009) A promise fulfilled- pyrethroid development and the benefits for agriculture and human health. Bayer Cropsci J 62:135–144
Husain R, Husain R, Adhami VM et al (1996) Behavioral, neurochemical, and neuromorphological effects of deltamethrin in adult rats. J Toxicol Environ Health 48:515–526
Kim KB, Anand SS, Kim HJ et al (2008) Toxicokinetics and tissue distribution of deltamethrin in adult Sprague dawley rats. Toxicol Sci 101:197–205
Kotila T, Yön ND (2015) The effects of permethrin on rat ovarian tissue morphology. Exp Toxicol Pathol 67:279–285
Kumar S, Thomas A, Pillai M (2011) Deltamethrin: promising mosquito control agent against adult stage of *Aedes aegypti* L. Asian Pac J Trop Med 4:430–435
Laskowski DA (2002) Physical and chemical properties of pyrethroids. Rev Environ Contam Toxicol 174:49–170
Lee S, Gan J, Kim J et al (2003) Microbial transformation of pyrethroid insecticides in aqueous and sediment phases. Environ Toxicol Chem 23:1–6
Lengeler C (2004) Insecticide-treated bed nets and curtains for preventing malaria. Cochrane Database Syst Rev 2:CD000363

Lewis RC, Cantonwine DE, Anzalota Del Toro LV et al (2014) Urinary biomarkers of exposure to insecticides, herbicides, and one insect repellent among pregnant women in Puerto Rico. Environ Health 13:97

Luo Y, Zhang M (2011) Environmental modeling and exposure assessment of sediment-associated pyrethroids in an agricultural watershed. PLoS One 6:e15794

Malik R, Ward MP, Seavers A et al (2010) Permethrin spot-on intoxication of cats literature review and survey of veterinary practitioners in Australia. J Feline Med Surg 12:5–14

Malik JK, Aggarwal M, Kalpana S et al (2017) Chlorinated hydrocarbons and pyrethrins/pyrethroids. In: Gupta RC (ed) Reproductive and developmental toxicology. Academic Press/Elsevier, Amsterdam, pp 633–655

Maloney SE, Maule A, Smith AR (1988) Microbial transformation of the pyrethroid insecticides: permethrin, deltamethrin, fastac, fenvalerate, and fluvalinate. Appl Environ Microbiol 54:2874–2876

Manna S, Bhattacharyya D, Mandal TK et al (2005) Neuropharmacological effects of alpha-cypermethrin in rats. Indian J Pharm 37:18–20

Marban E, Yamagishi T, Tomaselli GF (1989) Structure and function of voltage-gated sodium channels. J Physiol 508:647–657

Matsuo N, Ujiharah K, Shono Y et al (2005) Discovery and development of a novel pyrethroid insecticide'metofluthrin (SumiOne®,Eminence®)' (2005). Sumitomo Kagau 2:1–15

Maund SJ, Travis KZ, Hendley P et al (2001) Probabilistic risk assessment of cotton pyrethroids: V. Combining landscape-level exposures and ecotoxicological effects data to characterize risks. Environ Toxicol Chem 20:687–692

Mcdaniel KL, Moser VC (1993) Utility of a neurobehavioral screening battery for differentiating the effects of two pyrethroids, permethrin and cypermethrin. Neurotoxicol Teratol 15:71–83

Mueller-Beilschmidt D (1990) Toxicology and environmental fate of synthetic pyrethrois. J Pestic Reform 10:32–37

Mun JY, Lee WY, Han SS (2005) Effects of cypermethrin on the dopaminergic neurons in the progressive hemiparkinsonian rats. Toxicol Mech Methods 15:399–404

Omotoso GO, Onanuga IO, Ibrahim RB (2014) Histological effects of permethrin insecticide on the testis of adult wistar rats. J Med Biomed Sci 6:125–129

Outhlote Y, Bouchard M (2013) Urinary metabolities of organophosphate and pyrethroid pesticides and behavioral problems in Canadian children. Environ Health Perspect 121:1378–1384

Pap L, Bajomi D, Szekely I (1996) The pyrethroids, an overview. Int Pest Council 38:15–19

Peter JV, John G, Cherian AM (1996) Pyrethroid poisoning. J Assoc Physicians India 44:343–344

Raghavendra K, Barik TK, Reddy BPN et al (2011) Malaria vector control: from past to future. Parasitol Res 108:757–779

Rodriguez H, Tamayo C, Inostroza J et al (2009) Cypermethrin effects on the adult mice seminal glands. Ecotoxicol Environ Saf 72:658–662

Sanders HJ, Taff AW (1954) Staff industry collaborative report allethrin. Ind Eng Chem 46:414–426

Shafer TJ, Rijal SO, Gross GW (2008) Complete inhibition of spontaneous activity in neuronal networks in vitro by deltamethrin and permethrin. Neurotoxicology 29:203–212

Sharma P, Firdous S, Singh R (2014) Neurotixic effect of cypermethrin and protective role of resveratrol in wistar rats. Int J Nutr Pharmacol Neurol Dis 4:104–111

Singleton ST, Lein PJ, Farahat FM et al (2014) Characterization of α-cypermethrin exposure in Egyptian agricultural workers. Int J Hyg Environ Health 217:538–545

Skolarczyk J, Pekar J, Nieradko-Iwanicka B (2017) Immune disorders induced by exposure to pyrethroid insecticides. Postepy Hig Med Dosw 71:446–453

Soderlund DM, Bloomquist JR (1989) Neurotoxic actions of pyrethroid insecticides. Annu Rev Entomol 34:77–96

Song JH, Narahashi T (1996a) Differential effects of the pyrethroid tetramethrin on tetrodotoxin-sensitive and tetrodotoxin-resistant single sodium channels. Brain Res 712:258–264

Song JH, Narahashi T (1996b) Modulation of sodium channels of rat cerebellar purkinje neurons by the pyrethroid tetramethrin. J Pharmacol Exp Ther 277:445–453

Thatheyus AJ, Selvam ADG (2013) Synthetic Pyrethroids: toxicity and biodegradation. Appl Ecol Environ Sci 1:33–36

Vandame R, Meled M, Colin ME et al (1995) Alteration of the homingflight in the honey bee *Apis mellifera* L. exposed to sublethal dose of deltamethrin. Environ Toxicol Chem 14:855–860

Viel JF, Warembourg C, Le Mauer-Idrissi G et al (2015) Pyrethroid insecticide exposure and cognitive developmental disabilities in children: the PELAGIE mother-child cohort. Environ Int 82:69–75

Walker K (2000) Cost-comparison of DDT and alternative insecticides for malaria control. Med Vet Entomol 14:354–354

WHOPES (2005) Safety of pyrethroids for public health use. https://apps.who.int/iris/bitstream/handle/10665/69008/WHO_CDS_WHOPES_GCDPP_2005.10.pdf;jsessionid=E3B3DD50F6AFF1C15CCB063EC397E062?sequence=1. Accessed 22 May 2019

Wirtz K, Bala S, Amann A et al (2009) A promise extended-future role of pyrethroids in agriculture. Bayer Cropsci J 62:145–158

World Health Organization (WHO) (2014) Specifications and evaluations for public health pesticides alpha-cypermethrin long-lasting (incorporated into filaments) insecticidal net. https://www.who.int/whopes/quality/Alpha-cypermethrin_incorporated_LN_specs_eval_WHO_May_2014.pdf. Accessed 11 May 2019

Natural Products as Fungicide and Their Role in Crop Protection

9

Hiran Kanti Santra and Debdulal Banerjee

Abstract

Seeking solutions from nature for solving one and all problems is the age-old practice for mankind, and natural products are proved to be the most effective one for keeping up the balance of development as well as the "healthy, wealthy, and well" condition of mother nature. Fungal pathogens are proved to be a common and popular contaminant of agroecosystem that approximately causes 70–80% of total microbial crop loss. To meet the proper global increasing need of food products as a result of population explosion, managing agricultural system in an eco-friendly and profitable manner is the prime target; thus the word "sustainable agriculture" plays it part, and this package is highly effective when coupled with nature-derived fungicidal products that can minimize the event of fungal infections in agrarian ecosystem. Present study enlists the most common and effective natural products that might be of plant or microbial origin, their mode of action, day-by-day development of phytopathogenic resistance against the prevailing fungicides, and also their role in maintenance of sustainability of agricultural practices with special emphasis on their acceptance over the synthetic or chemical one. A large number of bioactive compounds ranging from direct plant (both cryptogams algae and moss and phanerogams)-derived natural extracts, essential oil of aromatic plants, and low-molecular-weight antimicrobial compounds known as phytoalexins to secondary metabolites that are both volatile and nonvolatile organic compounds of microbes (fungal and actinobacterial members) residing inside the host tissue, called endophyte, are widely used as agricultural bioweapons. The rhizospheric partners of plant, mycorrhizae, are also a prime agent of this chemical warfare and protect their green partners from fungal invaders and emphasize the concept of "sustainable agriculture."

H. K. Santra · D. Banerjee (✉)
Microbiology and Microbial Biotechnology Laboratory, Department of Botany and Forestry, Vidyasagar University, Midnapore, West Bengal, India
e-mail: db@mail.vidyasagar.ac.in

J. Singh, A. N. Yadav (eds.), *Natural Bioactive Products in Sustainable Agriculture*, https://doi.org/10.1007/978-981-15-3024-1_9

Keywords
Sustainable agriculture · Eco-friendly · Bioactive compounds · Algae · Moss · Phanerogams · Phytoalexins · Mycorrhizae · Chemical warfare

9.1 Introduction

Natural products are the best weapon for the survival of any type of problems regarding infection, pathogenesis, or protection from diseases. Due to their degradability in nature, they are the first options to be used by agriculturalists and plant biologist for combating fungal pathogenesis. The eukaryotic organism fungi have a separate kingdom in Whittaker's five kingdom classification and are prime member of this ecosystem as a potent decomposer. In spite of their heavy and multidimensional applications in agricultural, medicinal, and industrial field ranging from the production of life-saving medicines to food supplements, they are the cause of huge global crop loss each year and lead to economic exhaustion. Macro- and microscopic fungi-producing fruit bodies on different portions of a plant body (stem, leaf, fruit, root) lead to the death and decline of the crop species. Several methods have already been tried since the start of civilization for crop protection but because of plant and fungal coevolution, fungi have dominated on the green eukaryotes and caused significant reduction in the crop yield. Particularly in a country like India where the central GDP largely depends on agricultural output, the fungal pathogenesis has been a matter of grave concern for the agriculture department and policy makers. Huge amount of money and manpower is invested to fight against the fungal diseases for ensuring higher and qualitative yield, but still it has been a burning problem of today's conditions. The problem with chemical and synthetic tools for combating the parasitic infections includes their toxicity leading to quality deterioration and environmental pollution accompanied with side effects on human health. In a case study, it has been reported that the extreme use of antibiotics in agricultural field and their direct consumption by humans through their daily meal have led to resistance of those antibiotics in human fungal pathogens. So we are in search of bioactive agents that will be of biological origin, selective to their host, and produce no secondary symptoms with least negative impact. The problem with fungi is their secretion of various types of mycotoxins (aflatoxins, ochratoxins, patulin, fumonisin, zearalenone, deoxynivalenol, etc.) in the stored food products, causing postharvest loss of cereals, pulses, dry fruits, and spices. Mycotoxins are not only food spoilers but also potent disease-causing agents in humans leading to cancer, liver damage, kidney failure, and paralysis (Miller 1995). The severe effects of fungal crop loss are visible largely in tropical or subtropical regions where the temperature is moderately higher than the other parts of the world. Fungal devastation occurs in two phases: firstly, when the crops are growing on the field and, secondly, when they are stored for further transportation, postharvest loss. The third type of contamination occurs when the microscopic airborne pathogens like molds grow on cooked foods and lead to food spoilage. At each and every level, scientists have developed techniques to minimize fungal food loss. On a gross annual estimate, almost 25% of agricultural food items are of no use due to fungal contamination (Pittet 1998). The major issues with fungi-related crop loss are deterioration as a result of increase of fatty acid conditions, change of color and texture of food items, poor nutritional conditions, and poor germinability of stored seeds (Dhingra et al. 2001). Reports from Asia and Africa include death of humans and animals due to consumption of mycotoxin-contaminated foods (Reddy and Raghavender 2007). Fungal pathogens are sometimes dependent on more than one host for their successful completion of life cycle and disease development (*Puccinia graminis* var. *tritici*, causal agent of black stem rust of wheat that requires *Berberis aristata* for successful infection other than there main target wheat plant). So physical controls like eradication of secondary or collateral host and burning of the old livestocks and remnants of the field are the primary measures adopted by the farmers for disease-free crop production. So maintaining the sustainability along with less pathogenic infection is the deep ecological movement for crop maintenance. There are reports of resistance developed against the common and widely used antibiotics of agricultural importance. Blasticidin S, an antibiotic obtained from *Streptomyces* sp. (a type of actinobacteria predominantly present in soil samples), interacts with the protein synthesis and causes the death of the rice blast pathogens. Development of resistance of this antibiotic is reported to be present in some fungal pathogens that detoxify it by deamination (Dayan et al. 2009). Compounds of bacterial and fungal origin from both soil and endophytic sources are used as an alternative source over the chemical ones. Plant extracts especially essential oils from plant taxa of Lamiaceae family are of immense importance and are used as fungicidal or fungistatic. Most of the active ingredients act upon the fungal cell wall by either blocking the cellular processes like respiration, cell wall and cell membrane synthesis, ergosterol biosynthesis, protein synthesis, or DNA replication. Not only the secondary metabolites of plant and microbial origin but also the direct application of microorganisms in terms of biocontrol agent could be used as potent antifungals. Other than these, plants' own defense molecules, known as the phytoalexins, could provide a strong line of defense against mycorrhizae; the root symbionts of higher plants can physically, biologically, and biochemically protect the plant root from pathogenic invasion and provide an enhanced resistance conditions to their hosts. This study includes the role of these compounds as natural agents of antifungal property and their role in disease prevention.

9.2 Mycorrhizae as a Biocontrol Agent

9.2.1 Introduction

Mycorrhiza being the perfect example of symbiosis is known to be the oldest association between higher plant (both angiosperm and gymnosperm, monocot and dicot plants) and fungi and is an astonishing phenomenon of nature. The mycorrhizal association is one of nature's privileges for maintaining the sustainability of agriculture. In present day's changing environment, haphazard use of pesticides (fungicides) and chemicals poses a great risk to the existence and survival of mycorrhizal species in its complete biologically active form. There is a need to increase awareness in order to save mycorrhizal fungi from extinction.

Plants form beneficial association with other variants of life forms (animals, bacteria, or fungi) to complete their life processes, to fight against pathogenic microorganisms, and most importantly to thrive in adverse environmental situations. The plant root and its associated living microbial flora are together called "rhizosphere," particularly the area of mycorrhizal occurrence. The term mycorrhiza is derived from two Greek words: *mycos* which means fungus and *rhiza* which means roots. In nature, more than 80% of angiosperms and almost all of gymnosperms are known to have mycorrhizal associations. The common two types of mycorrhizal associations that exist in nature are endomycorrhizae, also called arbuscular mycorrhizae (AM), for example, *Endogone* sp. and *Rhizophagus* sp., and ectomycorrhizae (EM), for instance *Amanita muscaria* and *Laccaria bicolor*. Mycorrhizal associations support its host plants to survive in untimely soil conditions and drought situations by increasing the surface area of root and efficiency of mineral uptake. Environmental threats including problems of temperature increase, climate changing, drought, and infertility of soil are some of the major challenges in agriculture and have to be mitigated to ensure global food security. In this context, mycorrhiza-based crop production is one of the key components of sustainable agriculture practices.

9.2.2 Interaction Between AM Fungi and Plant Pathogens

In most of the cases, AM fungi–mediated suppression of root pathogenic fungi is achieved by either morphological, physiological, or biochemical alterations of the host. Several experiments on fungistatic activity of the mycorrhizal species have been done, and fruitful results are found against pathogenic fungi such as *Aphanomyces* spp., *Botrytis fabae, Chalara* (*Thielaviopsis*) *basicola, Dothiorella gregaria, Fusarium oxysporum, Gaeumannomyces graminis* var. *tritici, Ganoderma pseudoferreum, Pythium ultimum, P. splendens, Phytophthora parasitica, P. cactorum, P. vignae, Rhizoctonia solani, R. bataticola,* and *Sclerotium rolfsii* (Lioussanne et al. 2009; Bagyaraj 2006; Bagyaraj and Chawla 2012). The most common outcome of AM fungal colonization is seen as an increase in number of branches, resulting in a relatively larger proportion of higher-order roots in the root system. Thickening of the cell walls due to lignification and production of polysaccharides in mycorrhizal plants are the common mode of prevention of penetration and growth of pathogens like *Fusarium oxysporum* and *Phoma terrestris*. A huge percentage of AM-root pathogen interaction studies have been conducted in crop plants of agricultural and horticultural importance. But the information available on forest tree species is scanty. Mycorrhizal technology can thus play an important role in production of low-cost quality seedlings and provide plant protection. Like other methods of biological control, AM fungi are not able to offer complete immunity against the infection caused by plant pathogens. They could only impart a degree of resistance against soilborne plant pathogens. However, the possibility of biologically controlling soilborne plant pathogens looks promising.

9.2.3 Enhancement of Plant Defense Mechanism

AM fungi play a protective role for plants by activating the defense mechanisms for the better resistance of crop plants and thus may protect the host plant from further fungal pathogenic attack, thus working as a potent biocontrol agent. Researchers have proved that AM symbiosis triggers the activation of several defense-related genes and also expression of pathogenesis-related proteins. Evidences are drawn from modern techniques like molecular biology methods and immunological and histochemical analysis that strongly supports this concept. AM fungi first act as a biotrophic agent, and before entering the host plant's root cell, they cause a sharp change in endogenous salicylic acid that is reflected in quick accumulation of reactive oxygen species (ROS), a wide range of hydrolytic enzymes, and also the activation of phenylpropanoid biosynthetic pathway (Güimil et al. 2005, Paszkowski 2006, Roman et al. 2011). Research findings have proved that the amount of defense-related compounds (essential enzymes like PAL, phenylalanine ammonia-lyase, a product of phenylpropanoid pathway, enzymes needed for flavonoid or isoflavonoid biosynthesis like chalcone isomerase) that act for the protection of plant from fungal and bacterial pathogen is higher in the case of mycorrhiza-inoculated plant than in the uninoculated ones (Volpin et al. 1994, 1995). Host's physiological and biochemical processes are greatly influenced by the mycorrhizal association in terms of decreased root exudation, higher concentration of phenylalanine and serine contents, ortho-dihydroxy phenols, increased membrane phospholipid content, etc. (Smith et al. 1994). When the phospholipid contents are high, it reduces the chances of root pathogenic attack, and higher concentrations of ortho-dihydroxy phenols show inhibitory activity against root rot pathogen *Sclerotium rolfsii* (causal agent of southern blight), whereas the non-mycorrhizal plants are affected by the southern blight disease. Tomato plants when inoculated with *G. fasciculatum* show inhibitory activity against root knot nematodes. Host–AM association leads to the formation of defense-related compounds like phytoalexins, chitinases (CHI), β-1,3-glucanase (GLU) (enzyme related to hydrolysis of fungal cell wall), peroxidases (POX), hydroxyproline-rich glycoproteins, and phenolics (St Arnaud and Vujanovic 2007). Synergistic effect of PGPRs along with AM fungi is proved to be a system of better protection than the use of AM fungi alone (Linderman 1994; Bagyaraj and Chawla 2012). Fungal wilt of common medicinal plant Indian coleus (*Coleus forskohlii*) caused by *Fusarium chlamydosporium* could be minimized by the joint action of AM fungus and *Trichoderma viride* and cause a sharp increase in root yield and root forskolin concentration and may also reduce the severe disease conditions (Singh et al. 2012).

9.2.4 Change in Rhizospheric Microbial Population

AM causes a drastic change in the rhizospheric microbiota and intentionally either removes directly the pathogenic microorganisms or stimulates the accumulation of potent microbial partners especially fungus that are heavily antagonistic to the plant pathogenic ones. Plants with mycorrhizal association harbor higher population of rhizospheric microorganisms, thus making it impossible for the pathogen to compete and invade the root. In the case of *Phytophthora cinnamomi*, the numbers of sporangia and zoospores are found to be reduced when rhizospheric soil extracts of AM plants are applied; it means the AM fungi are able to alter the microbial population and particular functional groups of rhizospheric microorganisms (Meyer and Linderman 1986; Larsen and Bodker 2003). They cause qualitative and quantitative changes in the fungal community by several factors like changed exudation patterns; altered root size and architecture; different physiological and biochemical parameters like sugar, organic acids, and amino acids; and also putative direct AM fungal effects (Toljander et al. 2007; Ahmed et al. 2013; Vigo et al. 2000). Fungistatic siderophore (low-molecular-weight chelating agents having higher affinity for ferric ion)-producing microorganisms are found to be crowded in mycorrhiza-infected roots and rhizospheric regions. Mycorrhizal plants are to be reported with more actinomycetes and bacterial (Gram-positive *Paenibacillus* sp. against *Phytophthora parasitica*) flora antagonistic to soilborne root pathogens (Azcon-Aguilar and Barea 1996; Budi et al. 1999).

9.2.5 Change in Root Anatomy

Apart from providing biochemical and physiological defense strategies, arbuscular mycorrhizal species also exhibit physical barrier of defense by changing the root anatomy, morphology, and even architecture in terms of increased nutrient uptake, meristematic and nuclear activities of root cell, higher rate of growths, and branching patterns (Atkinson et al. 1994; Gamalero et al. 2010; Gutjahr and Paszkowski 2013). Thus responses of root morphology as a result from AFM colonization seem to depend on plant characters, tap root system, etc. More benefits are seen in tap root system than fibrous root system in terms of gained biomass and nutrient acquisition. Though there is a gap of knowledge in how increased root branching caused by mycorrhizal infection help the plant to defend fungal pathogenesis, synergism is seen as something that can balance the suppressed root growth caused by several root pathogens and restore the root health.

9.2.6 Enhanced Nutrient Uptake by the Host and Competition Between the Symbiotic Partner and Pathogenic One

Mycorrhiza-mediated strengthening of the vascular system allows the higher rate of flow of nutrients, increased mechanical strength, and also inactivation of vascular pathogens. In conditions of limited resource such as carbon requirement and space for inoculation, a competition between the symbiotic partner (mycorrhizae) and pathogenic fungi is very common and expected (Vos et al. 2014). In the direct warfare, mycorrhizae win over the pathogenic one and thus obtain higher amount of nutrients (almost 4–20% of total assimilated carbon by host plant) and occupy large areas of available root cortical cells (Jung et al. 2012; Vierheilig et al. 2008). Defeating the pathogenic fungi in terms of nutrient uptake and providing a little or no room for infection are probably the mightiest cause of biocontrol ability of AM fungus (Hammer et al. 2011). Output of AM and *Phytophthora* interaction indicates that the pathogen does not penetrate cortical arbuscular cells, suggesting that localized competition for infection site does occur between the pathogenic fungi and the AM fungus. Not only fungi but also plant-invading nematodes are in the queue for colonization and nutrient uptake (Smith 1988). The infection of southern root-knot nematode (*Meloidogyne incognita, M. exigua*) is reduced when the roots are priorly inoculated with symbiotic partners like in the case of coffee plants also (Alban et al. 2013; Dos et al. 2010). Reports have suggested that the number of infected sites is reduced within mycorrhizal root system than in the uninoculated one and thus strongly supports the mycorrhizal role as a biofungicide (Vigo et al. 2000). AM fungi can help the plant uptake of nutrients like phosphate, nitrogen, minerals, microelements (zinc), and water at a higher rate than the uninfected one (Baum et al. 2015; Parniske 2008), and as a result, they are provided with photosynthetic carbon (Smith and Smith 2011). The plants capable of absorbing higher amount of nutrients in terms of AM fungal association have the potential to tolerate pathogenic infections (Karagiannidis et al. 2002). Though the improved nutrition and increased tolerance are not involved in a cause–effect relationship, proofs are there that higher uptake of phosphate results in remarkable reduction in pathogenic infection in mycorrhizal plant but not in non-mycorrhizal plant (Bodker et al. 1998). Tomato plants already infected with *Rhizophagus irregularis* are not colonized by the pathogen *A. solani*, whereas non-mycorrhizal plant is affected by the pathogen (Fritz et al. 2006). Mixed action of arbuscular mycorrhizal fungi (AMF) *Glomus intraradices* and *Trichoderma harzianum* as a biocontrol agent significantly reduces the damping off disease caused by *Rhizoctonia solani* in the case of tomato seedlings (Amer and Seud 2008).

9.3 Phytoalexins as Plant Protectants

9.3.1 Introduction

In order to combat parasitic (fungal, bacterial, viral, nematoidal, and insectal) infection like mammalian cells, plant cells also develop defense systems that mediate the release of low-molecular-weight and short-lived (generally 72–96 h of existence) antimicrobial compounds or molecules known as the phytoalexins (Braga 1991; Echiverri et al. 2010; Paxton 1980). These secondary metabolites help the plant to withstand biotic and abiotic stress (Grayer and Kukubun 2001). Most of them being

Table 9.1 Mode of action of phytoalexin as an antifungal agent

Phytoalexin	Host plant	Mode of action	Target organism	References
Phaseolin or kievitone	*Phaseolus mungo*	Inhibition of glucose uptake of fungal cell	*Rhizoctonia solani*	VanEtten and Bateman (1971)
Rishitin, phytuberin, Anhydro-β-rotunol, solavetivone	*Solanum tuberosum*	Loss of motility and swelling of the zoospores, formation of cytoplasmic granules, and bursting of the cell membrane (leakage of electrolytes and metabolites)	*Phytophthora* sp.	Harris and Dennis (1977)
Stilbene, resveratrol, pterostilbene	*Vitis vinifera*	Disorganization of mitochondria Disruption of plasma membrane Uncoupler of ETS and blocker of photosynthesis	*Botrytis cinerea*	Pezet and Pont (1990), Adrian et al. (1997), Adrian and Jeandet (2012)
Camalexin	*Phaseolus vulgaris*	Induction of the fungal programmed cell death (PCD) by apoptotic mechanisms	*Botrytis cinerea*	Shlezinger et al. (2011)

lipophilic compounds can cross the plasma membrane and act inside the fungal cell causing cytoplasmic granulation of the infecting fungal cells, disorganization of the cellular components, rupture of the plasma membrane, and inhibition of the fungal enzymes and mycelial growth (Cavalcanti 2005). Mode of action of phytoalexins against fungal pathogenesis varies from species to species (Table 9.1). Metabolism of phytoalexin mediated by fungus involves the tendency for its increased polarity by addition of hydroxyl group (oxygenation), removal of methyl group (demethylation), etc. (Jeandet et al. 2014). Muller and Borger first enlightened the concept of phytoalexins almost 70 year ago (Muller and Borger 1940). The first reported case analyzed with the concept of phytoalexin was potato tuber infection by the different strains of causal organism of "late blight of potato," *Phytophthora infestans*. This pathogenic fungus initiated the hypersensitive reactions that lead to the formation of some "plant secondary metabolite" that inhibited further infection of the same plant when infected with another strain of the same genus of *Phytophthora*. Muller and his coworker named this "principle" as "phytoalexins" that have protected the plant from secondary infection (Deverall 1982). Accumulation of phytoalexins in the green plant tissue clearly indicates the presence of remarkable amount of fungal and bacterial infections in the host tissue (Stoessl 1980). Phytoalexins are naturally occurring products secreted and accumulated temporarily by plants in response to pathogenic attack or abiotic stress and agents like heavy metal toxicity, UV radiation, and wounds on tissue (Naoumkina et al. 2007). The inducer agent may be of two types, elicitor and elicitin. The elicitors are commonly the oligosaccharides from fungal cell origin (like hepatosaccharide from soja cell wall) (Sharp et al.

1984). The elicitin types of molecules are generally a type of glycoproteins secreted by the fungal cells (Cordelier et al. 2003). Reports on detailed investigations about phytoalexins have covered a very few families (Leguminosae and Solanaceae) of the green world (Ingham 1982; Kuc 1982). Though investigations on some selected number of species and genera are made from plant families including both monocotyledonous (Amaryllidaceae, Orchidaceae, Poaceae) and dicotyledonous plants (Apiaceae, Asteraceae, Convolvulaceae, Chenopodiaceae, Euphorbiaceae, Linaceae, Moraceae, Piperaceae, Rosaceae, Rutaceae) and even gymnospermic taxa (Ginkgoaceae) (Coxon 1982), cash crops like members of Poaceae (focusing on maize and rice), Vitaceae, and Malvaceae (cotton) have been studied for their phytoalexin production (Schmelz et al. 2014; Langcake and Pryce 1976; Jeandet et al. 2010; Sunilkumar et al. 2006). Though till date a lot of researches have already been performed regarding phytoalexins, a natural weapon against mycopathogens, but still to increase the fungitoxic effectivity of these stress metabolites, further advancement in design and genetic control is needed (Pont and Pezet 1990).

9.3.2 Biosynthetic Pathways and Regulatory Mechanism

Phytoalexin synthesis not only is dependent on pathogenic attack but also could be influenced by various abiotic factors such as temperature, humidity, and water availability (Fig. 9.1). There are evidences that different parts of the plant like leaves, flowers, stems, seeds, and root tubers are site of phytoalexin biosynthesis (Mikkelsen et al. 2003). Different biochemical pathways are used for producing various types of phytoalexins. The three most common pathways include (i) the

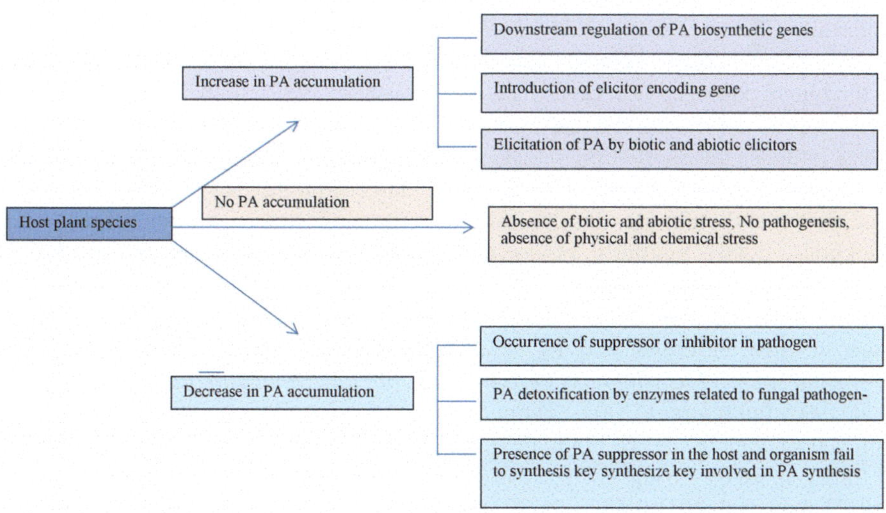

Fig. 9.1 PA production in host plant species is regulated by the interaction between the plant and pathogenic microbes

phenylpropanoic–polymalonic acid pathway, (ii) the methylerythritol phosphate (MEP) and geranylgeranyl diphosphate (GGDP) route, and (iii) the indole phytoalexin (IP) pathway (Jeandet et al. 2014). It is not always obvious that phytoalexins could be categorized not only by their chemical structure or biosynthetic pathway but also by their function and tissue specificity. Examples include the occurrence of momilactone A on different plant parts of rice plant (Lee et al. 1999; Cartwright 1981). Momilactone A is known to be residing in rice husks and rice stems constitutively, but they are also a phytoalexin of rice leaves. Further studies by Toyomasu and his coworkers conclude that momilactone A is constitutively synthesized and oozed out from root of rice plants. Still there is no sufficient data available to consider phytoalexins as ubiquitous throughout the whole plant kingdom. A lot of studies have revealed their complex biochemical synthetic machinery that involves their de novo synthesis, regulation, and mode of action (Jeandet et al. 2013 Ahuja et al. 2012). Regulatory mechanisms involve defense-related marker genes, calcium sensors, hormone signaling, phosphorylation cascades, and also their antipathogenic activity. There are reports on genetic engineering-mediated manipulation of phytoalexin production and increased disease resistance in the case of plants (Delaunois et al. 2009; Jeandet et al. 2012, 2013).

9.3.3 Fungal Pathogenesis: A Stimulus for Phytoalexins Production

Phytoalexins are secondary or stress metabolites that are produced when the host plant is infected with pathogenic fungus. Phytoalexin-mediated defense response includes the expression of lytic enzymes such as chitinases and glucanases and a number of pathogenesis-related (PR) proteins, oxidizing agents, and lignification of cell walls (Dixon and Lamb 1999). Mode of action of phytoalexin involves the coordinated synergism between several defense factors for the effective inhibition of the fungal pathogen (Purkayastha 2017; Mansfield 1999). In the case of *Sorghum* plant, significant infection caused by *Fusarium proliferatum* and *Fusarium thapsinum* stimulates the production of 3-deoxyanthocyanidin, apigeninidin, and luteolinidin and also the concentration of defense-related proteins like peroxidases, β-1,3 glucanases, and chitinases that help to fight the pathogenic infection (Huang and Backhouse 2004). Similarly, phytocassanes A, B, C, D, and E are produced as a result of *Magnaporthe oryzae*, *Rhizoctonia solani*, and *Phytophthora infestans* infections on rice plants. Remarkably phytocassane E, and momilactones A and B exert in vitro antifungal activity against *Magnaporthe oryzae*, *Botrytis cinerea, Fusarium solani, Fusarium oxysporum,* and *Colletotrichum gloeosporioides* (Koga et al. 1997; Fukuta et al. 2007). There are several ways of blocking the fungal infection in host plant tissues by phytoalexin-mediated response. That includes inhibition of fungal spore on the leaf surface and inhibition during and after penetration to host cell (Usman et al. 2018). The occurrence of fungal germ tube on the leaf surface and diffusion of fungal metabolites through the leaf cells cause the accumulation of phytoalexins by the underlying cells and provide the first line of induced chemical defense (VanWees et al. 2003). Phytoalexins may be located on papillae or cell walls, thereby producing a localized, fungitoxic barrier to penetration (Friend 2016). Examples include occurrence of fungitoxic (against *Erysiphe graminis*) flavonoid (thought to be phytoalexin) on papillae of resistant barley leaves.

9.3.4 Fungal and Green Plant Extracts: In Vivo Inducers of Phytoalexins Production

Phytoalexins are known to be solely produced as a result of induction or stimulus by external agents. Fruitful evidences could be drawn regarding this fact. Induction of disease resistance in plants is developed through the direct and indirect involvement of elicitors. Extracts of fungal basidiocarp, essential oils of aromatic plants (Walters et al. 2013), and also synthetic chemicals like aminobutyric acid, salicylic acid, jasmonic acid, and acibenzolar-S-methyl (Garcia-Mier et al. 2013) are the inducers of phytoalexins production. Preparations of horse tail pteridophytic genus *Equisetum* sp. induce the production of glyceolin in soybean plant (*Glycine max*) cotyledons and significantly reduce the *Rhizoctonia solani* infection (Guimaraes et al. 2015). Further studies include the effect of aqueous extracts of basidiocarps of *Agaricus blazei*, *Lentinula edodes*, and *Pycnoporus sanguineus* (Arruda et al. 2012) on the production of glyceollins. Deoxyanthocyanidins and glyceolins are also synthesized by the tinctures of medicinal plants like *Ruta graveolens*, *Origanum majorana*, *Baccharis trimera* (Matiello and Bonaldo 2013), *Hymenolobium petraeum*, *Qualea albiflora*, and *Corymbia citriodora* (Matiello et al. 2016) that act as the elicitors of deoxyanthocyanidins and glyceolin production. Homeopathic preparations of species of Calcarea (*C. citriodora* and *Calcarea carbonica*), essential oils of *Eucalyptus globulus* (Telaxka et al. 2014; Oliveira et al. 2014), and mild concentrations of salicylic acid (Durango et al. 2013) are major elicitins of pistain production and accumulation in cotyledons of common bean (*Phaseolus vulgaris*). Silicon-mediated enhancement of disease resistance by peroxidase (POX), polyphenol oxidase (PPO), chitinases (CHI), β-1,3-glucanases (GLU), and phenylalanine ammonia-lyase (PAL) is found in the case of leaf spot of cotton plant caused by *Ramularia areola* (Curvêlo et al. 2013).

9.3.5 Modern Approaches Involving Amphibians' Extract as Defense Inducers of Plants

Southern Amazonian amphibian family Bufonidae represents the true toads, and their cutaneous secretions are of diverse source of bioactive compounds which can be fruitful as new chemical weapons for agrochemical development. Use of elicitors in the case of crop protection nowadays is becoming a very popular method of inducing response which are proved to be durable and broad-spectrum disease control mechanism where the plant's own resistance is used. A group of seven Brazilian scientists (Deice et al. 2019) evaluated the possibilities of methanolic extracts of cutaneous secretions of two species of Bufonidae, *Rhaebo guttatus* and *Rhinella marina*, on synthesis of phytoalexins named glyceolin (soybean plant), deoxyanthocyanidins (*Sorghum* plants), and phaseolin (mung plant) in soybean cotyledons, sorghum mesocotyls, and bean hypocotyls, respectively. There is a direct relationship between the phytoalexins production and defense ability of the host plant against the fungal pathogenesis. Studies reveal that when the phytoalexin glyceolin is produced in higher amount in the soybean plant (cultivar TMG 132 RR) as a result of methanolic extracts of amphibian's (*R. guttatus*) cutaneous secretion (at a concentration of 0.2 mg/mL), stimulates the enzyme β-1,3-glucanase that can cause the hydrolysis of the fungal cell wall along with other defense-related enzymes (chitinase) is also produced in higher amount, but when suppression of glyceolin occurs, that particular enzyme is also not produced. There are evidences in the case of *Glycine max* that the effectivity of phytoalexins varies from cultivar to cultivar. Application of *R. marina* (amphibian) methanolic extracts induced glyceolin production in TMG 132 RR and Monsoy 8372 cultivars IPRO but did not induce TMG 132 RR cultivars to synthesize these defense-related compounds.

9.3.6 Phytoalexins Versus Phytopathogenic Fungi: A Direct Chemical Warfare

Less toxicity of phytoalexins than chemical fungicides is the reason for their universal acceptance. For over 75 years, phytoalexins have been a detailed area of study for its antimicrobial activity, especially antifungal properties. Several investigations include the in vivo bioeffectivity of the phytoalexins against serious plant pathogenic fungi (Table 9.2). Phytoalexin synthesizing genes have also been genetically modified to cope up with the pathogenic evolution. Still reports are there that include examples of cruciferous phytoalexins detoxification by fungal enzymes (Pedras and Abdoli 2017). Modification of pathogen to overcome phytoalexin-mediated damage includes curved germ tubes as a result of asymmetric growth of the germ tube. Phytoalexins are natural products of diverse chemical nature, for example, alkaloids, coumarins, isoflavonoid (coumestans, isoflavans, isoflavones, isoflavanones, pterocarpans, pterocarpenes), lignans, polyacetylenes, pterocarpons (pisatin, phaseolin, glyceollin, medicarpin, and maackiain), terpenes, and non-isoflavonoid compounds (furanoacetylenes and stilbenes) (Fig. 9.2) (Grayer and Kokubun 2001). Both in vitro and in vivo fungicidal activity are shown by sakuranetin (rice phytoalexins) against the blast fungus (Hasegawa et al. 2014). Reduction of green mold (caused by *Penicillium digitatum*) infections is achieved by the action of coumarin type of phytoalexin (scopoletin) of orange (Sanzani et al. 2014). The loss of apples production caused by *Penicillium expansum* and accumulation of patulin is minimized by the action of phenolic phytoalexins like resveratrol, scopoletin, scoparone, and umbelliferone (Sanzani et al. 2009). In the case of *Medicago sativa* (alfalfa), the isoflavonoid 7-O-methyltransferase provides increased resistance against *Phoma medicaginis* by synthesizing *maiackiain* (He and Dixon 2000). For soybean plants, transformation of resveratrol to pterostilbene

Table 9.2 Production of phytoalexin by host plant species in response to pathogenic fungal infection

Fungal pathogen	Name of phytoalexin	Host	References
Phytophthora drechsleri	*Trans-trans*-3,11-tridecadiene-5,7,9-triyne-1,2-diol	*Carthamus tinctorius* (safflower) (Asteraceae)	Allen and Thomas (1971)
Helminthosporium turcicum	Sativin, vesitol	Alfalfa (*Medicago sativa*), bird's-foot trefoil (*Lotus corniculatus*) (Nymphaeaceae)	Bonde et al. (1973)
Ceratocystis fimbriata	Ipomeamarone Xanthotoxin Polyacetylenes/falcarinol Phenolics: xanthotoxin and 6-methoxymellein	Sweet potato (*Ipomoea batatas*) (Chenopodiaceae) *Pastinaca sativa (parsnip root)*	Johnson (1973)
Cercospora beticola	2′,5-Dimethoxy-6,7-methylenedioxyflavanone, 2′-hydroxy-5-methoxy-6,7-methylenedioxyisoflavone, betagarin, isoflavones, betavulgarin	*Beta vulgaris* (Chenopodiaceae)	Geigert (1973)
Hendersonula sp., *Phytophthora* sp.	Xanthoxylin	*Citrus limon* (Rutaceae)	Hartmann and Niehaus (1974)
Rhizopus stolonifer, Aspergillus niger, Fusarium moniliforme	Diterpenes, casbane	*Ricinus communis* (Euphorbiaceae)	Sitton and West (1975)
Phytophthora infestans	Dihydrophenanthrenes (loroglossol)	*Loroglossum hircinum* (Orchidaceae)	Ward (1975)
Melampsora lini	Coniferyl alcohol, coniferyl aldehyde	*Linum usitatissimum* (Linaceae)	Keen and Littlefield (1975)
Botrytis cinerea	Flavans, 7-hydroxyflavan, 7,4′-dihydroxyflavan, 7,4′-dihydroxy-8-methylflavan, trans-resveratrol (4,3′,5′-trihydroxy stilbene), 6-methoxymellein, *p*-hydroxybenzoic acid, polyacetylene falcarinol	Daffodil (*Narcissus* sp.) (Amaryllidaceae) *Vitis vinifera*, grape vine (Vitaceae) Carrot root	Langcake and Pryce (1976), Coxon (1980), and Harding and Heale (1981)

(continued)

Table 9.2 (continued)

Fungal pathogen	Name of phytoalexin	Host	References
Fusarium solani	Furanopterocarpans, moracins E, F, G, and H	*Morus alba (Moraceae)*	Takasugi (1979)
Bipolaris leersiae	Brassicanal A Brassicanal C Brassinin Cyclobrassinin Dehydro-4-methoxycyclobrassinin, dioxibrassinin 1-Methoxybrassenin 4-Methoxybrassinin 1-Methoxyspirobrassinol Methyl ether	*Brassica oleracea* *B. rapa* *B. napus* *B. carinata* (Brassicaceae)	Takasugi et al. (1986), (1988); Monde et al. (1990a, b), (1991), (1994); and Gross et al. (1994)
Phoma lingam	Brassicanal A Brassilexin Camalexin Cyclobrassinin Methyl 1-methoxyindole-3-carboxylate Spirobrassinin	*B. rapa, B. campestris* *B. carinata* *Capsella bursa-pastoris* *B. napus* *R. sativus* (Brassicaceae)	Devys et al. (1988), Dahiya and Rimmer (1988), Browne et al. (1991), Conn et al. (1994), Pedras and Khan (1996), Pedras et al. (1997), and Pedras and Sorensen (1998)
Sclerotinia sclerotiorum	Cyclobrassinin	*B. napus* (Brassicaceae)	Dahiya and Rimmer (1988)
Cladosporium sp.	Camalexin, Cyclobrassininsulfoxide 1-Methoxycamalexin 6-Methoxycamalexin	*A. thaliana* *B. carinata* (Brassicaceae)	Devys and Barbier (1990)
Alternaria brassicae	Camalexin 1-Methoxybrassinin 6-Methoxycamalexin	*Arabidopsis thaliana* *B. napus* *B. oleracea* *C. bursa-pastoris* *Camelina sativa* (Brassicaceae)	Browne et al. (1991), Tsuji et al. (1992), and Gross et al. (1994)
Rhizoctonia solani	Camalexin Cyclobrassinin	*C. bursa-pastoris* *B. napus* (Brassicaceae)	Browne et al. (1991)
Pythium ultimum	Cyclobrassinin 1-Methoxybrassinin	*B. napus* *B. oleracea* (Brassicaceae)	Conn et al. (1994)

(continued)

Table 9.2 (continued)

Fungal pathogen	Name of phytoalexin	Host	References
Cladosporium cucumerinum	1-methoxybrassinin, 1-methoxyspirobrassinin, cyclobrassinone, sinalexin, spirobrassinin	*B. oleracea* *B. carinata* *B. juncea* (Brassicaceae)	Gross et al. (1994)
Phomopsis perniciosa	Biphenyls, auarperin, dibenzofurans, cotonefurans	*Aronia* sp., *Chaenomeles* sp., *Eriobotrya* sp., *Malus sp.* (Rosaceae)	Kokubun and Harborne (1995)
Pyricularia oryzae	Brassinin Spirobrassinin	*B. napus* *B. carinata* (Brassicaceae)	Storck and Sacristan (1995)
Erwinia carotovora	Camalexin	*C. bursa-pastoris* (Brassicaceae)	Jimenez et al. (1997)
Fusarium oxysporum	Camalexin	*Arabis lyrata* (Brassicaceae)	Zook et al. (1998)
Bipolaris maydis	*3-Deoxyanthocyanidin, apigeninidin, luteolinidin, and apigeninidin 5-O-arabinoside*	*Sorghum bicolor* (Poaceae)	Wharton and Nicholson (2000)
Fusarium proliferatum, F. thapsinum	*Apigeninidin, luteolinidin*	*Sorghum* sp. (Poaceae)	Huang and Backhouse (2004)
Puccinia coronata	Avenalumins I, II, III	Barley (Poaceae)	Chaube and Pundhir (2005)
Monilinia fructicola	Trifolirhizin	Red clover (*Trifolium pretense*) (Fabaceae)	Chaube and Pundhir (2005)
Penicillium expansum	Resveratrol, scopoletin, scoparone, umbelliferone	Apple (*Malus* sp.) (Rosaceae)	Sanzani et al. (2009)
Mycosphaerella fijiensis (black Sigatoka disease)	Irenolone, musanolones	Banana (*Musa* sp.) (Musaceae)	Echeverri et al. (2012)
Phytophthora megasperma	Glyceollin	Soybean (*Glycine max*) (Fabaceae)	Ng et al. (2011)
Fusarium graminearum Cochliobolus heterostrophus, Rhizopus microsporus, Colletotrichum sublineolum, Aspergillus flavus, Aspergillus sojae, Ustilago maydis	Zealexins (sesquiterpinoid) Zealexins A1, A2, A3, B1	*Zea mays* (maize) (Poaceae)	Huffaker et al. (2011)

(continued)

Table 9.2 (continued)

Fungal pathogen	Name of phytoalexin	Host	References
Cochliobolus victoriae *Fusarium* sp.	*Diterpenes (solenoids), momilactones A, B, phytocassane A–E, flavonoids (5,4-dihydroxy-7-methoxyflavanone0*	*Oryza sativa* (rice) (Poaceae)	Ahuja et al. (2012)
Botrytis fabae	Wyerone	Pea (*Pisum sativum*) (Fabaceae)	Slusarenko et al. (2012)
Penicillium digitatum (green mold symptoms)	Scopoletin	Orange (*Citrus sinensis*) (Rutaceae)	Sanzani et al. (2014)
Rhizoctonia solani	Pterostilbene	Soybean (*Glycine max*) (Fabaceae)	Zernova et al. (2014)

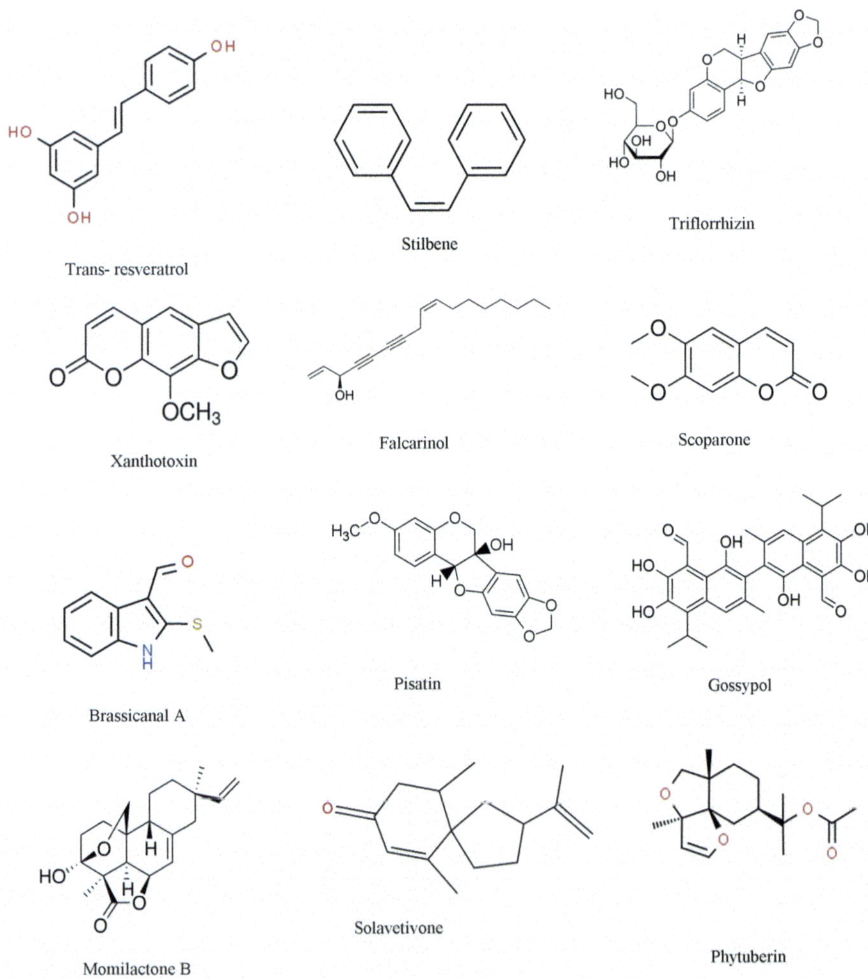

Fig. 9.2 Different types of phytoalexins having major role in plant protection

includes protection against *Rhizoctonia solani* (Zernova et al. 2014). Scientists have proven that not only fungal infection acts as the stimulus for phytoalexin synthesis but also the hormone levels; phosphorylation cascades play a major role in this purpose. Cytokinin overexpression in *Nicotiana tabacum* is directly associated with its resistance against *P. syringe* by higher concentration of capsidiol and scopoletin (Grosskinsky et al. 2011.) The fungitoxicity of the phytoalexin could be enhanced by methylation or presence of electron-attracting groups on aromatic rings that is directly involved in affinity with membrane proteins being an uncoupler of ETS system.

9.4 Endophytes: An Untapped Source of Biofungicides

Endophytes are a type of hidden beneficial microorganisms that reside within the host plant causing no visible disease symptoms and syndrome and promote the plant to maintain its existence in typical harsh conditions. Sometimes they could be latent pathogens at a very distant path of the host's life cycle but are simply a unique area of research where plant science and their microbial association get new definition. Endophytes have been a constant and reliable source of exploration of bioactive compounds, but extensive search has not been performed till date, and that has given the endophyte biologists a great opportunity to search endophytic fungal and actinobacterial flora for the establishment of novel bioactive compounds. Selection of plant for endophytic isolation is the most vital part of this study. Exploitation of the proper isolates accelerates this search and opens up new angle of research. The search for uncommon products of agrochemical importance is a common demand of todays' world. The safer the antifungal agents become, the more it is well accepted in the scientific community as well as agricultural market. In general, the screening of thousands of natural products ends up giving only one commercial product. So indeed it's a tough job to end the search of new antibiotics with a hopeful result. A total of 6 out of 20 of the popular prescribed medicines are of fungal origin, and it is a fact that 5% of the fungi have been described till date (Hawksworth 1991, 2001). So fungi serve as a continuous dependable source of new natural products. The intelligent screening procedure includes the selection of fungal flora of endophytic sources to open up the untapped potential of secondary metabolites synthesized by fungi.

9.4.1 Process of Screening of Antifungal Metabolites from Endophytic Origin

Microorganisms grown in the petri plates or culture broth constitute minimal growth medium needed for their survival. Any kind of stress or transfer of microorganisms on selective media acts as a stimulation for production of their secondary bioactive compounds. These secondary metabolites are produced for their survival in odd environments and strictly act as the selection force for the expression of their antimicrobial-producing genes. These crude by-products of microbial cultures are filtered and purified for their industrial, medicinal, and agricultural exploitation. Soil microorganisms have been exploited for a long time for production of antibiotics, but microorganisms inhabiting plants are a new source in that respect.

Plants are selected usually with potent medicinal applications. Here the knowledge of ethnobotany and folk taxonomy contributes a lot in this selection procedure. A strong literature survey supports the plant selection. The plants are surface sterilized and plated in nutrient-less solid plates. The fungi emerge out from different explants using the decaying plant parts as their primary growth substance. The isolates are identified by microscopic structures focusing on their conidial morphology, spore sculpturing, and colony characters. Confirmatory identification includes 18s rRNA analysis. Endophytic fungi are tested for their antifungal activity against phytopathogenic fungi by one-to-one inhibition assay or antagonistic test (Fig. 9.3). Two agar portions containing fungal hypha of endophyte and pathogen are placed on opposite sides of the plate. If the growth of pathogenic one is arrested partially or completely, that endophytic isolate is marked as antifungal agent and selected for further studies. Another way of screening includes separating the agar plate into two equal halves, and two fungi are placed on two separate sides of the discontinuous plate. This test aims to screen the endophytes that produce volatile antifungals. If the isolate is potent enough to produce volatile organic compounds with fungi static or cidal activity, this will cease the growth of the pathogenic strain. Then that isolate would be qualitatively and quantitatively measured for their volatile emissions using GC-MS as the master equipment (Fig. 9.4). Liquid extracts of endophytic fungi are also tested for antifungal potentials by agar well-diffusion method. The fungal extract having antibiotic property shows clear zone of inhibition of growth of the pathogenic fungus surrounding the area of application of that fungal liquid. The potent isolate will be mass cultured, and the bioactive molecules will be extracted using organic solvents like ethyl acetate, ethyl ether, and n-hexane. Steps include purification of that fungal extract by column chromatography, detection of the compound by thin-layer chromatography, and analysis of compounds by HPLC and mass chromatography. Field experiment includes the synergistic effect of a pure compound with mixture of natural compounds, and the effectivity of a newly applied antifungal agent strictly depends on the host plant and pathogenic microorganism's interaction, environmental condition, and development of drug resistance by that organism. Application of pellets soaked in fungal extracts also is a method of determination of antifungal activity.

9.4.2 Diversity of Antifungal Metabolites

Secondary metabolites are itself diverse in nature. A variety of bioactive secondary metabolites are produced at significant concentrations by the endophytic microbial flora. The major components include quinones, phenols, phenolic esters, steroids, terpenoids, cytochalasins, benzopyranone, alkaloids, isocoumarins, and chromones.

Fig. 9.3 Different approaches for the screening of endophytic fungi with antifungal activity

Till date, a large number of plants have been studied for their endophytic flora as antifungal agents (Table 9.3).

9.4.2.1 Alkaloids

Alkaloid was the first ever reported insecticidal bioactive product. Cryptocin was isolated from endophyte of *Tripterygium wilfordii*, a plant of Celastraceae family. The inner barks of the stem were used as explant, and *Cryptosporiopsis* cf. *quercina* was isolated as a potent endophyte active against *Pyricularia oryzae* and some other phytopathogenic fungi (Li et al. 2000). *Colletotrichum* sp. produces 6-isoprenylindole-3-corboxylic acid having inhibitory action against *Phytophthora*

Confirmation by olfactory test by different persons Or
Inhibition of pathogenic fungal growth in split plate method.

↓

Direct extraction of the volatile emission by SPME fiber from the petri plate Or
Growth of the selected fungus on GC-vials

↓

Detection by SPME method or head space technique using GC-MS as the tool

↓

Detection of volatiles by proper method and generation of complete chromatogram and
report development by the help of NIST library

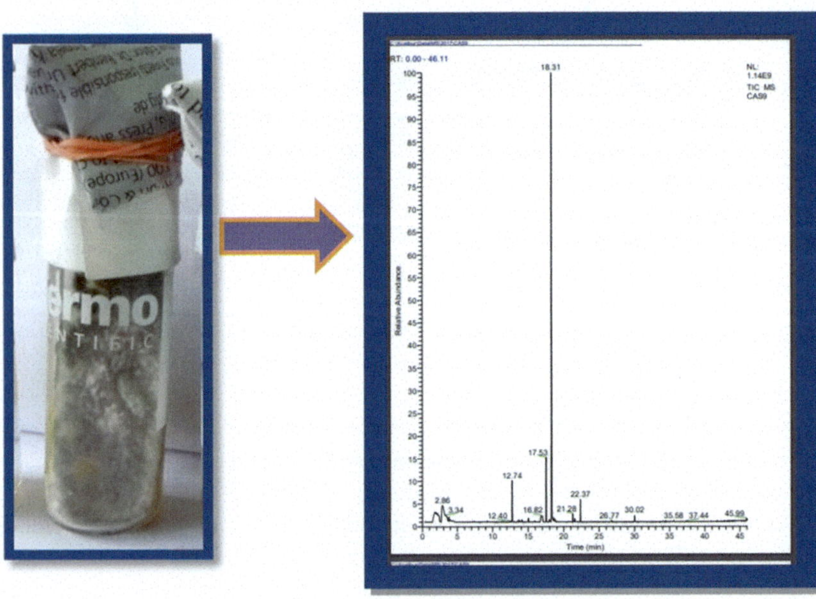

Fig. 9.4 Detection of volatile organic compounds of endophytic origin

capsici, a pathogen of Cucurbitaceae, Fabaceae, and Solanaceae, and also other phytopathogens *Rhizoctonia cerealis* and *Gaeumannomyces graminis* var. *tritici*, a common pathogen of Poaceae family (Lu et al. 2000). Epoxycytochalasin H and cytochalasins N and H were isolated as chloroform and methanolic extracts of

Table 9.3 Antifungal activity of endophytic fungi

Name of the endophytic isolate	Source plant	Antifungal against	References
Ovulariopsis sp. and *Alternaria* sp.	*Datura stramonium*	*Aspergillus niger*, *Colletotrichum gloeosporioides*, *Fusarium* sp., *Phytophthora nicotianae*, *Scopulariopsis* sp., *Trichoderma* viride, *Verticillium* sp.	Li et al. (2005a, b)
Beauveria bassiana, *Trichoderma koningii*, *Alternaria alternata*, *Phoma* sp., *Acremonium strictum*	*Zea mays* (maize) (roots)	*Fusarium oxysporum*, *Fusarium pallidoroseum*, *Fusarium verticillioides*, *Cladosporium herbarum*	Orole and Adejumo (2009)
Alternaria sp., *Chaetomium* sp., Dothideomycetes sp., *Thielavia subthermophila*	*Tylophora indica*	*Sclerotinia sclerotiorum*, *Fusarium oxysporum*, *Fusarium oxysporum*	Kumar et al. (2011)
Chaetomium globosum	*Withania somnifera*	*S. sclerotiorum*	Kumar et al. (2013)
Nigrospora oryzae, *Fusarium proliferatum*, *Guignardia cammillae*, *Alternaria destruens*, *Chaetomium* sp.	*Jatropha curcas*	*S. sclerotiorum*	Kumar and Kaushik(2013)
Phytophthora infestans, *Fusarium oxysporum*	*Triticum durum*	*Alternaria* sp., *Cladosporium* sp., *Penicillium* sp., *Aspergillus* sp., *Chaetomium* sp., *Phoma* sp.	Sadrati et al. (2013)
Cladosporium sp., *Curvularia* sp., *Penicillium* sp.	Moso bamboo (*Phyllostachys edulis*) (seeds)	*Curvularia eragrostidis*, *Pleospora herbarum*, *Arthrinium sacchari*, *Arthrinium phaeospermum*	Shen et al. (2014)
Trichothecium sp.	*Phyllanthus amarus*	*Penicillium expansum* (blue mold of apples)	Taware et al. (2014)
Alternaria sp., *Biscogniauxia mediterranea*, *Cladosporium funiculosum*, *Paraconiothyrium* sp.	*Opuntia humifusa*	*Colletotrichum fragariae*, *C. gloeosporioides*, *C. acutatum*	Silva-Hughesa et al. (2015)
Rhexocercosporidium sp., *F. solani*	*Sophora tonkinensis* Gapnep	*Alternaria panax*, *F. solani*, *C. gloeosporioides*	Yao et al. (2017)

(continued)

Table 9.3 (continued)

Name of the endophytic isolate	Source plant	Antifungal against	References
Glomerella cingulate, *Colletotrichum gloeosporioides*, *C. truncatum*, *Lasiodiplodia pseudotheobromae*, Dothideomycetes sp.	*Houttuynia cordata* Thunb.	*F. oxysporum*, *S. rolfsii*, *T. harzianum*, *Rhizoctonia sp.*, *A. brassicicola*, *P. palmivora*	Aramsirirujiwet et al. (2016)
C. boninense, *F. chlamydosporum*, *C. aeria*, *M. yucatanensis*, *Cladosporium* sp.	*Monarda citriodora* (leaf, roots, and flowers)	*F. solani*, *Sclerotinia* sp., *Colletotrichum capsici*, *A. flavus*, *A. fumigatus*	Katoch and Pull (2017)
Trichoderma longibrachiatum strain BHU-BOT-RYRL17, *Syncephalastrum racemosum* strain AQGSS 12, *Trichoderma longibrachiatum* voucher 50	*Markhamia tomentosa*	*Fusarium oxysporum*, *Sclerotinia sclerotiorum*, *Rhizoctonia solani*, *Botrytis cinerea*	Ibrahima et al. (2017)
Penicillium simplicissimum, *Leptosphaeria* sp., *Talaromyces flavus*, *Acremonium* sp.	Cotton roots (*Gossypium hirsutum*)	*V. dahlia* (Verticillium wilt disease)	Yuan et al. (2017)
Aspergillus sp., *Xylaria* sp., *Fusarium* sp., *Trichothecium* sp., *Oidium* sp.	*Camellia oleifera*	*Camellia oleifera* anthracnose pathogen	Yu et al. (2018)
Penicillium sp. (ARDS-2.3), *Aspergillus oryzae* (ARHS-1.1)	*Asparagus racemosus* Willd	*Botrytis cinerea* (gray mold), *Sclerotinia sclerotiorum* (stem rot), *Rhizoctonia solani* (root rot), *Fusarium oxysporum* (wilt)	Chowdhary and Kaushik (2018)
Aspergillus sp., Curvularia sp., *Fusarium oxysporum*	*Dendrobium lindleyi*	*Fusarium* sp., *Sclerotium* sp., *Colletotrichum* sp., *Curvularia* sp., *Phytophthora* sp.	Bungtongdee (2019)

Phomopsis sp., an endophyte of *Gossypium hirsutum*. It showed potent antifungal activity against species of *Bipolaris* (*B. sorokiniana*, *B. maydi*), *Botrytis* (*B. cinerea*), *Sclerotinia* (*Sclerotinia sclerotiorum*), *Rhizoctonia* (*R. cerealis*), and *Fusarium* (*Fusarium oxysporum*) (Fu et al. 2011). A lot of endophytes have been explored for their antifungal production, but only a few of them were positive for antifungal metabolites categorizing in alkaloids. The common alkaloids acting as the antifungal agents of endophytic fungal origin are gliotoxin, cryptocanadin, tyrocidine A, fumigaclavine C, fumitremorgin C, 1-N-methyl albonoursin, and phomapsichalasin.

9.4.2.2 Terpene Derivatives

The terpenoids, usually called isoprenoids, are large and diverse group of naturally occurring organic compounds derived from terpenes that are multicyclic. Sixty percent of all the known natural products are terpenoids in nature. Some endophytic fungicidal products are of terpenes by their native chemical structure. Endophytic isolates (*Hormonema* sp.) of gymnospermous plant *Juniperus communis* were reported to be antifungal producers of a triterpene glycoside enfumafungin (Pelaez et al. 2000). Known antifungal sterols of endophytic origin are 3β-hydroxy-ergosta-5-ene, 3-oxoergosta-4,6,8,22-tetraene, etc. The sterols are strong inhibitors of *Helminthosporium sativum* (present name: *Bipolaris sorokiniana*), the asexual stage of *Cochliobolus sativus*, a common root rot pathogen of wheat and barley crops which also infects leaf and stems of Poaceae plants (Lu et al. 2000). Sesquiterpenes are reported to be the growth inhibitors of *Cladosporium phlei* (causal organism of leaf spot disease of timothy grass, *Phleum pratense*). This is a unique example where the host plant (*Phleum pratense*) itself harbors the endophyte (*Epichloe typhina*) that inhibits the growth of its leaf spot pathogen (*Cladosporium phlei*).

9.4.2.3 Isocoumarins

From the point of view of organic chemistry, isocoumarins are defined as the isomer of coumarin where the orientation of the lactone is reversely arranged. Zhang and his coworkers in the year 2008 isolated an endophytic fungus named *Microdochium bolleyi* from *Fagonia cretica* (also known as virgin's mantle of Zygophyllaceae family), a herb of semiarid regions of Gomera. Isocoumarins were identified as the active compounds having antifungal activity against *Microbotryum violaceum* (previously known as *Ustilago violacea*), an obligate parasite of Basidiomycete group and a common infectant of members of Caryophyllaceae causing smut of anther. The four isolated and identified isocoumarins are monocerin, 12-oxo epimers of monocerin, and open-ring derivative compounds of monocerin. The compounds are obtained as mixtures by column chromatography followed by Sephadex LH-20 chromatography techniques. Preparative TLC further differentiated the four compounds. Monocerin and its analogues were previously reported as antifungal compounds from fungal sources of *Drechslera monoceras*, *Exserohilum monoceras*, *Helminthosporium monoceras*, *Exserohilum turcum*, and *Fusarium larvarum* (Aldridge and Turner 1970; Robeson and Strobel 1982; Grove and Pople 1979; Claydon et al. 1979). These secondary metabolites act on pathogens by interfering stages of divisional phases of cell cycle. The second isocoumarin was colorless oil. The third and fourth one are represented by the empirical formula of $C_{16}H_{20}O_7$ and $C_{16}H_{22}O_7$. The fourth one is structurally correlated to fusarentin 6,7-dimethyl ether. Both the compounds are of heptaketide in their origin, and it is revealed that fusarentins are the probable precursors of the active compounds monocerins (Scott et al. 1984; Axford et al. 2004). Dihydroisocoumarins, mellein (an isocoumarin derivative), (R)-7-hydroxymellein, and fonsecinone were reported from species of *Xylaria* (endophyte of *Piper aduncum*), *Pezicula*, *Penicillium* (*Alibertia macrophylla*), and

Aspergillus (*Cynodon dactylon*), respectively (Oliveira et al. 2011; Schulz et al. 1995; Song et al. 2004).

9.4.2.4 Phenolics as the Most Potent Antifungal from Endophytic Source

Phenols (popularly known as phenolics) represent a class of chemical compounds characterized with a hydroxyl group attached to an aromatic hydrocarbon group. Phenol (or carbolic acid) is a colorless crystalline solid, aromatic compound having benzene rings. They are predominantly found in the plant kingdom as a response to stress and are of utmost importance. Endophytic culture extracts are also known to be rich sources of phenolics; usually they are directly proportional to the antioxidative property of any fungal isolate, but in some particular cases, they are characterized with their antifungal potentials against phytopathogenic fungus. Usually the liquid culture extracts of the endophytic isolates are subjected to solvent extraction using ethyl acetate, n-hexane, ethyl ether, etc. Those organic solvents are believed to extract the phenolics from the water-based culture broth. Those extracted compounds are further screened for their antifungal efficiency. Ethyl acetate extracts of endophytic *Phoma* sp. are reported to contain tetralone metabolites (derivatives of α-tetralone, 3,6,7-trihydroxy-α-tetralone) inhibiting the growth of two common broad phytopathogenic fungus *Fusarium oxysporum* and *Rhizoctonia solani*. Griseofulvin is known to be the first antifungal compound isolated from *Penicillium griseofulvum*. Later it is isolated from several species of fungi including endophytic *Penicillium canescens* and *Xylaria* sp. (member of Xylariaceae family). Griseofulvin from endophytic *P. canescens* of popular Chinese medicinal plant *Polygonatum cyrtonema* (Polygonaceae) showed strong inhibitory effectivity against phytopathogenic *Botrytis cinerea*, *Sclerotinia sclerotiorum*, *Colletotrichum orbiculare*, and *Didymella bryoniae* (Wang et al. 2010). Other than *Penicillium*, endophytic *Xylaria* sp. isolated as an endophyte of *Abies holophylla* yields griseofulvin and dechlorogriseofulvin for in vitro and in vivo effectivity against pathogenic *Magnaporthe grisea*, *Corticium sasakii*, *Blumeria graminis* (Park et al. 2005). The ascomycete fungus *Pestalotiopsis* is known to be a common plant pathogen but also has been reported many times because of their endophytic existence in the host plants. The two common species *Pestalotiopsis microspora* (host: tropical plant *Terminalia morobensis*) and *P. fici* are reported to be producing antifungal metabolites isopestacin and pestalofones D–E (Harper et al. 2003; Liu et al. 2009a, b). Chlorogenic acid and colletotric acids are antifungal phenolics of *Colletotrichum gloeosporioides* and Sordariomycetes sp., respectively (Chen et al. 2010; Zou et al. 2000). They were isolated from medicinal plants of China (*Artemisia mongolica* and *Eucommia ulmoides*) and effective against fungi imperfecti *Helminthosporium sativum*. Orcinol is used for the production of a dye called orcein used randomly for the staining of cells and chromosomes. Orcinol is popularly known for its antifungal activity too and has been isolated as a product of endophytic origin of *Penicillium* sp. from *Alibertia macrophylla* (a plant of Rubiaceae) showing bioactivity against *Cladosporium cladosporioides* and *Cladosporium sphaerospermum* (Oliveira et al. 2014). Endophytic *Phomopsis* sp., *Dothiorella* sp., and *Diaporthe* sp. have also been tested for their antifungal production and antifungal compounds that were detected (Brady et al. 2000; Xu et al. 2004; Huang et al. 2008).

9.4.3 Volatile Organic Compounds

Volatile organic compounds (VOCs) are said to be a type of organic low-molecular-weight carbon-containing small compounds (up to C20) that have a high vapor pressure with low molecular mass (100–500 daltons) at room temperature. The high vapor pressure results from a low boiling point of that chemical compound, which causes a huge quantity of molecules to evaporate from the liquid, solid, or semisolid form of the compound and gets released into the surrounding environment. The endophytes are unique in their volatile emissions. The term mycofumigation that is very much popular with the treatment of agricultural phytopathogens is actually the output of VOCs that originated from endophytic isolates. The first ever reported volatile antibiotic producer was *Muscodor albus* (Xylariaceae family), an endophyte of *Guazuma ulmifolia* (a plant of Sterculiaceae family collected from tropical forest of SW Ecuador), isolated by Gary Strobel and his co-workers (Strobel et al. 2007). The major compounds isolated by GCMS are known to be involved in antifungal, antibacterial activity. Compounds include butanoic acid, 2-methyl-; butanoic acid, 3-methyl-; 2-butenal, 2-methyl-; butanoic acid, 3-methylbutyl ester; 3-buten-1-ol, 3-methyl; guaiol; 1-octene, 3-ethyl-; formamide, N-(1-methylpropyl); azulene and naphthalene derivatives; caryophyllene; phenylethyl alcohol; acetic acid, 2-phenylethyl ester; bulnesene; and various propanoic acid, 2-methyl- derivatives. These compounds were tested against a number of phytopathogenic fungi (*Botrytis cinerea*, *Mycosphaerella fijiensis*, *Pythium ultimum*, *Phytophthora cinnamomi*) showing partial or complete death or growth inhibition of those pathogens after 2 or 4 days of incubation. *Muscodor albus* was reported from a diverse type of host plants, i.e., *Myristica fragrans*, *Terminalia prostrata, Cinnamomum zeylanicum,* and *Ginkgo biloba,* by several workers (Worapong et al. 2001; Sopalun et al. 2003; Ezra and Strobel 2003; Mercier et al. 2004; Ezra et al. 2004a, b; Atmosukarto et al. 2005; Lacey and Naven 2006; Lacey et al. 2009; Strobel et al. 2007; Banerjee et al. 2010a, b; Corcuff et al. 2011; Alpha et al. 2015). The mycofumigants are effective against pathogen *Fusarium culmorum*, causal agent of seedling blight, foot rot, ear blight, stalk rot, and common rot of cereals. Sexual stage (teleomorph) of *Glomerella cingulata,* a fungus of Glomerellaceae, is a potent pathogen causing anthracnose-like symptoms of water-soaked, sunken spots and necrotic lesions on fruits of forest trees. This phytopathogen is strictly inhibited by the volatile emissions of this novel endophyte. Banerjee et al. (2010a, b) first reported *Muscodor albus* strain GBA from the USA as an isolate of *Ginkgo biloba* (first isolate of *M. albus* from *G. biloba*) and tested the biological efficacy of its volatile mixtures against agricultural pathogens and also evaluated its promises to be used as a commercial mycofumigant agent for controlling the fungal diseases in storage fruits and vegetables, that is, agricultural productions and during food transportation. The strain GBA in comparison to other strains of *Muscodor* E6 and CZ620 completely inhibits and potentially kills the member of Phycomycetes, *Pythium ultimum* after 2 days of exposure of the mixture of volatiles. The organic compounds include alcohols, acids, esters, ketones, and lipids as their active components. 1-butanol, 3-methyl-, acetate was found in significant quantities. Vitrine, a terpenoid, was first isolated from *Muscodor albus* strain GAB. The volatile mixture is artificially produced by the mixture of the pure compounds, and that mixture is again checked for antifungal activity. A positive mycocidal or mycostatic effect similar to the effect of endophyte's volatile emission will confirm establishment of the endophyte and its mixture as the biocontrol or antifungal agent. *Myrothecium inundatum*, an endophyte of herbaceous *Acalypha indica* (Euphorbiacea member collected from north-eastern part of India), produces unique mixture of volatile components having 3-octanone, 3-octanol, 7-octen-4-ol, sesquiterpenes, organic acids, methyl esters, naphthalene, 2-octanoic acid, heptanoic acid, etc. This endophyte produces foam in its liquid culture predominant with long-chain carbon compounds like octane, 1,4-cyclohexadiene, 1-methyl- and cyclohexane, and 1-ethylpropyl. The presence of this type of compounds emphasizes the concept of "mycodiesel" Several other species of *Muscodor*, for example, *Muscodor heveae*, *Muscodor ghoomensis, M. indica, M. camphora, M. suthepensis, Muscodor tigerii, Muscodor darjeelingensis, Muscodor strobelii, Muscodor kashayum, Muscodor musae, Muscodor sutura, Muscodor cinnamomi, Muscodor crispans, Muscodor vitigenus*, and *M. roseus*, were isolated from different parts of the world and from various types of hosts, for example, *Hevea brasiliensis, Cinnamomum camphora, Cinnamomum bejolghota, C. zeylanicum, Aegle marmelos, Musa acuminate, Prestonia trifidi, Grevillea pteridifolia, Erythrophleum chlorostachys, Paullinia paullinioides*, and *Hevea brasiliensis* (Meshram et al. 2013, 2015, 2017; Suwannarach et al. 2015; Saxena et al. 2015; Suwannarach et al. 2010, 2012, 2013; Kudalkar et al. 2012; Mitchell et al. 2010; Worapong et al. 2002; Daisy et al. 2002; Siri-udom et al. 2016), showing antifungal activity against a large number of phytopathogens including *A. fumigatus, Botrytis cinerea, Colletotrichum lagenarium, Ceratocystis ulmi, Cercospora beticola, Geotrichum candidum, Mycosphaerella fijiensis, Phytophthora cinnamomi, Phytophthora palmivora, Pythium ultimum, Rhizoctonia solani, Sclerotinia sclerotiorum,* and *Verticillium dahlia.*

9.4.3.1 VOC Producers as the Postharvest Biocontrol Agents and Mycofumigants

Postharvest fungal disease is one of the prime causes of agricultural loss of crops. Use of biological agent to minimize this loss is one of the vital targets of agriculturalists, horticulturalists, and plant biologists. Several chemical agents have been already tested for practical applications, but endophytes are less explored organisms in this arena. Volatiles from endophytic source open up new scope of utilization of unique mixtures of chemicals to be used as mycofumigant agents. A large number of endophytes have already been screened for their postharvest disease management ability (Table 9.4). The volatiles of the endophyte could be considered as the natural fungicides. *Muscodor vitigenus*, an endophyte of *Hevea brasiliensis*, was analyzed in GC-MS for their volatile production. The isolates produce a unique mixture of

Table 9.4 VOCs produced by endophytic fungi with antifungal activity

Endophytic fungi	Source	VOCs emitted	Antifungal activity against	References
Muscodor albus	*Guazuma ulmifolia, Myristica fragrans*, unidentified small vine, *Terminalia prostrata, Cinnamomum zeylanicum, Ginkgo biloba,*	1-Butanol, 3-methyl-acetate, citrine (terpenoid), naphthalene, tetrahydrofuran, 2-methyl furan, 2-butanone, aciphyllene, azulene derivative. Germacrene B; acetic acid; methyl ester; 1-butanol; benzeneethanol; acetate; pyrrolidine; 2-heptanoic acid, 4-cyclopropyl-; bicyclo[3.1.1]heptane, 6-methyl-2-; propanoic acid, 2-methyl-	*Aphanomyces cochlioides, Aspergillus fumigatus, A. ochraceus, Fusarium culmorum, Glomerella cingulate*	Worapong et al. (2001), Sopalun et al. (2003), Ezra and Strobel (2003), Mercier and Jimenez (2004), Ezra et al. (2004a, b), Atmosukarto et al. (2005), Lacey and Naven (2006), Lacey et al. (2009), Strobel et al. (2007), Banerjee et al. (2010a, b), Corcuff et al. (2011), and Alpha et al. (2015)
Gliocladium sp.	*Eucryphia cordifolia*	1-Butanol; 3-methyl-octene, 1-propanol, 2-methyl-; 1-butanol, 2-methyl-; propanoic acid; octanone; 1,3,5,7-cyclooctatetraene (azulene); acetic acid; 2-phenylethyl ester; phenylethyl alcohol	*Pythium ultimum, Verticillium dahliae*	Stinson et al. (2003)
Edenia gomezpompae	*Callicarpa acuminata*	Naphthoquinone spiroketal, preussomerin EG 1–3	*Colletotrichum* sp., *Alternaria solani, Phytophthora capsici, Phytophthora parasitica, Fusarium oxysporum*	Macias et al. (2008)
Gliocladium roseum (NRRL 50072)	*Eucryphia cordifolia*	Undecane, 2,6-dimethyl; decane, 3,3,5-trimethyl; cyclohexene, 4-methyl; decane, 3,3,6-trimethyl; and undecane, 4,4-dimethyl	*Pythium ultimum*	Strobel et al. (2008a, b)

(continued)

Table 9.4 (continued)

Endophytic fungi	Source	VOCs emitted	Antifungal activity against	References
Nodulisporium sp.	*Lomatia fraseri*, *Olearia argophylla*	Isobutanol; origanene; myrcene; ß-pinene; 1,3,8-p-menthatriene; 6-isopropenyl-3-methoxymethoxy-3-methyl-cyclohexene; eucalyptol; benzene, 1-methyl-4-(1-methylethenyl)-; 2-cyclohexen-1-one, 2-(2-methyl-2-propenyl)-; ß-elemene; caryophyllene; α-farnesene	*Rhizoctonia fragariae*, *Fusarium oxysporum*, *Sclerotium rolfsii*, *Verticillium dahlia*, *Colletotrichum acutatum*	Mann et al. (2008)
Oidium sp.	*Terminalia catappa*	Esters of propanoic acid, 2-methyl-; butanoic acid, 2-methyl-; and butanoic acid, 3-methyl-	*Pythium ultimum*	Strobel et al. (2008a, b)
Oxyporus latemarginatus	*Capsicum annum*	5-Pentyl-2-furaldehyde	Controls postharvest apple decay and *Rhizoctonia* sp. root rot of moth orchid	Lee et al. (2009)
Botrytis sp. BTF 21, *Cladosporium* sp. MIF01, *Penicillium* sp. BTF08		Butane-2-methyl; β-butyrolactone; 2-butenedinitrile; 1-propanol, 2-methyl-; 1-butanol, 3-methyl-; 2-butenedinitrile	Biocontrol potential against phytopathogen *Fusarium oxysporum*	Ting et al. (2010)
Hypoxylon sp.	*Persea indica*	1,8-Cineole, 1-methyl-1,4-cyclohexadiene, (+)-alpha-methylene-alpha-fenchocamphoron (monoterpene)	*Botrytis cinerea*, *Phytophthora cinnamomi*, *Cercospora beticola*, *Sclerotinia sclerotiorum*	Tomsheck et al. (2010)
Muscodor crispans	*Ananas ananassoides*	Propanoic acid, 2-methyl-; methyl ester; propanoic acid, 2-methyl-; 1-butanol, 3-methyl-; 1-butanol, 3-methyl-, acetate; propanoic acid, 2-methyl-, 2-methylbutyl ester; and ethanol	*Pythium ultimum*, *Phytophthora cinnamomi*, *Sclerotinia sclerotiorum*, *Mycosphaerella fijiensis* (the black Sigatoka pathogen of bananas)	Mitchell et al. (2010)
Muscodor cinnamomi	*Cinnamomum bejolghota*	Azulene; butanoic acid, 2-methyl-, methyl ester; propanoic acid, 2-methyl-, methyl ester	*Rhizoctonia solani*	Suwannarach et al. (2010, 2012)

Muscodor fengyangensis	*Pseudotaxus chienii, Actinidia chinensis, Abies beshanzuensis*	Naphthalene derivatives; a-phellandrene; β-phellandrene; 2-cyclohexen; propionic acid, 2-methyl-; and methyl ester	*Botrytis cinerea, Aspergillus clavatus, Colletotrichum fragariae, Didymella bryoniae, Fusarium oxysporum, Magnaporthe oryzae, Pythium ultimum, Rhizoctonia solani, Sclerotium rolfsii, Verticillum dahlia, Penicillium digitatum*	Zhang et al. (2010)
Myrothecium inundatum	*Acalypha indica*	1,4-Cyclohexadiene, 1-methyl- and cyclohexane, (1-ethylpropyl)3-octanone, 3-octanol, 7-octen-4-ol	*Pythium ultimum, Sclerotinia sclerotiorum*	Banerjee et al. (2010a, b)
Nodulisporium sp. *CF016*	*Cinnamomum loureirii*	β-Elemene, β-selinene, α-selinene, 1-methyl-1,4-cyclohexadiene	Postharvest biocontrol agent of apple and a potent biofumigant agent against *P. ultimum, R. solani, F. oxysporum, Phytophthora capsici, Sclerotinia sclerotiorum, Colletotrichum coccodes, Magnaporthe oryzae, Alternaria panax, Botrytis cinerea, Penicillium expansum*	Park et al. (2010)
Phoma sp.	*Larrea tridentata*	Alpha-humulene (sesquiterpene), alcohols, reduced naphthalene derivatives, trans-caryophyllene	*Verticillium, Ceratocystis, Cercospora, Sclerotinia Trichoderma, Colletotrichum, Aspergillus* sp.	Strobel et al. (2011)
Phomopsis sp.	*Odontoglossum* sp.	1-Butanol, 3-methyl-; benzeneethanol; 1-propanol, 2-methyl-; and 2-propanone	*Pythium ultimum, Phytophthora palmivora, Sclerotinia sclerotiorum, Rhizoctonia solani, Fusarium solani, Botrytis cinerea, Verticillium dahliae,* and *Colletotrichum lagenarium*	Singh et al. (2011)

(continued)

Table 9.4 (continued)

Endophytic fungi	Source	VOCs emitted	Antifungal activity against	References
Candida intermedia	*Strawberry*	1,3,5,7-cyclooctatetraene; 3-methyl-1-butanol; 2-nonanone; pentanoic acid, 4-methyl-; ethyl ester; 3-methyl-1-butanol, acetate; acetic acid, pentyl ester	Postharvest control agent of *Botrytis* fruit rot in strawberry plants	Huang et al. (2011)
Muscodor sutura	*Prestonia trifidi*	Thujopsene, chamigrene, isocaryophyllene, butanoic acid, 2-methyl-	Antifungal activity against 12 pathogenic fungi, e.g., *A. fumigatus*, *Botrytis cinerea*, *Colletotrichum lagenarium*, *Ceratocystis ulmi*, *Cercospora beticola*, *Mycosphaerella fijiensis*, *Phytophthora cinnamomi*, *Phytophthora palmivora*, *Pythium ultimum*, *Rhizoctonia solani*, *Sclerotinia sclerotiorum*, *Verticillium dahlia*	Kudalkar et al. (2012)
Nodulisporium sp.	*Myroxylon balsamum*	1, 4-Cyclohexadiene, 1-methyl-; 1,4-pentadiene and cyclohexene, 1-methyl-4-(1-methylethenyl)-; alkyl alcohols starting with 1-butanol-3-methyl, 1-propanol-2-methyl, 1-pentanol, 1-hexanol, 1-heptanol, 1-octanol, 1-nonanol along with phenylethyl alcohol, secondary alkyl alcohols, esters, ketones, benzene derivatives, terpenoids	*Aspergillus fumigatus*, *Rhizoctonia solani*, *Phytophthora cinnamomi*, *Sclerotinia sclerotiorum*	Mends et al. (2012)
Muscodor musae	*Musa acuminata*	Ethyl 2-methylpropanoate, methyl 2-methylbutanoate, 2-methylpropyl acetate, methyl 3-methylbutanoate, 2-methylpropyl propanoate, 2-methylpropan-1-ol, 3-methylbutanoyl acetate, 2-methylbutyl 2-methylpropanoate, 3-methylbutan-1-ol, 3-methyl-3-buten-1-ol, ethyl 2-hydroxy-2-methylpropanoate, 3-hydroxy-2-butanone	*Alternaria porri*, *Alternaria solani*, *Botrytis cinerea*, *Colletotrichum capsici*, *Colletotrichum gloeosporioides*, *Colletotrichum musae*, *Fusarium oxysporum*, *Fusarium solani*, *Nigrospora oryzae*, *Penicillium digitatum*	Suwannarach et al. (2013)

Muscodor equiseti	*Equisetum debile*, *Hevea brasiliensis*	Isobutyric acid, $C_4H_8O_2$	Pathogens (filamentous fungi)	Suwannarach et al. (2013) and Siri-udom et al. (2016)
Muscodor kashayum	*Aegle marmelos*	3-Cyclohexen-1-ol, 1-(1,5-dimethyl-4-hexenyl)-4-methyl; 1,6-dioxacyclododecane-7,12-dione; 2,6-bis(1,1-dimethylethyl)-4-(1-oxopropyl) phenol; 2,4-di-tert-butylthiophenol and 4-octadecylmorpholine	Antifungal and mycofumigant agent	Meshram et al. (2013)
Nodulisporium sp.	*Thelypteris angustifolia*	Series of ketones, including acetone; 2-pentanone; 3-hexanone, 4-methyl-; 3-hexanone, 2,4-dimethyl; 2-hexanone, 4-methyl and 5-hepten-2-one; 1,8-cineole; 1-butanol, 2-methyl and phenethyl alcohol; cyclohexane, propyl	*Phytophthora palmivora*, *Rhizoctonia solani*, *Sclerotinia sclerotiorum*, *Phytophthora cinnamomi*	UL-Hassan et al. (2013)
Muscodor strobelii	*Cinnamomum zeylanicum*	Viridiflorol, tetraoxapropellan, terpinolene, octadec-9-enoic acid, aspidofractinine-3-methanol	*Penicillium citreonigrum*, *Aspergillus japonicas*	Meshram et al. (2014)
Muscodor darjeelingensis	*Cinnamomum camphora*	2,6-Bis(1,1-dimethylethyl)-4-(1-oxopropyl) phenol, 1, 6-dioxacyclododecane-7, 12-dione and 4-octadecylmorpholine	Species of *Candida*	Saxena et al. (2014)
Muscodor tigerii	*Cinnamomum camphora*	4-Octadecylmorpholine, 1-tetradecanamine, N,N-dimethyl and 1,2-benzenedicarboxylic acid, mono(2-ethylhexyl) ester	Inhibits the growth of fungal and bacterial pathogens like *Cercospora beticola*, *Penicillium marneffei*, *Rhizoctonia solani*	Saxena et al. (2015)

(continued)

Table 9.4 (continued)

Endophytic fungi	Source	VOCs emitted	Antifungal activity against	References
M. suthepensis	*Cinnamomum bejolghota*	Ethyl 2-methylpropanoate, methyl 2-methylbutanoate, 2-methylpropylacetate, methyl 3-methylbutanoate, methylpropyl propanoate, methylpropan-1-ol, 3-methylbutanoylacetate, 2-methylbutyl 2-methylpropanoate, 3-methylbutan-1-ol, 3-hydroxy-2-butanone, 3-ethyl-2-methylpentane, 6-methyl-5-hepten-2-one, 3-methylhexane, 4-dimethyl-1-heptene, 2-ethylhexylacetate, 4,5-dimethyl-1,3-cyclopentanedione, 2-methylpropanoicacid	A potent biofumigant agent against *Penicillium digitatum* to control tangerine fruit rot	Suwannarach et al. (2015)
Muscodor ghoomensis *M. indica* *M. camphora*	*Cinnamomum camphora*	Tetracontane; 4-octadecylmorpholine; N,Ndimethyl-1-pentadecanamine and cis-9-hexadecena; 4-octadecylmorpholine; 1, 6-dioxacyclododecane-7, 12-dione; 1,4-dimethyl-7-prop-1-en-2-yl-2,3,3a,5,6,7,8,8a-octahydro-1H-azulen-4-ol	*Alternaria alternate, Arthrinium phaeospermum, Aspergillus flavus, Botrytis cinerea, Cercospora beticola, Colletotrichum gloeosporioides, Fusarium solani, Muscodor albus, Penicillium marneffei, Rhizoctonia solani*	Meshram et al. (2015, 2017)
Nodulisporium sp. GS4d2Il♀a	*Gliricidia sepium*	Mono- and sesquiterpenes, especially eucalyptol and limonene, amines, 2-pentylfuran, α-phellandrene, α-myrcene, 3-carene, butyl isocyanatoacetate, tetrahydro-3-methyl-furan, 1,8-nonadiyne	*Phytophthora capsici, Pythium aphanidermatum, Phytophthora cinnamomi, Phytophthora parasitica*	Fernández et al. (2016)
Xylaria sp. *PB3f3*	*Haematoxylum brasiletto*	3-Methyl-1-butanol, thujopsene, 2-methyl-1-butanol, 2-methyl-1-propanol	*Pythium aphanidermatum, Phytophthora capsici, Alternaria solani, Fusarium oxysporum*	Sanchez et al. (2016)

Daldinia cf. *concentrica*	*Olea europaea*	Germacrene A, α-bulnesene, α-selinene, terpenes, β-elemene, phenylethyl alcohol, 4-heptyn-2-ol, isoamyl acetate, 3-methyl-1-butanol, 2-methyl-1-butanol, 4-heptanone, isoamyl acetate and trans-2-octenal, 3-methoxy-2-napthol	*Alternaria alternata* pathotype tangelo, *Alternaria alternate, Aspergillus niger, Botrytis cinerea, Colletotrichum* sp., *Coniella* sp., *Fusarium euwallaceae, Fusarium mangiferae, Fusarium oxysporum, Lasiodiplodia theobromae, Neoscytalidium dimidiatum, Penicillium digitatum, Phoma tracheiphila, Pythium aphanidermatum, Pythium ultimum, Rhizoctonia solani, Sclerotinia sclerotiorum*	Liarzi et al. (2016)
Hypoxylon anthochroum strain Blaci	*Bursera lancifolia*	Phenylethyl alcohol and eucalyptol	Anti phytopathogenic activity	Ulloa-Benitez et al. (2016)
Muscodor vitigenus	*Hevea brasiliensis*	Naphthalene, azulene, 3-methylbutan-1-ol, 3-methylbutyl acetate	Potent biocontrol agent	Siri-udom et al. (2016)
Muscodor heveae	*Hevea brasiliensis*	3-Methylbutan-1-ol, 3-methylbutyl acetate, 2-methylpropanoic acid, and azulene derivatives	*Phellinus noxius, Rigidoporus microporus* (causal organism of root rot disease in rubber)	Siri-udom et al. (2016)
Nodulisporium sp. *strain GS4d2II1*	*Solanum lycopersicum*	Caryophyllene, 4-methyl-2,6-di-tert-butylphenol, alcohols' mixture, phenylethyl alcohol, 2-methyl-1-butanol, 3-methyl-1-butanol, eucalyptol, ocimene, terpinolene	Potent postharvest biocontrol agent for tomato against *Fusarium oxysporum*	Romero et al. (2017)

naphthalene, azulene, 3-methylbutan-1-ol, and 3-methylbutyl acetate that partially or completely inhibits the growth of the phytopathogenic fungal species. *Colletotrichum gloeosporioides*, a commercially significant plant pathogen, usually acts as a secondary invader of injured tissue or saprophyte and causes bitter root in variety of crops; tropical fruits like yams, papaya, avocado, coffee, sweet pepper, tomato; and also perennial grasses. *Rhizoctonia solani*, a commercially significant plant pathogen of Basidiomycotina, causes symptoms of brown patch on turf grass, damping off of soybean seedlings, black scurf of potatoes, root rot of sugar beet, belly rot of cucumber, sheath blight of rice, etc. Another phytopathogen named *Fusarium oxysporum*, the causal agent of *Fusarium* wilt or koa wilt, and *Rigidoporus microporus*, the causal agent of white root rot disease on tropical crops like cacao, were suppressed in terms of their growth upon exposure to the volatiles of *M. vitigenus*. *Phytophthora parasitica* (oomycetous fungi) causes destructive diseases of a wide range of crop plants (*Arabidopsis thaliana* and *Medicago truncatula*, the two guinea pigs of plant science), nursery and ornamental plants, and forest ecosystems. Another phytopathogenic species of *Ganoderma austral* that forms white heart rot in *Tilia* trees, *Quercus* sp. (oaks), *Fagus* sp. (beech), and *Betula* sp. (birch) is found to be inactive when exposed to the volatiles. *Phellinus noxius* is reported to be a serious threat to almost 200 plants covering 59 families of tropical forests of Asia, Africa, Japan, Taiwan, and the Pacific Islands, which is inhibited significantly by *M. vitigenus* volatiles. So *M. vitigenus* opens up scope for the biocontrol of all these seven harmful phytopathogens and ensures the complete protection of host from fungal attack. This could be concluded as natural fungicides. So mycofumigation by these volatiles could minimize the loss caused by these pathogens. Stinson and his co-workers (2003) isolated endophytic *Gliocladium* sp. (Hypocreaceae) from *Eucryphia cordifolia* and reported multiple number of volatile compounds, for example, 1-butanol, 3 methyl; octene; 1-propanol, 2-methyl-; 1-butanol, 2-methyl-; propanoic acid; octanone; 1,3,5,7-cyclooctatetraene (azulene); acetic acid; 2-phenylethyl ester; and phenylethyl alcohol. These volatiles are lethal to major phytopathogens, for example, *Pythium ultimum* and *Verticillium dahliae*. *Pythium ultimum* is known to infect plants causing damping off and root rot disease of corn, soybean, potato, and wheat. *Verticillium dahlia* causes *Verticillium* wilt resulting in curled and discolored appearance of leaves of almost 350 species of eudicots of temperate regions. So using *Gliocladium* sp. as a biocontrol agent in tropics may reduce the agricultural loss to some extent. *Edenia gomezpompae*, a member of Pleosporaceae isolated as an endophyte of *Callicarpa acuminata*, was reported to produce naphthoquinone spiroketal showing antifungal activity against *Colletotrichum* sp., *Alternaria solani*, and *Phytophthora capsici* (Macias et al. 2008). Lee et al. (2009) performed mycofumigation with *Oxyporus latemarginatus* EF069 volatiles for control of postharvest apple (*Malus pumila*) decay and *Rhizoctonia* root rot infection on moth orchid (*Phalaenopsis* sp.). Apple is an important economic fruit, and biocontrol of apple fungal pathogens by volatile emissions of endophytic fungi is completely an innovative way of treatment. Suwannarach and his coworkers in the year 2010 isolated a new species of Muscodor, *Muscodor cinnamomi* CMU-Cib 461, from a member of Lauraceae named

Cinnamomum bejolghota. This isolate was known to produce azulene, a new compound detected first from any *Muscodor* species. This species was tested in vitro and in vivo for antifungal activity against a common worldwide devastating pathogen *Rhizoctonia solani* (causal agent of damping off). The VOCs produced by this fungi include (S)-(+)-5-methyl-1-heptanol; ethyl acetate; propanoic acid, 2-methyl-, methyl ester; cis-2,4-dimethylthiane; S,S-dioxide; cyclopentane; butanoic acid, 2-methyl-, methyl ester; 1-butanol, 3-methyl-, acetate; β-humulene; azulene, 1,2,3,5,6,7,8,8a-octahydro-1,4-dimethyl-7-(1-methylethenyl)-; 1S-(1.α., 7.α., 8a.β); and eudosma-4(14),11-diene 1,1,1,5,7,7,7-heptamethyl-3,3-bis(trimethylsiloxy) tetrasiloxane. *Rhizoctonia solani*–infected seedlings were treated with volatile mixtures to assess the mycofumigation property. In vivo experiment was conducted on four seedlings of bird pepper, bush bean, garden pea, and tomato. It was concluded that 30 gm of *Muscodor cinnamomi* prepared on rye grain solid media is the minimum dose required for inhibition of *Rhizoctonia* infection and total control and elimination of damping off symptoms. *Muscodor cinnamomi*-infected soil does not show any seed germination inhibition in comparison to *Rhizoctonia solani*–infected soil. So it is a type of pioneer study of using endophytic species as potent agents of fumigation and biocontrol. *Candida intermedia* strain C410 (Saccharomycetaceae) was isolated as an endophyte of strawberry (*Fragaria ananassa*), and the volatile emission was known to be a mixture of 49 organic compounds including esters, alcohols, alkenes, alkanes, alkynes, organic acids, ketones, and aldehydes of which 1, 3, 5, 7-cyclooctatetraene and 3-methyl-1-butanol were the most dominant (Huang et al. 2011). Volatiles of strawberry endophyte were itself useful as postharvest control agent for the host plant against *Botrytis* fruit rot. Other compounds include 1,3,5,7-cyclooctatetraene; 3-methyl-1-butanol; 2-nonanone; pentanoic acid, 4-methyl-, ethyl ester; 3-methyl-1-butanol, acetate; acetic acid, pentyl ester; and hexanoic acid, ethyl ester that were found to be extremely inhibitory to conidial germination (reproductive growth) and also vegetative (mycelial) proliferation of *B. cinerea*. When the fruits are exposed to *C. intermedia* synthetic volatiles or itself to the fungus, the incidence of *Botrytis* fruit rot reduces significantly. Strawberry fruits inoculated directly with the endophyte also remain disease-free. So mixtures of *Candida intermedia* C410, the unique natural products, are useful as mycofumigation technique or for postharvest disease management by biological control policies.

Tangerine fruit (*Citrus tangerine*), the commercial citrus crop of Northern Thailand, faces huge postharvest losses due to pathogenesis of green mold (*Penicillium digitatum*). The pathogen is the prime cause of worldwide deterioration of tangerine fruits by mycopathogenesis. Out of 32 detected compounds, the most predominant were 2-methylpropanoic acid and 3-methylbutan-1-ol. Other compounds include carbitol, octanoyl chloride, azulene, 3-methylhexane, 2-methylpropan-1-ol, 2,3-butanediol, caryophyllene, 2-methylbutyric acid, ethyl 2-hydroxyproponate, etc. The pathogen was treated both in vitro and in vivo for their inhibition by endophytic volatile components. In both cases, pathogen growth was restricted. During transportation of the fruits, fungus causes huge crop loss by infecting the fruits; when fruits were inoculated with 30 gm of rye grain culture of

M. suthepensis (1 month old), the disease development is ceased. So it is a classic example of mycofumigation by the biocontrol agent of tangerine fruit for the control of rot lesions caused by *P. digitatum* infection. The in vivo application requires the proper surface sterilization (using sodium hypochlorite) of the targeted parts where the inoculation is going to be done, for example, fruit, stem, root, and leaf. Usually infection on fruits for assessing the biocontrol potential is the most common and popular method. The seeds will be washed in distilled water, and using sterile needle, uniformly the whole area would be done, and the whole area would be infected or inoculated with the endophytic liquid extracts containing spore suspensions. *Muscodor albus* VOCs are potent enough to cause a significant reduction of in vitro spore germination of the *Tilletia* species *T. horrida*, *T. indica*, and *T. tritici*. Endophytic *Nodulisporium* spp., *Trichoderma* spp., *Phomopsis* spp., and *Oxyporus latemarginatus* are reported to produce VOCs that inhibit mycelial growth of phytopathogenic fungi (Lee et al. 2009; Park et al. 2010; Ajith and Lakhsmidevi 2010; Amin et al. 2010). Black sigatoka disease (also known as leaf spot or black leaf streak disease) of banana (*Musa paradisiaca*) is caused by *Mycosphaerella fijiensis* (ascomycete fungus). This phytopathogen is inhibited by the volatile emissions of *Muscodor sutura*, an endophytic isolate of *Prestonia trifidi*. The volatiles are effective also against *Ceratocystis ulmi*, the causal agent of Dutch elm disease of American elm (*Ulmus americana*). The volatile mixtures include thujopsene, chamigrene, isocaryophyllene, and butanoic acid, 2-methyl- that are potent inhibitors of the common anthracnose pathogen of cucumber, muskmelon, and watermelon (members of cucurbits), *Colletotrichum lagenarium*. So this unique endophyte and its chemical mixtures are potent mycofumigants and ensure crop protections against destructive pathogens like *C. ulmi* and *C. lagenarium*, *Sclerotinia sclerotiorum* (causing white mold, cottony rot, water soft rot, stem rot, drop, crown rot, and blossom blight diseases of the host), and also *Phytophthora palmivora* (oomycete fungi), the causal agent of bud root of palms and areca nut predominantly occurring in regions of South India (Kudalkar et al. 2012). Liarzi and his coworkers tested the biological control efficacy of the endophytic *Daldinia* cf. *concentrica*, isolated from olive tree (*Olea europaea* L.) of Israel against 18 phytopathogens, and the unique mixtures of 27 volatile were effective against the phytopathogenic mycelial growths. The mixtures include a variety of organic compounds: 3-methyl-1-butanol, 2-methyl-1-butanol, 1-methyl-1,3-cyclohexadiene, 1-methyl-1,4-cyclohexadiene, 4-heptanone, isoamyl acetate, 4-heptyn-2-ol, 2-octenal, octanal, β-elemene, α-guaiene, β-selinene, α-selinene, α-bulnesene, germacrene A, etc. The unique mixtures having broad-spectrum antifungal property could be used for fumigation for eliminating the pathogenic infections of *Aspergillus niger* (mold-causing organism on fruits of economic importance). So the endophytic *D.* cf. *concentrica* opens up opportunities for fungal disease control in food and agricultural industries (Liarzi et al. 2016). *Nodulisporium sp.* strain GS4d2II1 (*Hypoxylon anthochroum*) and *Hypoxylon anthochroum* strain Blaci are potent enough to be used as biopesticide against *Fusarium oxysporum*, a common contaminant of *Solanum lycopersicum* var. *cerasiforme* (cherry tomato) causing a great percentage of crop loss globally. Six VOCs of alcohols' mixture, phenylethyl alcohol, 2-methyl-1-butanol, 3-methyl-1-butanol, eucalyptol, ocimene, and terpinolene, were detected and applied together with synergistic effect and individually both in vitro and in vivo. Inoculation of pathogen on the cherry tomato fruits yields significant reduction in *Fusarium* contamination. Both agar dilution techniques and gas test were done to assess the in vitro antifungal activity, and the endophytic volatile mixtures were effective in both the cases. Volatiles kill the pathogens probably by interfering cell membrane permeability, hyphal morphology, and respiratory activity of the pathogenic *Fusarium oxysporum*. So it is a great opportunity to use the unique mixture of volatile organic compounds of the endophytic isolate to reduce the crop loss caused by the pathogenic infection on the commercially valuable plant of cherry tomato worldwide. Endophytic *Phoma* sp. (Didymellaceae) and *Phomopsis* sp. (Valsaceae) were isolated from *Larrea tridentata* and *Odontoglossum* sp. (Strobel et al. 2011; Singh et al. 2011). The volatiles detected are effective against phytopathogens *Verticillum* sp., *Ceratocystis* sp., *Cercospora* sp., *Sclerotinia* sp. *Sclerotinia* sp., and *Botrytis* sp.

9.5 Seaweeds as Natural Fungicides

Algae are diverse group of autotrophs and the leading producers of O_2 in the ecosystem. They range from prokaryotic unicellular to eukaryotic complex multicellular forms involved in the marine and terrestrial food chain. Antifungal activity of the seaweed (members of Phaeophyceae and Rhodophyceae) is a major weapon for natural fungicides along with their antibacterial, anti-protozoan, and antiviral activities. Algal seaweeds are potent holders of large number of secondary metabolites including phenolics, terpenes, alkaloids, and lectins which are not directly involved in photosynthesis and reproduction and thus fall under the category of secondary metabolites. They are common antimicrobial of algal origin that act on the target organisms by altering the microbial cell permeability accompanied with the loss of internal macromolecules or sometimes interfere with the membrane function causing cellular disintegrity ultimately leading to cell death (Abu-Ghannam and Rajauria 2013). Several studies include antifungal activity of algal members against human pathogens; a very few studies include their efficacy against plant pathogens (Cheung et al. 2014; Singh et al. 2007; Stirk et al. 2007; Padmakumar and Ayyakkannu 1997; Ismail et al. 2014; Genovese et al. 2013; Lopes et al. 2015). Padmakumar and Ayyakkannu tested 80 species of algae against a variety of bacterial and fungal pathogens. Out of the all screened organisms, 70% exhibited antibacterial efficiency, and only 27.5% inhibited fungal growth. Polysaccharides found in the cell wall and deposited in terms of storage food from red and brown algal sources include ulvans (obtained from *Ulva* sp.), alginates and fucans (from *Fucus* sp.), laminarin (*Laminaria* sp.), and carrageenans that can induce defense responses in plants against phytopathogens by pathogen-associated molecular patterns (MAPs) and are capable of inducing plant resistance (Vera et al. 2011). Polysaccharides stimulate regular cellular changes associated with pathogen perception and defense activation by change in Ca^{2+} concentration and burst due to oxidative stress activation of salicylate, ethylene, and jasmonate biosynthetic pathways and by activating pathogenesis-related proteins (PRPs) (Jaulneau et al. 2010; Zhao et al. 2012). As a result of the depolymerization of the polysaccharides, the obtained oligosaccharides induce protection against a variety of fungal, viral, and bacterial diseases by accumulation of the antimicrobial compounds in the cell. Algal polysaccharides as an alternative weapon over the synthetic agricultural drugs for controlling plant disease have been widely studied (Stadnik and Freitas 2014; Hahn et al. 2008).

9.5.1 Laminarin: Defense-Inducing Polysaccharide from Brown Algae

Brown algae *Laminaria digitata*, a genus of Phaeophyceae, is commonly called seaweeds and known to be the potent producers of kelp, an iodine-rich substance needed for the normal functioning of thyroid gland. *Laminaria* produces laminarin, glucan polysaccharide–containing 1,3-linked β-d-glucose moiety, a reserve food material found on the vacuoles of the vegetative cells of this genus. β-Glucans are involved as a major part of daily diet and obtained from the brands of common cereals. They are involved in the defense responses of agricultural crops like tomato (*Lycopersicon esculentum*), eggplant (*Solanum melongena*), pepper (*Piper nigrum*), watermelon (*Citrullus lanatus*), grape (*Vitis vinifera*), apple (*Malus* sp.), and pear (*Pyrus communis*). Elicitation of defense response by laminarin against causal agents of gray mold (*Botrytis cinerea*) and downy mildew (*Plasmopara viticola*) in grapevine plants remarkably suppresses their infection up to 55% and 75%, respectively (Copping et al. 2004). So, natural product from brown algae *Laminaria* sp. known as laminarin or laminaran can act as the biofungicide or biocontrol compounds. Use of laminarin significantly reduces the mycelial growth and aflatoxin production in *Aspergillus flavus* and ensures its use as a fungicide (Liangbin et al. 2012). The advantage of using laminarin over other products is that as it breaks down finally to glucose molecules, it has no maximum residue limit (MRL) on the plant treated with this product. So, there is no need of preharvest interval constraint. This has been a prime cause why laminarin has substituted five popular fungicides involved in the treatment of apple scab (*Venturia inaequalis*) in France (Mery et al. 2013). This phyto-pharmaceutical is used widely in France and some countries of Europe in the name of Vacciplant (major active constituent is laminarin). Laminarin has broad-spectrum applicability on fire blight of apples and pears in Greece, France, Belgium, Switzerland, Portugal, and also Morocco. It is effective for apple scab disease in France and Belgium and for curing storage diseases of apples caused by *Gloeosporium* sp. in Belgium. Laminarin comes out as a fungicide of natural origin after being eligible in 33 tests between 2001 and 2011 in several parts of Europe, for example, France, Belgium, Italy, and Poland, on natural contamination of orchards on several sensitive strains of scab fungus including Golden Delicious, Golden Smoothie, Read Cheaf, Galaxy, Gala, and Pink Lady. Laminarin is applied widely against secondary scab (to minimize secondary scab during summer and up to harvest) as a result of its uniqueness in its mode of action. It does not involve cell death of the host plant or hypersensitivity induction in the host organism but rather stimulates plants' natural resistance (Klarzynski et al. 2000). Aziz et al. (2003) reported its effectiveness in tobacco plants, wheat, strawberries, apples, and vines. The application of laminarin and alginate reduced the development of wilt symptoms caused by *Verticillium dahliae* on olive twigs, stimulating its phenolic metabolism (Salah et al. 2018). Moreover, alginates reduced pathogen growth in vitro. Laminarin induces the release of H_2O_2in cells of tobacco plants and leads to the increase in PAL activity (phenylalanine ammonia-lyase) and causes the accumulation of PR-1, PR-2 (glucanase), PR-3 (chitinase), and PR-5. Concerning red algae polysaccharides, carrageenans induced protection against a broad range of pathogens such as tobacco mosaic virus (TMV), *B. cinerea*, and *E. carotovora* on tobacco (Vera et al. 2011). Again on tobacco, Mercier et al. (2001) showed that carrageenan infiltrated the leaves and increased the expression of genes coding for a sesquiterpene cyclase involved in the synthesis of the antimicrobial terpenoid capsidiol, PR-3 proteins (basic chitinases), and proteinase inhibitor with antipathogenic activity. An adequate percentage of growth and spore germination inhibition of *Botrytis cinerea* was mediated by the hexane extracts of *Laminaria digitata* and *Undaria pinnatifida. Porphyra umbilicalis*, laverbread, is an edible seaweed (Corato et al. 2017). Other than B-glucan polysaccharides (laminarin) of *Laminaria*, other secondary metabolites (phenols, terpenes) of phaeophycean algae (*Sargassum* sp.) showed effectivity against common pathogens *Fusarium solani*, *Rhizoctonia solani*, *Aspergillus* spp., *Fusarium oxysporum*, *Penicillium* spp., and *Botrytis cinerea* (Khallil et al. 2015; Ibraheem et al. 2017; Mabrouk et al. 1985; Liu et al. 2014).

9.5.2 Cyanobacterial Polysaccharides Versus Fungal Infection

Cyanophycean blue-green algae are abundant all over the world and ranging from pond ecosystem to oceanic system. Though they have been reported to produce a large number of toxins and involved in death and disease of cattle and human being, they are of serious interest from the point of view of natural fungicidal products. Drawing the similarities with bacteria, they are characterized with a mucilaginous or gelatinous sheath composed of polysaccharides which are the weapon against fungal pathogenesis. Cyanobacterial polysaccharides (POL) show higher disease resistance against *B. cinerea* when they are applied on the intact fruit (preharvest conditions when fruit is attached to the plant) rather than the fruit detached (postharvest conditions) from the plant (Zheng et al. 2011; Feliziani et al. 2015; Yao and Tian 2005). Polysaccharides are involved in elicitation as elicitors for development of local and systemic disease resistance and expression of defense enzyme synthesis, for example, chitinases and glucanases that are involved directly in antifungal responses (Paulert et al. 2009; Reymond and Farmer 1998; Sharma et al. 2014). Water extracts of common BGA *Anabaena* sp., *Ecklonia* sp. (common edible marine algae of Japan and Korea), and *Corallina* sp. (hard seaweed of Corallinaceae family) exhibit antifungal activity against *Podosphaera xanthii* (causal agent of powdery mildew of cucurbits) on zucchini plant, *Cucurbita pepo*, of Cucurbitaceae

(Roberti et al. 2015, 2016). In the recent past, fungi inhibitory ability of algal members has been reported by several workers (Righini et al. 2018; Corato et al. 2017; Khallil et al. 2015; Ibraheem et al. 2017). In vitro growth inhibition of *Aspergillus oryzae* and *Penicillium notatum* has been seen by cyanophycean *Anabaena laxa* (Frankmölle et al. 1992). Devastating plant pathogens *Pythium* sp., *Fusarium* sp., and *Rhizoctonia* sp. were restricted by extracts of *Anabaena* sp. (Moon et al. 1992; Manjunath et al. 2010). The use of BGA extract as the growth inhibitor of pathogenic *Chaetomium globosum, Cunninghamella blakesleeana, Aspergillus oryzae, Rhizoctonia solani, Fusarium* sp., *Pythium* sp., and *Sclerotinia sclerotiorum* is reported. The extracts of *Phormidium fragile* and *Nostoc muscorum* (Rizk 2006) are effective control agents of sugar beet pathogens (*Sclerotium rolfsii, Rhizoctonia solani, Fusarium verticillioides*). The common root rot pathogen *Rhizoctonia solani* is inhibited by the extracts of cyanophycean algae *Nostoc entophytum* and *N. muscurum* (Osman et al. 2011). The recent investigations of Dukare et al. (2011) included the inhibitory activity of three strains of BGA (C4, C8, C12) against *F. solani, F. oxysporum, F. oxysporum* f. sp. *lycopersici, F. moniliforme, P. debaryanum*, and *R. solani*. Righini et al. (2018) reported for the first time about the resistance-inducing ability of the polysaccharides of the brown (*Ecklonia* sp.), red (*Jania* sp.), and cyanophycean algae (*Anabaena* sp.) and their possible role in disease control ability. Study on zucchini plant revealed the enhancement of defense-related enzyme on the plant mediated by the treatment of *Anabaena* extract (Roberti et al. 2015). Water extracts of *Ecklonia* sp., *Anabaena* sp., and *Corallina* sp. are potent antifungals against *Podospora xanthii* on zucchini plant (Roberti et al. 2016).

9.6 Plant Amphibians as an Alternative Source

Bryophytes, the simplest member of the broad umbrella of Embryophyta, are situated between algae and pteridophytes, are known to be plant amphibians growing in the marshy or shady habitat, and require water for their fertilization and for the perfect swimming motility of their sperms. They have been evaluated for their antimicrobial activity for a long time. It has been proved that these cryptograms are rich source of bioactive secondary metabolites and can easily be exploited as an alternative source of fungicidal compounds. As they grow in marshy habitats and can protect themselves from biotic (ultraviolet rays, heat stress, and predation) and abiotic stress (fungal or bacterial attack), they are store house of diverse bioactive chemicals (Xie and Lou 2008). Members of Hepaticopsida and mosses (the evolved members of bryophytes) are known to possess antifungal activity and are rich source of flavonoids, terpenoids, bibenzyls, and fatty acids of therapeutic importance (Krzaczkowski et al. 2008). Bryophytes are known to possess antibiotic property (Banerjee and Sen 1979; Banerjee 2000; Singh et al. 2007; Shirzadian et al. 2009; Savaroglu et al. 2011). Their antibiosis has been evaluated against a large number of plant and human pathogenic fungus (Mekuria et al. 2005). Antimicrobial compounds from bryophyte can cure the problems of conventional antibiotic resistance (Vanden Bossche et al. 1998). The antifungal efficacy is tested by disc diffusion

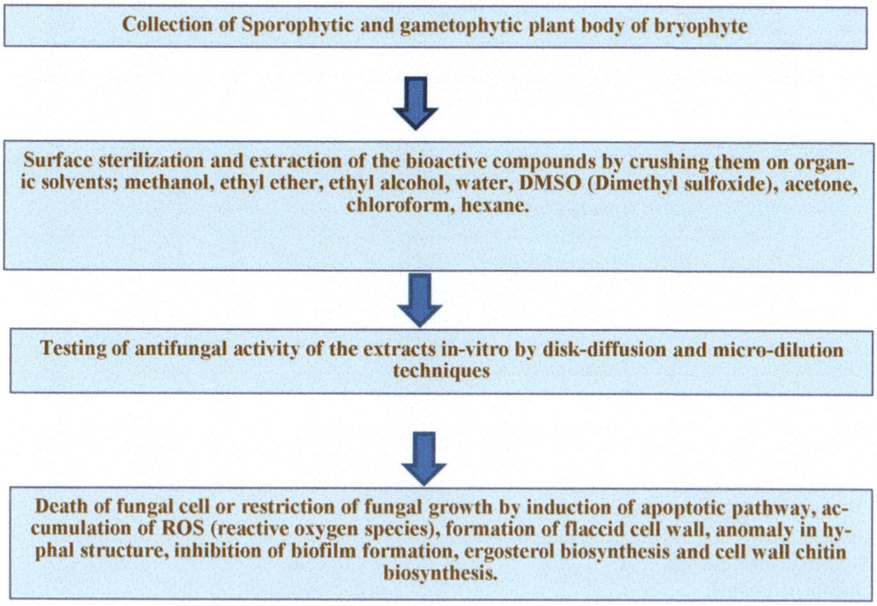

Fig. 9.5 Pharmacological extraction of bioactive compounds from bryophytes

assay and microdilution method (Fig. 9.5). Different concentrations of the extracts are prepared and checked for their antifungal efficacy against phytopathogenic fungi. They may be fungicidal or fungistatic in nature, interfering at cellular, genetic level and creating blockage at metabolic pathways. Extracts are made on several organic solvents or water extractions and also mixture of one or two organic solvents. The solvents popularly used are ethanol, methanol, chloroform, ether, dimethyl sulfoxide (DMSO), acetone, chloroform, and hexane (Table 9.5). Sporophytes and gametophytes of different bryophytes at different stages of growth and at a different amount are first surface sterilized and then crushed on the organic solvents and used as antifungals in vitro against the fungal pathogens (Wolters 1964). A large number of phytopathogenic fungi (*A. niger, R. bataticola, F. moniliforme, Penicillium funiculosum, T. viride, P. ochrochloron, A. versicolor, A. fumigatus Trichoderma viride, Aspergillus niger, A. flavus, P. funiculosum, Tilletia indica, Sclerotium rolfsii, R. solani, Penicillium ochrochloron, Alternaria alternate, Botrytis cinerea, Botryodiplodia theobromae, F. oxysporum* f. sp. *gladioli, Penicillium expansum, P. chrysogenum, Trichoderma viride*) are reported to be partially or completely inhibited by the bryophyte extracts of *Marchantia polymorpha*, *Atrichum undulatum*, *Physcomitrella patens*, *Rhodobryum ontariense*, *Ctenidium molluscum*, *Ptilidium pulcherrimum*, *Hypnum cupressiforme*, *Fontinalis antipyretica* var. *pyretica*, *Plagiochasma appendiculatum*, and *Dumortiera hirsuta* (Sabovljevic et al. 2011; Pejin et al. 2012; Veljic et al. 2009; Gahotri and Chaturvedi 2011; Alam et al. 2011; Deora and Jain 2008; Dey and De 2011; Deora and Suhalka 2017).

Table 9.5 Antifungal activity of extracts of bryophytes

Extractions	Bryophyte genus	Antifungal against	References
Organic solvent extracts	*Scleropodium purum, Sphagnum fimbriatum, S. nemoreum, S. subsecundum, Pogonatum aloides, P. urnigerum, Polytrichum commune, P. formosum, Plagiothecium denticulatum, Mnium hornum, Oligotrichum hercynicum, Atrichum undulatum, F. antipyretica*	*C. cerebella, B. alli, P. oryzae, P. versicolor, F. bulbigenum, Rhizoctonia solani,*	Wolters et al. (1964) and Savaroglu et al. (2011)
A-Herbertenol, β-herbertenol, α-formylherbertenol, β-bromoherbertenol	*Herbertus aduncus*	*Botrytis cinerea, Rhizoctonia solani*	Matuso et al. (1986)
5- and 7-Hydroxycalamenenes, drimenol, drimenal, viridiflorol, gymnomitrol, bisbenzyls	*Bazzania trilobata*	*Botrytis cinerea, Cladosporium cucumerinum, P. infestans, Pyricularia oryzae, Septoria tritici*	Scher et al. (2004)
Trans-β-methylthioacrylate	*Balantiopsis cancellata*	*Cladosporium herbarum*	Labbe et al. (2005)
Ether, alcohol, and hexane extract	*Pallavicinia lyellii, Scapania verrucosa*	*Aspergillus niger, F. oxysporum, P. oryzae*	Subhisha and Subramoniam (2005), Guo et al. (2008)
Methanol and ethanol extracts	*Pleurozium schreberi, Palustriella commutata, Homalothecium philippeanum, Anomodon attenuatus, Rhytidium rugosum, Hylocomium splendens, Dicranum scoparium, Leucobryum glaucum*	Variety of phytopathogens *A. niger, P. ochrochloron*	Sabovljevic et al. (2006), Veljic et al. (2009)
Aqueous extracts	*Plagiochasma appendiculatum, Dumortiera hirsute*	*Alternaria alternate, A. niger, Botrytis cinerea, Botryodiplodia theobromae, F. oxysporum f. sp. gladioli, Penicillium expansum, P. chrysogenum, Trichoderma viride*	Deora and Jain (2008) and Alam et al. (2011)

(continued)

Table 9.5 (continued)

Extractions	Bryophyte genus	Antifungal against	References
Acetone, ethanol, chloroform, and distilled water extracts	*Thuidium delicatulum*, *Plagiochasma appendiculatum*, *Bryum argenteum*, *B. cellulare*	*A. niger*, *R. bataticola*, *F. moniliforme*	Bodade et al. (2008)
Methanolic and chloroform extracts	*Ctenidium molluscum*, *Ptilidium pulcherrimum*, *Marchantia polymorpha*, *Hypnum cupressiforme*, *Fontinalis antipyretica* var. *pyretica*	*Trichoderma viride*, *Aspergillus niger*, *A. flavus*, *P. funiculosum*, *Tilletia indica*, *Sclerotium rolfsii*, *R. solani*, *Penicillium ochrochloron*	Veljic et al. (2009), Gahotri and Chaturvedi (2011)
DMSO extracts	*Marchantia polymorpha*, *Atrichum undulatum*, *Physcomitrella patens*, *Rhodobryum ontariense*	*Penicillium funiculosum*, *T. viride*, *P. ochrochloron*, *A. versicolor*, *A. fumigatus*	Sabovljevic et al. (2011), Pejin et al. (2012)
Acetone and methanol extracts	*Riccia gangetica*	*Curvularia lunata*	Deora and Suhalka (2017), Deora and Guhil (2015, 2016)

9.7 Ray Fungi–Based Antifungal Activities

Actinobacteria are a group of gram-positive filamentous bacteria that are called as the branched bacteria or ray fungi (from Greek *actis*, ray beam, and *mykes*, fungus) and are characterized with the high G + C content occurring in mostly aerobic conditions but occasionally being anaerobes (Ludwig and Klenk 2005; Olanrewaju and Babalola 2019). Their morphology varies from forming branching filaments or mycelial growth to external spores. They are ubiquitous in nature ranging their distribution from soil and human microbiota to plant and even animal kingdom. They are predominant in aquatic as well as terrestrial ecosystem playing a major part in mineralization and recycling of organic matters leading to soil formation (Sharma et al. 2014). They are not only free-living members of the ecosystem but also a plant symbiont or endophyte, contributing to the plants' survival in extreme conditions and pursuing several bioactivities in vivo and in vitro. Actinomycetes produce a diverse range of secondary metabolites, for example, antibiotics, antitumor, insect-repellent, and immunosuppressive agents, and plant growth-promoting regulators (PGPRs) that are of immense pharmaceutical and agricultural importance. They are the prime producers of diverse antibiotics after the landmark discovery of penicillin in the year 1928. The single genus of *Streptomyces* sp. itself produces 76% of the total known bioactive (10,000 are produced by actinobacteria out of 23,000 produced by microorganisms, almost 45%) compounds from actinobacterial and bacterial source (Berdy 2012) and is known to be the prime organism in the pharmaceutical world. They are equally profitable when isolated from plant source and designated as endophytic actinomycetes. So exploitation of the actinobacterial novel bioactive compounds both from endophytic and non-endophytic source is the ultimate way to fight against human and plant diseases. Here we focus only on actinobacterial compounds' antifungal activity and role in plant protection from deadly diseases caused by severe phytopathogens leading to irreparable crop loss and economic breakdown of agricultural sectors.

9.7.1 Isolation, Identification, and Detection of Bioactive Compounds with Anti-phytopathogenic Activity

Actinomycetes from soil source are selected based on the enrichment culture technique and are plated on selective media for isolation. Antifungal agents, for example, nystatin and cycloheximide, are supplemented for the inhibition of fungal contamination. For isolation of endophytic actinobacteria from plant source, plants are first selected and surface sterilized for the elimination of the epiphytic contaminants and finally plated on the selective growth media like starch casein nitrate agar (SCNA), chitin-vitamin B, tap water yeast extract agar (TWYA), soybean, humic acid-vitamin B (HV), yeast extract casamino acid (YECA), modified Gausse, and glycine-glycerol (Ivantiskaya et al. 1978; Küster 1959; Küster and Williams 1964; Williams and Davies 1965; Hayakawa and Nonomura 1987; Crawford et al. 1993). International Streptomyces Project (ISP) medium is also popular media used for isolation, and they are supplemented with amino acids (L-asparagine for ISP 5, tryptone for ISP 1), inorganic trace salts, starch or carbohydrate sources (malt extract for ISP 4), and agar as solidifying agent. pH set at near to optimum or slightly basic is mandatory for proper isolation techniques using ISP medium. The actinomycete isolates are grown in solid or liquid medium for their antifungal bioactivity detection. Antagonistic activities of the potent isolates are tested by growing them on both sides of the fungal hyphae, and isolate having anti-phytopathogenic activity will inhibit the growth of the pathogens. Actinobacterial aqueous- or solvent-based extracts will be evaluated for either fungistatic or fungicidal activity by agar well-diffusion techniques. Soluble bioactive compounds of antifungal importance will be extracted using wide range of organic solvents followed by purification by column and thin-layer chromatographic techniques. HPLC analysis will be the most useful method for the detection of the purity of the compound, and further NMR studies are needed for the proper identification of the bioactive compound. Cell line studies are made with the coupling of bioinformatics tools for the proper knowledge about their mode of action.

9.7.2 Role in Plant Protection

9.7.2.1 Actinobacterial Flora Producing Antibiotics of Agricultural Importance

Actinobacteria can be a part of plant as endophyte, rhizospheric soil as symbiont for plant growth-promoting substance producer, and organisms' normal microbial flora as gut microorganism. So they are ubiquitous in their distribution. Out of several biologically potent compound produced from the actinobacterial source, antibiotics are the major contribution of these microorganisms toward human civilization. All the known antibiotics (blasticidin, mildiomycin, natamycin, validamycin, kasugamycin) are of actinobacterial (most of them are the *Streptomyces* sp.) source showing protective activity for the plants against agricultural fungal pathogens (Tables 9.6 and 9.7).

9.7.2.2 Endophytic Actinomycetes

As human are dependent completely on nature and more particularly natural components of agricultural origin and importance, dependence on agricultural crops is of a known fact. But the problem arises when the crops are affected most by the fungal pathogens leading to huge crop loss, and thus the search for novel antibiotics is on, and the search has shifted to actinobacterial source, and endophyte plays an important role in this respect. There are significant reports of antifungal compounds from bacterial origin, but now the focus has shifted to microbes of endophytic origin (Table 9.8). Till date, a huge number of antibiotics are already reported and have minimized the crop loss to a notable amount (Fig. 9.6). Antibiotics and other antifungal compounds include munumbicins A, B, C, D, E-4, and E-5, vanillin, saadamycin, 5,7-dimethoxy-4-p-methoxyphenyl coumarin, coronamycin, and fistupyrone isolated from different strains of *Streptomyces* sp. (*Streptomyces aureofaciens* CMU Ac 130, *Streptomyces* sp. NRRL 30562, *Streptomyces* sp. Hedaya48, *Streptomyces* sp. MSU-2110, *Streptomyces* sp. TP-A0569) and from different parts (stem, leaf, inflorescence, root, fruit, internal healthy tissues) of diverse medicinal plants, for example, *Kennedia nigricans* (Fabaceae), *Aplysina fistularis* (yellow-green candle sponge or yellow tube sponge), *Monstera* sp. (Monsteraceae), *Allium fistulosum* (Liliaceae), and *Zingiber officinale* (Zingiberaceae) of different regions all over the world. The bioactive compounds are effective against a wide range of phytopathogens including *Magnaporthe oryzae, Fusarium graminearum, F. oxysporum* f. sp. *lycopersici, F. oxysporum* f. sp. *cubense, F. verticillioides, Colletotrichum* sp., *Pestalotiopsis* sp., *Diaporthe* sp., *Xylaria* sp., *P. aphanidermatum, Pythium oligandrum, Pythium ultimum, Pythium aphanidermatum, Phytophthora cypogea, Phytophthora cactorum, Phytophthora infesta, Fusarium solani, Aspergillus fumigatus, Mycosphaerella fijiensis, Sclerotinia sclerotiorum, Rhizoctonia solani,* and *Phytophthora erythroseptica* (Shan et al. 2018; Costa et al. 2013; Igarashi et al. 2002; Tian et al. 2004; Zin et al. 2007) and are protecting a large number of cereals and other important cash crops from being affected by these common contaminants. Endophytic actinobacteria directly counteract with fungal plant pathogens not only by producing bioactive compounds but also by enhancing the plant's growth through

Table 9.6 Mode of action of antibiotics against fungal crop pathogens

Active component	Trade name	Source	Effective against	Mode of action	References
Kasugamin ™	Kasugamycin	*Streptomyces kasugaensis*	Rice blast and other diseases in Japan (*Magnaporthe oryzae*)	Inhibition of protein synthesis by t-RNA and ribosome assembly interference	Tanaka et al. (1966) and Zaker (2016)
Polyoxin B and polyoxorim	Polyoxin Z, Polyoxin AL	*Streptomyces cacaoi*	Fungal pathogens	Protein synthesis inhibitor	Isono and Suzuki (1979)
Validamycin	Validacin, Valimun, Sheathmar, Mycin	*S. hygroscopicus*	Rhizoctonia root rot	Knockout of an enzyme necessary for glucose generation on hyphal tips	Karneda et al. (1987) and Dayan et al. (2009)
Mildiomycin	Mildiomycin	*Streptoverticillium rimofaciens* (soil actinomycete)	Powdery mildews in Japan	Inhibition of protein synthesis by blockage of peptidyl transferase	Om et al. (1984)
Blasticidin S	Bla-S	*S. griseochromogenes* (soil actinomycete)	Rice blast disease in Eastern Asia	Inhibition of protein synthesis	Kimura and Yamaguchi (1996)
Natamycin	Delvolan (amphoteric macrolide)	*S. natalensis* or *S. chattanoogenesis*	Fungal disease of ornamental plants	Membrane dysfunction by binding to membrane ergosterol	te Welscher et al. (2008)

Table 9.7 Antifungal activity of different species of *Streptomyces*

Actinomycetes	Antifungal against	References
Streptomyces griseus	*Rhizoctonia solani*	Merriman et al. (1974)
Streptomyces kasugaensis	*Fusarium* sp.	De Vasconcellos and Cardoso (2009)
Streptomyces sp.	*Alternaria brassicae, Colletotrichum gloeosporioides, Rhizoctonia solani, Phytophthora capsici*	Srividya et al. (2012)
Streptomyces sp.	*Sclerotium rolfsii*	Gholami et al. (2014)
Streptomyces griseus E44G	*Fusarium oxysporum* f. sp. *lycopersici*	Al-Askar et al. (2015)
Streptomyces sanglieri	*Ganoderma boninense*	Azura et al. (2016)
APA2 *Streptomyces longisporoflavus*, AAH53 *Streptomyces mutabilis*, APC70 *Streptomyces griseus*	*Alternaria solani, Colletotrichum coccodes, Fusarium oxysporum*	Dávila et al. (2016)
	Wilt disease of banana caused by *Fusarium oxysporum* f. sp. *cubense* (FOC)	Dewi et al. (2017)
Streptomyces strain KX852460	*Rhizoctonia solani* AG-3 KX852461	Ahsan et al. (2017)
Streptomyces 12-09-4 and 12-09-11	*Botrytis cinerea, Fusarium oxysporum* f. sp. *cucumerinum*	Wang et al. (2018)

the production of plant growth promoters and making the plant less susceptible to pathogenic invasion. They are efficient agent of reducing the symptoms that arise due to exposure to environmental stress (Shimizu 2011). Enhanced production of indole acetic acid (IAA) was mediated by *Streptomyces* sp. (isolated from *Centella asiatica*) and *Nocardiopsis* sp. (Dochhil et al. 2013; Shutsrirung et al. 2014; Gangwar et al. 2014). Experimental trials on cucumber indicate positive result as the isolates *Actinoplanes campanulatus*, *Micromonospora chalcea*, and *Streptomyces spiralis* enhanced plant growth and improved yield conditions (El-Tarabily et al. 2010). Other than auxin, auxin-like similarly functioning molecules named as pteridic acids A and B are found to be inducers of adventitious root proliferation in kidney bean plants at very minute concentrations of 1 mM (Igarashi et al. 2002).

9.7.2.3 Siderophores and Chitinases

Chitin is a major fungal cell wall polysaccharide (the second most abundant polysaccharide in nature after cellulose) component and is the first line of defense of fungal cells. Actinobacteria antagonize the fungal cell by producing chitinases (an enzyme capable of hydrolyzing fungal cell wall) and break the glycosidic bonds in chitin and lead to the death of the pathogenic cell. Endophytic *Kitasatosporia* sp.

Table 9.8 Antifungal compounds from bacterial origin and their common targets

Bacillomycin D (*B. subtilis* AU195), bacillomycin and fengycin (*B. amyloliquefaciens* FZB42), Zwittermicin A (*B. cereus* UW85), geldanamycin (*S. hygroscopicus var. geldanus*), chaetomin (*Chaetomium globosum*), gliotoxin (*T. virens*), 2,4-diacetylphloroglucinol (*Pseudomonas fluorescens* F113)	*Aspergillus flavus*, *Fusarium oxysporum*, *Phytophthora medicaginis*, *R. solani*, *P. ultimum*, *Pyricularia oryzae*	DeBoer et al. (1970), Shanahan et al. (1992), Smith et al. (1993), Moyne et al. (2001), Wilhite et al. (2001), Koumoutsi et al. (2004) and Anitha Murugesan (2005)
Pyrrolnitrin, pyoluteorin, pseudane (*Burkholderia cepacia*, *P. fluorescens* Pf-5), phenazines (*P. fluorescens* 2–79 and 30–84)	Damping off (*Phytophthora medicaginis*, *P. aphanidermatum*), damping off and rice blast (*R. solani*, *Pyricularia oryzae*, *Pythium ultimum*), take-all (*Gaeumannomyces graminis* var. *tritici*)	Howell and Stipanovic (1980), Homma et al. (1989), Thomashow et al. (2002) and Smith et al. (1993)
Harpin proteins (*Erwinia amylovora*), trade name: Harpin αβ (ProAct)	Induction of systemic acquired resistance (SAR) and less susceptibility to fungal and bacterial disease	Wei et al. (1992)
Strobilurin and oudemansin (members of basidiomycete grows on dead wood) Commercial synthetic analogues: azoxystrobin and kresoxim-methyl	Fungal pathogens (blockage of mitochondrial respirations by blocking of ubiquinone receptor)	Kraiczy et al. (1996)
Mycosubtilin (*B. subtilis* BBG100), iturin A (*B. subtilis* QST713), herbicolin (*Pantoea agglomerans* C9-1), xanthobaccin A (*Lysobacter* sp. strain SB-K88),	Damping off (*Pythium aphanidermatum*, *Botrytis cinerea*), root rots (*R. solani*), fire blight (*Erwinia amylovora)*, damping off (*Aphanomyces cochlioides*)	Sandra et al. (2001), Kloepper et al. (2004), Leclere et al. (2005), Islam et al. (2005) and Paulitz and Belanger (2001)

(isolate of *Catharanthus roseus*) and *Kibdelosporangium* sp. (isolate of *Achillea fragrantissima*) are reported to be chitinase producers (El-Shatoury et al. 2009; Mini Priya 2012). *Actinoplanes missouriensis* isolated from *Lupinus* sp., a member of Fabaceae family, produces chitinase causing hyphal cell lysis and reducing the conidial germination rate and protects the plant from pathogenic attack of *Plectosporium tabacinum*, the causal agent of lupin root rot in Egypt (El-Tarabily 2003; El-Tarabily and Sivasithamparam 2006). Siderophores are soluble, small, high-affinity iron carriers produced by bacterial or fungal members and are involved in the transportation of iron (Fe^{3+}) across the cell membrane. They have caught sudden attention due to their involvement in plant growth promotion as well as antagonistic ability against phytopathogens (Cao et al. 2005; Tan et al. 2006; Rungin et al. 2012). Endophytic actinobacteria from *Aloe vera*, *Mentha arvensis*, and *Ocimum sanctum* are known to be producers of hydroxymate type and catechol type of siderophores, and the isolate *Saccharopolyspora* O9 is known to be the potent inhibitor

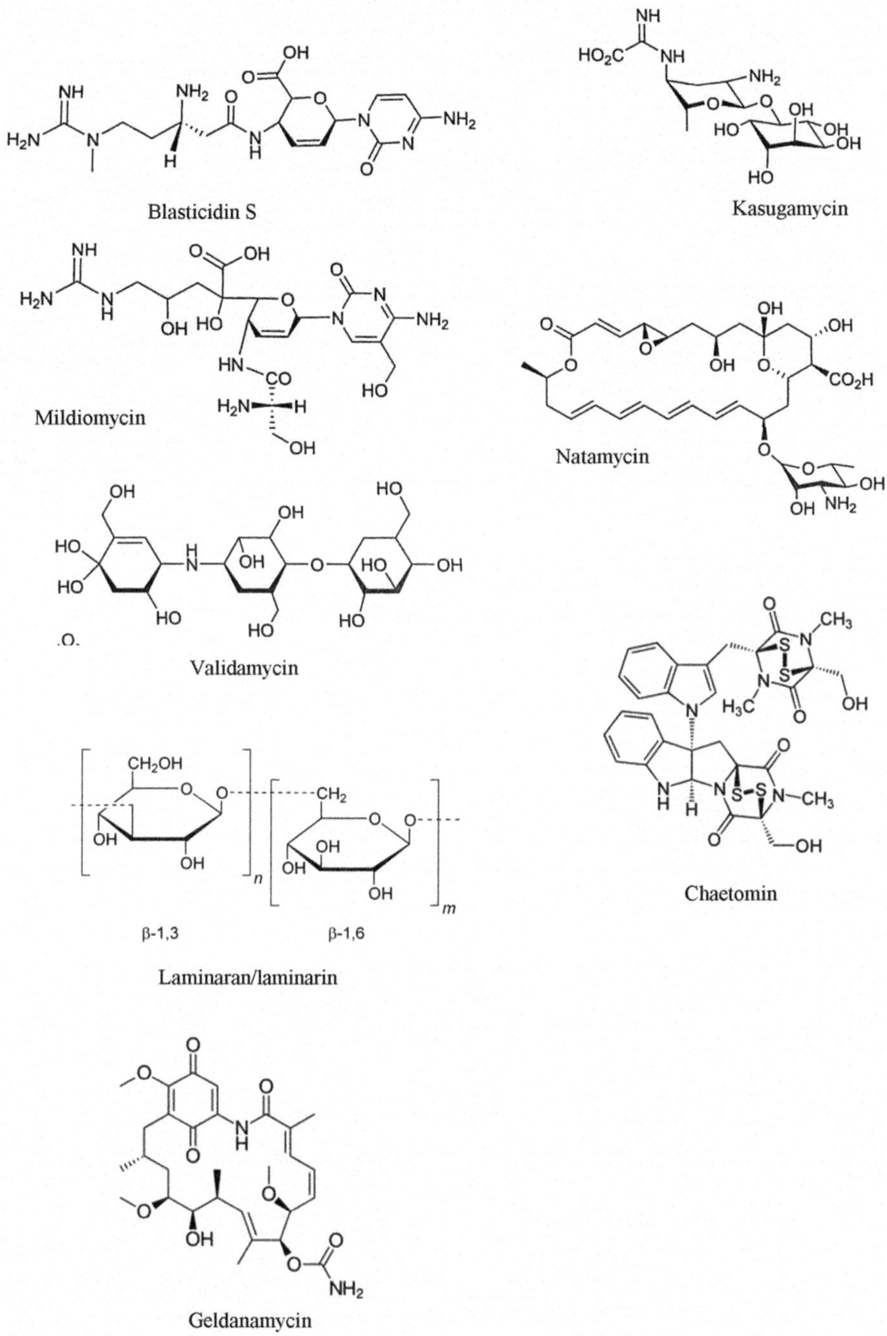

Fig. 9.6 Antibiotics and polysaccharides with plant protective ability

of *Alternaria brassicicola*, *Botrytis cinerea*, and *Fusarium oxysporum* (Gangwar et al. 2014; El-Shatoury et al. 2009).

9.7.2.4 Biocontrol Activity

Endophytic isolates of *Cucumis sativus* (cucumber), identified as *Actinoplanes campanulatus*, *Micromonospora chalcae*, and *Streptomyces spiralis*, are reported to control the growth and development of damping off, crown rot, and root rot pathogen *Pythium aphanidermatum*. They are known to promote plant growth and to protect seedlings and mature plants. A novel bioactive compound identified as 6-prenylindole was isolated from endophytic *Streptomyces* sp. showing strong antifungal activity against a broad range of phytopathogens: *Alternaria brassicicola* and *Fusarium oxysporum* (Igarashi 2004). Another new prenylated indole derivative from endophytic actinobacterial source inhibited the growth of *Colletotrichum orbiculare*, *Phytophthora capsici*, *Corynespora cassiicola*, and *Fusarium oxysporum* (Zhang et al. 2014). Naphthomycins A and K isolated from *Streptomyces* sp. CS have antifungal activity against *Penicillium avellaneum* (Lu and Shen 2003, 2007). Biocontrol ability of fistupyrone has made it a useful tool to minimize the crop loss of *Brassica* due to black leaf spot disease caused by *Alternaria brassicicola* (Igarashi 2004). Interest on actinomycetes of endophytic origin as an alternative tool for antifungal agent is increasing day by day (Table 9.9).

9.8 Plant Extracts as the Prime Source for Antifungals

Since the beginning of human civilization, whenever human race has faced any turbulence in its path of existence, they have rushed to their green friends, trees, for the ultimate solution. Search for bioactive products of medical importance has been a thirst area from time immemorial. Whether it is a concern of human or plant health, trees have given answers in all aspects. In the recent past, phytopathogenic infection has pushed the agricultural productive parameters to a real challenge, and plant extracts in its crude and purified form are applied as biocontrol methods (Table 9.10). The existing synthetic chemicals are facing problem of immediate or delayed drug resistance and also issues of nephrotoxicity (the gold standard; amphotericin B), biomagnification, or quality assurance of the food products and thus are inconsistent in their business (Goa and Barradell 1995; Cuenca-Estrella et al. 2000). So green plant extracts are the novel, safest, and the best effective treatment tool in this arena. Plants are mysterious in their chemical nature and in respect to their secondary metabolite production. The faith is consistent on green plants due to the fact that plants protect themselves from fungal or bacterial diseases specially for the taxa that occur in marshy shady or water-logged or stress conditions (Gurgel et al. 2005). So the search is primarily made on the wild native taxa or invasive species that have higher potential of antimicrobial production. The knowledge of ethnobotany comes in this context, and tribal people are imitated for the gathering of crude knowledge. The problem of fungal pathogenesis is mainly faced by plants of economic importance, that is, cash crops. A single event of pathogenic attack can

Table 9.9 Endophytic actinomycetes as antifungal agents

Actinomycetes	Host plant	Antifungal against	References
Streptomyces sp.	Rhododendron	*Phytophthora cinnamomi, Pestalotiopsis sydowiana*	Shimizu et al. (2000)
Streptomyces NRRL 30562	*Snake vine medicinal plant* (*Kennedia nigriscans*)	*Pythium ultimum, Rhizoctonia solani, Phytophthora cinnamomi, Geotrichum candidum, Sclerotinia sclerotiorum*	Castillo et al. (2002), (2006)
Streptomyces sp.	*Allium fistulosum*	*Alternaria brassicicola* on Chinese cabbage seedlings	Igarashi et al. (2002)
Streptomyces sp. AOK-30	Mountain laurel (*Kalmia latifolia*)	*Pestalotia rhododendri*	Nisimura et al. (2002)
Streptomyces sp. CS	*Maytenus hookeri*	*Penicillium avellaneum* UC-4376	Lu and Shen (2003), (2007)
Streptomyces sp. TP-A0595, *Streptomyces* sp. TP-A0569	*Allium tuberosum*	*Alternaria brassicicola*	Igarashi (2004)
Streptomyces sp. MSU-2110	*Monstera* sp.	*P. ultimum, Phytophthora cinnamomi, Geotrichum candidum, F. solani, Rhizoctonia solani*	Ezra et al. (2004a, b)
Streptomyces griseofuscus, Streptomyces hygroscopicus, Streptomyces globisporus, Streptomyces aureus, Streptomyces albosporus	Rice (*Oryza sativa*)	*Magnaporthe grisea, Rhizoctonia solani, Xanthomonas oryzae, Fusarium moniliforme*	Tian et al. (2004)
Streptomyces aureofaciens CMU Ac130, *Streptomyces* sp. Tc022	*Zingiber officinale*	*Fusarium oxysporum, Colletotrichum musae*	Taechowisan et al. (2006), (2007)
Streptomyces sp., *Streptoverticillium* sp., *Streptosporangium* sp.	*Musa paradisiaca*	*Fusarium* sp. wilt pathogen	Cao et al. (2005)

(continued)

Table 9.9 (continued)

Actinomycetes	Host plant	Antifungal against	References
Streptomyces fulvoviolaceus, *Streptomyces caelestis*, *Streptomyces coelicolor*	*Thottea grandiflora*, *Mapania* spp., *Polyalthia* spp.	*Fusarium solani*, *Aspergillus fumigatus*, *Mycosphaerella fijiensis*, *Pythium ultimum*, *Sclerotinia sclerotiorum*, *Rhizoctonia solani*, *Phytophthora erythroseptica*	Zin et al. (2007)
Microbispora sp., *Nonomurae* sp., *Streptomyces* sp.	*Solanum lycopersicum*	*Rhizoctonia solani*, *Pythium irregulare*, *A. solani*, *P. parasitica*	Inderiati and Franco (2008)
Streptomyces sp., *Nocardia* sp., *Streptosporangium* sp., *Streptoverticillium* sp.	*Azadirachta indica*	*Pythium oligandrum*, *Pythium ultimum*, *Pythium aphanidermatum*, *Phytophthora cypogea*, *Phytophthora cactorum*, *Phytophthora infestans*	Verma et al. (2009)
Promicromonospora cymbopogonis, *Nonomuraea roseola*, *Micromonospora chokoriensis*, *Streptomyces ochraceiscleroticus*, *S. aurantiacus*, *S. griseocameus*, *S. chryseus*, *S. albogriseolus*	*Juncus effusus L.*, *Ainsliaea henryi*, *Stellera chamaejasme*, *Salvia miltiorrhiza*, *Lysimachia fortune*, *Senecio declouxii*, *Potentilla discolor*, *Achyranthes aspera*, *Cynanchum auriculatum*	*Verticillium dahliae*, *Fusarium oxysporum* f. sp. *vasinfectum*, *Aspergillus niger*, *Fusarium oxysporum* f. sp. *niveum*, *Colletotrichum orbiculare*, *Fusarium graminearum*, *Exserohilum turcicum*, *Curvularia lunata*, *Botrytis cinerea*	Zhao et al. (2011)
Streptomyces sp. Hedaya-48	*Aplysina fistularis*	*F. oxysporum*	El-Gendy and El-Bondkly (2010)
Actinoplanes campanulatus, *Micromonospora chalcea*, *Streptomyces spiralis*	*Cucumber (Cucumis sativa)*	*Pythium aphanidermatum*	El-Tarabily et al. (2010)
Cr-12 and Cr-20 unidentified isolates	*Catharanthus roseus*	*Curvularia lunata*, *Fusarium oxysporum*, *Fusarium solani*, *Rhizoctonia solani*	Kafur and Khan (2011)
Streptomyces coelicolor	*Rhizophora apiculata*, *Avicennia marina*	*A. niger*, *A. flavus*, *Penicillium* sp., *A. fumigatus*	Gayathri and Muralikrishnan (2013)

(continued)

Table 9.9 (continued)

Actinomycetes	Host plant	Antifungal against	References
Streptomyces sp. 16R3B	*Zea mays*	*P. aphanidermatum*, causal agent of damping off in cucumber (*Cucumis sativa*)	Costa et al. (2013)
Streptomyces sp. neau-D50	Soybean (*Glycine max*)	*Phytophthora capsici*, *Corynespora cassiicola*, *Fusarium oxysporum*, *Colletotrichum orbiculare*	Zhang et al. (2014)
Streptomyces cinereus AR16	*Emblica officinalis*	*Fusarium oxysporum*, *Rhizoctonia solani*, *Aspergillus niger*, *Alternaria brassicicola*, *Phytophthora dresclea*	Gangwar et al. (2015)
Streptomyces sp., *Leifsonia xyli*, *Microbacterium* sp., *Streptomyces* sp., *Brevibacterium* sp.	*Mirabilis jalapa*, *Clerodendrum colebrookianum*	*Rhizoctonia solani*, *Fusarium graminearum*, *Fusarium oxysporum*, *Fusarium proliferatum*, *Fusarium oxysporum* f. sp. *ciceris*, *Colletotrichum capsici*	Passari et al. (2015)
Streptomyces sp.	*Schima wallichii*	*Colletotrichum* sp., *Alternaria* sp., *F. oxysporum* f. sp. *ciceris*, *F. proliferatum* *F. culmorum*, *F. graminearum*	Passari et al. (2016)
Melia toosendan (chinaberry)	*Rhodococcus* sp., *Tomitella* sp.	*Colletotrichum orbiculare*, *Fusarium oxysporum*, *Alternaria solani*, *Magnaporthe grisea*, *Curvularia lunata*, *Gibberella saubinetii*	Zhao et al. (2018)
Streptomyces levis NBRC 15423(T), *Streptomyces gilvifuscus* T113(T), *Micromonospora olivasterospora* DSM 43868(T), *Actinomadura geliboluensis* A8036(T), *Streptomyces djakartensis* NBRC 15409(T), *Streptomyces griseoaurantiacus* NBRC 15440(T), *Nocardiopsis dassonvillei* NBRC 13392(T)	*Camellia sinensis*	*Magnaporthe oryzae*, *Fusarium graminearum*, *F. oxysporum* f. sp. *lycopersici*, *F. oxysporum* f. sp. *cubense*, *F. verticillioides*, *Colletotrichum* sp., *Pestalotiopsis* sp., *Diaporthe* sp., *Xylaria* sp.	Shan et al. (2018)

Table 9.10 Bioactive products with their common sources and targets

Bioactive product	Source	Effective against	References
Carvone	Dill and caraway seed	Inhibit the growth of storage pathogen	Moezelaar et al. (1999)
Gamma aminobutyric acid (GABA)	Any type of plant and animal source	Prevent powdery mildew on grapes, brown rot and shot hole of stone fruit; enhances growth of almond, broccoli, onions	Copping (2004)
Cinnamaldehyde from cinnamon leaf essential oil (mixture of α-methyl cinnamaldehyde, (E)-2-methylcinnamic acid, eugenol, isoeugenol, citral, geraniol oil), trade name: Vertigo and Cinnacure	*Cinnamomum osmophloeum* Alternate source: seeds of the weed *Cassia obtusifolia*	Dry bubble (caused by *Verticillium fungicola*), dollar spot (*Sclerotinia homoeocarpa*), pitch canker (*Fusarium moniliforme* var. *subglutinans*), wood rot fungi (*Coriolus versicolor* and *Laetiporus sulphureus*, *Aspergillus fumigatus* and *Trichophyton rubrum*)	Copping (2004), Wang et al. (2005), Cheng et al. (2008) and Khan and Ahmad (2011)
Jojoba oil (vegetable oil)	*Simmondsia chinensis* (Simmondsiaceae)	Powdery mildew for ornamental plants and grapes and applied as ground spray	Copping (2004)
Milsana (ethanolic extracts of plant)	*Reynoutria sachalinensis* (Polygonaceae)	Powdery mildew of wheat and grape (*Uncinula necator*, *Sphaerotheca fuliginea*) by induced resistance, used as a spray	Copping (2004)
Pink plume poppy blood root	*Macleaya cordata* *Sanguinaria Canadensis* (Papaveraceae)	Alkaloids; sanguinarine effective against *Rhizoctonia solani* by systemic acquired resistance (SAR)-mediated accumulation of endogenous phenolics	Liu et al. (2009a, b)

(continued)

Table 9.10 (continued)

Bioactive product		Source	Effective against	References
Sporan		Rosemary oil (*Rosmarinus officinalis*)	*Botrytis cinerea*	Zaker (2016)
Promax		Thyme oil (*Thymus vulgaris*)	*Botrytis cinerea*	Zaker (2016)
Trilogy		Neem oil (*Azadirachta indica*)	*F. oxysporum*, *R. solani*	Zaker (2016)
GC-3™		Cottonseed (*Gossypium hirsutum*) and garlic (*Allium sativum*)	–	Zaker (2016)
Fatty Acids	Polyacetylenic acids, octadeca-9,11,13-triynoic acid, trans-octadec-13-ene-9,11-diynoic acid	*Prunella vulgaris*	*M. oryzae*, *R. solani*, *P. infestans*, *S. sclerotiorum*, *F. oxysporum*, and *P. capsici*. Inhibits rice blast, tomato late blight, wheat leaf rust, and red pepper anthracnose	Yoon et al. (2010)
	Lipopeptides	*Bacillus* XT1 CECT 8661	*Botrytis cinerea*	Toral et al. (2018)
	Ginger oleoresin (GO)	Ginger	*Pestalotiopsis microspora*; dominant pathogenic fungi causing rotten disease in harvested Chinese olive (*Canarium album* Lour.) fruits	Chen et al. (2018)

(continued)

Table 9.10 (continued)

Bioactive product		Source	Effective against	References
Alkaloids	Steroidal alkaloids; verazine type (veramitaline, stenophylline B, veramiline) and jerveratrum type (jervine)	*Veratrum taliense*	*Phytophthora capsici*	Zhou et al. (2003)
	D-Calycanthin, L-folicanthine	*Chimonanthus praecox* (from seeds)	*Exserohilum turcicum, B. maydis, A. solani, S. sderotiorum, F. oxysporum*	Zhang et al. (2006)
	Securinine	*Phyllanthus amarus*	*Alternaria alternate, A. brassicae, A. brassicicola, Curvularia lunata, C. maculans, C. pallenscens, C. musae, Helminthosporium echinoclova, H. spiciferum*	Singh et al. (2008)
	Pipernonaline, a piperidine alkaloid	*Piper longum*	*M. oryzae, R. solani, B. cinerea, P. infestans, P. recondite, B. graminis*	Yoon et al. (2013)
Glycosides	Hemoiedemosides A, B (sulfated triterpene),	Patagonian sea cucumber (*Hemoiedema spectabilis*)	*Cladosporium cucumerinum*	Chludil et al. (2002)
	Triterpenic saponins,	*Sapindus mukorossi, Diploknema butyracea*		Saha et al. (2010)
	Pregnane (caudatin glycosides)	*Cynanchum wilfordii* (roots)	Barley powdery mildews and strawberry powdery mildew (*Sphaerotheca humuli*)	Yoon et al. (2011)

affect seriously the demand and supply ratio; thus the sustainability is lost, and restoring the good health of crops is a basic need of agricultural sectors but in an efficient way not hampering the soil health, ecosystem characters, and human health and also should be budget friendly. The search is strictly focused on plants of ethno-medicinal importance as history indicates the ability of medicinal plant extracts in human and animal mycoses and antifungal ability (Mathias-Mundy and McCorkle

1995). The statistics of the World Health Organization (WHO) states that 80% of the worlds' population in underdeveloped or developing countries depend on plants of ethnomedicinal importance for their primary healthcare issues.

9.8.1 Antifungal Properties of Secondary Metabolites of Plants

Secondary metabolites are plants' best weapon against phytopathogenic invasion. Several plant extracts have been assessed for their antifungal activity against a variety of phytopathogens of serious agricultural threats (Table 9.11). The metabolites are divided into terpenoids, saponins, phenolic compounds, flavones, flavonoids, flavonols, alkaloids, and coumarins (Table 9.12). Plant extracts are primarily tested for antifungal efficacy and further are purified by solvent extraction and chromatographic procedures leading to discovery of new antifungal agents. Terpenoids, also called as isoprenoids (under the chemical subclass of prenyllipids), are known to be the oldest group of widespread molecular compounds produced by plants. Scher et al. (2004) reported a variety of six sesquiterpenes of antifungal importance against the causal organisms of bunch rot (*Botrytis cinerea*) on grapes, scab of cucurbits (*Cladosporium cucumerinum*), potato blight (*Phytophthora infestans*), rice blast (*Pyricularia oryzae*), and blotch of wheat (*Septoria tritici*). Sesquiterpene isolated from *Polygonum punctatum* (dotted knotweed of knotweed family Polygonaceae) named after the chemical polygodial is an effective control agent of *Zygosaccharomyces bailii* (a common food spoilage yeast). Scab of cucurbits is a common and devastating fungal pathogenic disease in agricultural fields, and this disease is to some extent prevented by the use of clerodane diterpenes extracted from *Detarium microcarpum*, a plant of Leguminosae family (Cavin et al. 2006). Skaltsa (2000) reported fungi inhibitory (*Cunninghamella echinulata*) activity of costunolide and eudesmane derivatives isolated from *Centaurea* plants. Other than terpenes, saponins (triterpene ad steroidal saponins) are also effective antifungals reported from plant sources. Tea is one of the most vital cash crops in terms of foreign money earning and the most popular beverage having antioxidative properties. Pathogenic infection by *Pestalotia longiseta* causes a huge loss of tea production. Nagata et al. in the year 1985 isolated triterpenoid saponins camelids I and II from the leaves of *Camellia japonica* (Japanese camellia) that inhibited the tea pathogen *P. longiseta*. Cucurbitacins I, A, B, Q, and E isolated from cucurbitacins (*Ecballium elaterium*) have antifungal activity against *Botrytis cinerea* (Har-Nun and Meyer 1990). Phenolics are odorous compounds having antifungal compounds and are also responsible for the plant pigment production. Phenolics cover a large number of chemical compounds, for example, alkylated phenols, anthraquinones, coumarins, phenolic acid, phenols, phenylpropanoids, quinines, xanthones, hydroxycinnamic acid, p-coumaric acid, ferulic acid, and chlorogenic acid. Phenol derivatives like crassinervic acid (*P. crassinervium*), aduncumene (*P. aduncum*), hostmaniane (*P. hostamannianum*), and gaudichaudanic acid (*P. gaudichaudianum*) are effective against strawberry blossom blight pathogen *Cladosporium cladosporioides* (Lago et al. 2004). 3-Acetyl-4-acetoxyacetophenone showed antifungal activity against

Table 9.11 Bioactive phytochemicals against common plant diseases of crop plants

Name of the disease	Plant taxa and chemical class of the compound	References
Scab of cucumber (*Colletotrichum cucumerinum*)	Napthoxirenes from bark of *Sesamum angolense* of Pedaliaceae family, xanthone from roots of *Polygala nicaeensis* (Polygonaceae), sakurasosaponin from leaves of *Rapanea melanophloeos*, 5-methylcoumarins, mutisicoumarones C and D from *Mutisia friesiana* (Asteraceae), mollugenol A from *Mollugo pentaphylla* (Molluginaceae), flavone and flavonol of *Helichrysum decumbens* (Asteraceae), triterpenoid saponin from roots of *Dolichos kilimandscharicus* (Leguminosae), Clerodane diterpene from *Detarium microcarpum*, E-triticine, P-triticine, puroindoline from *Triticum aestivum* (Poaceae)	Spendley et al. (1982), Potterat et al. (1987), Martson et al. (1988), Marston et al. (1993),Viturro et al. (2004), Cavin et al. (2006a), and Dhatwalia et al. (2009)
Disease of woody plant (*Melampsora medusae*)	Pinocembrin from leaves of *Populus deltoides* (Salicaceae)	Shain and Miller (1982), Hoof et al. (2008)
Cladosporium fruit and leaf rot and bitter root (*Cladosporium gloeosporioides*)	Long-chain alcohol from peels of young fruit of *Persea americana* from Lauraceae, methylripariochromene A from roots of *Eupatorium riparium* (Asteraceae)	Prusky et al. (1983), Ratnayake Bandara et al. (1992)
Pine needle pathogen (*Dothistroma pini*)	Stearic acid from needles of *Pinus radiata* (Pinaceae)	Franich et al. (1983)
Pathogen of corn, sorghum, apple (*Helminthosporium carbonum*)	Luteone and wighteone from leaf surface of *Lupinus albus* (Leguminosae)	Ingham et al. (1983)
Black and brown spot of banana (*Colletotrichum musae*)	Dopamine from unripe banana fruit (*Musa* sp.)	Muirhead and Deverall (1984)
Powdery mildew of grains (*Erysiphe graminis*)	Gramine from leaves of *Hordeum vulgare* (Poaceae)	Wippich and Wink (1985)
Leaf spot, rots, and blights (*Alternaria alternata*), disease of cereal (*Penicillium verrucosum*)	Alizarin and emodin from root of *Rubia tinctorum* of Rubiaceae, alkylated phenols of peel and pulp of *Mangifera indica* (Anacardiaceae)	Cojocaru et al. (1986), Manojlovic et al. (2005)
Maize rot (*Fusarium moniliforme*), epidemic outbreak of glume and kernel discoloration (*Curvularia lunata*)	Flavan-4-ols of root bark of *Sorghum* cultivars of Poaceae	Jambunathan et al. (1986)

(continued)

Table 9.11 (continued)

Name of the disease	Plant taxa and chemical class of the compound	References
Rice blast disease (*Pyricularia oryzae*)	Jasmonic acid, hydroxybenzoic acid from leaves of *Oryza officinalis*; pisiferic acid from leaves of *Chamaecyparis pisifera* (Cupressaceae); gingerenones A, B and C; isogingerenone B from *Zingiber officinale*	Kobayashi et al. (1987), Endo et al. (1990), Cho et al. (1998)
Blue mold of tobacco (*Peronospora tabacina*)	Diterpenoids from *Nicotiana tabacum* of Solanaceae	Reuveni et al. (1987)
Leaf and fruit pathogen (*Cladosporium cladosporioides*)	Canaliculatol from bark of *Stemonoporus canaliculatus* and long-chain alcohol from *Persea americana*, phenylethanone from *Euodia lunuankenda*, sinharine and methylsinharine from *Glycosmis cyanocarpa*, Illukumbin from *Glycosmis mauritiana* (Rutaceae), phenylethanone from *Euodia lunuankenda* (Lauraceae), benzoquinone from *Croton lacciferus* (Euphorbiaceae)	Bokel et al. (1988), Ratnayake Bandara and Wimalasiri (1988), Kumar et al. (1990), Greger et al. (1992), Pacher et al. (2001), Springob and Kutchan (2009),
Butt rot on conifers; Douglas fir, spruce, fir, hemlock, pine, and larch (*Phaeolus schweinitzii*)	Astringin and rhaponticin from *Picea sirchellsis* (Pinaceae)	Woodward and Pearce (1988)
Sudden death of soybean (SDS) pathogen, damping off, corn rot, root rot, wilting, and necrotic spots (*Fusarium solani*)	Tomatine from *Lycopersicum esculentum* (Solanaceae)	Defago and Kern (1988)
Gray mold of grapes (*Botrytis cinerea*)	Surangin B from *Mammea longifolia* (Clusiaceae), cucurbitacin I of *Ecballium elaterium* (Cucurbitaceae), chalcone from *Bauhinia manea* (Leguminosae)	Achenbach et al. (1988), Har-Nun and Meyer (1990), Deng and Nicholson (2005)
Witches' broom of cocoa tree (*Crinipellis perniciosa*)	Polymeric procyanidin from *Theobroma cacao* of Sterculiaceae	Brownlee et al. (1990), Duke (2004)
Fusarium wilt of pathogen (*Fusarium oxysporum* f. sp. *lycopersici*)	Jodrellin B and oerodin from *Scutellaria woronowii and S. violacea* of Lamiaceae	Cole et al. (1991)
Collar rot, root rot, damping off, wire stem, primarily the pathogen of herbs (*Rhizoctonia solani*)	Dihydrochalcone from twigs and leaves of *Psidium acutangulum* (Myrtaceae), coumarin from *Mammea longifolia* (Clusiaceae), dolichin of *Dolichos lablab* (Fabaceae), methylquercetin from leaf of *Wedelia biflora*	Miles et al. (1991), Miles et al. (1993), Lee et al. (2003), Deng and Nicholson (2005) and Yoganandam et al. (2009)

(continued)

Table 9.11 (continued)

Name of the disease	Plant taxa and chemical class of the compound	References
Spoilage of fruit and vegetables (*Cladosporium sphaerospermum*)	Nonglycosidic iridoid from *Alibertia macrophylla* (Rubiaceae)	Young et al. (1992)
Late blight of potato or tomato (*Phytophthora infestans*), bbrown rust (*Puccinia recondita*), rice blast disease (*Pyricularia grisea*)	Emodin from *Cassia tora* (Leguminosae)	Kim et al. (2004)
Pathogen of Rhododendron (*Pestalotia guepinii*), cortical stem rot (*Fusarium avenaceum*), *Drechslera* leaf spot (*Drechslera* spp.)	Trifolin and hyperoside from *Camptotheca acuminata* (Cornaceae)	Li et al. (2005a, b)
Strawberry pathogen (*Cladosporium fragariae*) and anthracnose of *Lupin* sp., *Cladosporium acutatum*	Alkaloid (findersine, anhydroevoxine, haplamine) from *Haplophyllum sieversii* (Rutaceae)	Cantrell et al. (2005)
Pre- and postharvest fungal disease of cereal grains, legumes, and tree nuts (*Aspergillus flavus*)	Alkaloids from *Datura metel*	Dabur et al. (2005)
Early blight of potato and rice (*Alternaria solani*)	Meliacin-type of nortriterpenoid from *Chisocheton paniculatus* (Meliaceae)	Yang et al. (2009)
Take-all fungus in wheat, barley, rye, oat, turf grass (*Gaeumannomyces graminis*)	Avenacins from *Avena sativa* (Poaceae)	De Bertoldi et al. (2009)

Sclerotinia sp. Phenolic structures when contain a carbonyl group are known to be flavones, and the addition of an extra 3-hydroxyl group indicates flavonol. Flavonoids are also known to hydroxylated phenolics but occurring as a C6–C3 unit linked to aromatic ring. Not only plant samples directly but also plant derivatives like porpolis (galangin isolated from the bee glue or resinous mixture produced as a result of the mixture of tree buds, sap, botanical extracts, and bee exudates) are shown to be antifungal against green rot or mold of tangerine (pathogens *Penicillium digitatum*, *P. italicum*) and also control postharvest disease of cereal grains, legumes, and tree nuts caused by *A. flavus* (Afolayan and Meyer 1997). Flavones (6,7,4′-trihydroxy-3,5′-dimethoxyflavone, 5,5′-dihydroxy-8,2′,4′-trimethoxyflavone) from *Artemisia giraldi* are effective against *A. flavus* infections (Cowan 1999). Leaf wax of *Arrabidaea brachypoda* (Brazilian medicinal plant from Bignoniaceae) contains

Table 9.12 Direct extracts of plants tested for antifungal potency

Plant species	*Fungal pathogen*	References
Reynoutria sachalinensis	*Uncinula necator (the causal pathogen of grapevine (powdery mildew)*	Herger et al. (1988) and Abdu-Allah and Elyousr (2017)
Cassia tora (dealcoholized extract of leaves)	*Aspergillus niger*	Mukherjee et al. (1996)
Thymol, carvacrol, citronellol, geraniol, citral, perillyl, menthol, eugenol, 1,8-cineole, Y-terpinene, p-cymene, anethole from several plants like *Thymus vulgaris*, *T. spicata*, *T. pulegioides*, *Cymbopogon citratus*, *Cymbopogon martini*	*Fusarium moniliforme*, *Rhizoctonia solani*, *Phytophthora capsici*, *Monilinia fructicola*, *Botrytis cinerea*, *Curvularia lunata*	Mueller-Riebau et al. (1995), Krishna Kishore et al. (2007)
Phlomis fruticosa (Jerusalem sage)	*Aspergillus niger, Penicillium ochrochloron*, *Trichoderma viride*, *Fusarium tricinctum*, *Phomopsis helianthi*, *Cladosporium cladosporioides*, *Aspergillus ochraceus*	Ristic et al. (2000)
Vernonia tenoreana	*Aspergillus niger, A. flavus*	Ogundare et al. (2006)
Tagetes erecta	*Fusarium wilt* (*Fusarium oxysporum* f. sp. *niveum)*	Du et al. (2017)
Garlic *(Allium sativum L.)*, neem *(Azadirachta indica L.)*	Narrow brown leaf spot (*Cercospora oryzae*), sheath blight (*Rhizoctonia solani*), sheath rot (*Sarocladium oryzae*), false smut (*Ustilaginoidea virens*)	Mahmud et al. (2018)
Neem seed and eucalyptus extract	*Fusarium oxysporum* f. sp. *lycopersici* (FOL) tomato crop causing wilt disease	Poussio et al. (2018)
Stilbene and resveratrol	*Plasmopara viticola*	Krzyzaniak et al. (2018)
Zingiber officinale, Piper nigrum, Azadirachta indica, Nicotiana tabacum, Carica papaya	*P. expansum* pathogenic of yam tubers (*Dioscorea rotundata*)	Gwa et al. (2018)
Neem (*Azadirachta indica*), lantana (*Lantana camara*), and eucalyptus (*Eucalyptus globulus*)	*Pyricularia oryzae*	Wasimfiroz et al. (2018)
Onion (*Allium cepa*), garlic (*Allium sativum*), lantana (*Lantana camara*), marigold (*Tagetes erecta*), dathura (*Datura stramonium*), tulasi *Ocimum sanctum*), eupatorium (*Eupatorium rugosum*), parthenium (*Parthenium hysterophorus*), neem (*Azadirachta indica)*	*Botrytis oryzae* (brown leaf spot of paddy)	Channakeshava and Pankaja (2018)

(continued)

Table 9.12 (continued)

Plant species	*Fungal pathogen*	References
Methanol extracts of *Eucalyptus tereticornis, Ammi visnaga, Azadirachta indica, Rheum Palmatum, Adansonia digitata*	*Rhizoctonia solani (*root rot of maize)	Rashad et al. (2018)
Lawsonia inermis (henna), *Acalypha wilkesiana* (acalypha), *Melia azedarach* (chinaberry), *Punica granatum* (pomegranate), *Lantana camara* (lantana)	*Puccinia triticina* (leaf rust fungus)	Draz et al. (2019)
Azadirachta indica	*Sclerospora graminicola* (pearl millet downy Mildew)	Atri et al. (2019)

cirsiliol, cirsimaritin, and hispidulin and is showed to be effective against *Cladosporium sphaerospermum* (Alcerito et al. 2002). Galeotti et al. (2008) studied the antifungal activity of kaempferol 3-O-β-d-glucopyranosyl(1-2)-O-β-d-glucopyranosyl(1-2)-O-[α-1-rhamnopyranosyl-(1-6)]-β-d-glucopyranoside isolated from *Dianthus caryophyllus* (carnation, a member of Caryophyllaceae family) against *Fusarium oxysporum* f. sp. *dianthi* (Galeotti et al. 2008). *Fusarium culmorum*, a serous pathogen of seedling blight, foot rot, ear blight, stalk rot, and common rot of cereals and grasses, is found to be inhibited by six commercial coumarins: bergapten, herniarin, umbelliferone, xanthotoxin, and scopoletin. *Tithonia diversifolia*, the source of tithoniamarin, is effective against the anther smut fungus *Microbotryum violaceum*, earlier known as *Ustilago violacea* (Yemele-Bouberte et al. 2006). Berberine and jatrorrhizine (alkaloids) are isolated from *Mahonia aquifolium* (a plant of Berberidaceae family commonly called as Oregon grape and native to western North America) and are effective against human pathogenic *Candida* species. Pathogens of mango (*C. gloeosporioides*), anthracnose of lupin species, postbloom fruit drop of citrus, Valencia and navel oranges in Florida (caused by *C. acutatum*), and strawberry (caused by *Colletotrichum fragariae*) are inhibited by findersine, anhydroevoxine, and haplamine (Cantrell et al. 2005). Roots of *Cyathobasis fruticulosa* are source of beta-carboline, tryptamine, and phenylethylamine-derived alkaloids and are antifungal in nature (Bahceevli et al. 2005).

9.8.2 Essential Oil

Essential oils (EOs) of aromatic and medicinal plant origin are reported to possess antifungal properties and are of wide spectrum in their application for the control of agricultural pathogen (Table 9.13). EOs are mainly categorized under the plants' secondary metabolites and may fall under the category of terpenes, ketones, esters, aromatic phenols, ethers, alcohols, oxides, etc. (Fig. 9.7). They act by inhibiting the fungal hyphal growth either by accumulating in the fungal cell membrane or by

Table 9.13 Common sources of popular essential oils with antifungal property

Sl. no.	Name of the plant	Family
1	*Anethum graveolens*	Apiaceae
2	*Aniba rosaeodora*	Lauraceae
3	*Artemisia absinthium*	Asteraceae
4	*Boswellia thurifera*	Buseraceae
5	*Brassica nigra*	Brassicaceae
6	*Bunium persicum*	Apiaceae
7	*Cestrum nocturnum*	Solanaceae
8	*Calocedrus macrolepis* var. *formosana*	Cupressaceae
9	*Cananga odorata*	Annonaceae
10	*Carum carvi*	Apiaceae
11	*Cedrus deodara*	Pinaceae
12	*Chenopodium ambrosioides*	Amaranthaceae
13	*Cinnamomum jensenianum*	Lauraceae
14	*Cinnamomum zeylanicum*	Lauraceae
15	*Cicuta virosa*	Apiaceae
16	*Citrus limon*	Rutaceae
17	*Cuminum cyminum*	Apiaceae
18	*Cymbopogon citratus*	Poaceae
19	*Cymbopogon martini*	Poaceae
20	*Cymbopogon winterianus*	Poaceae
21	*Daucus carota*	Apiaceae
22	*Echinophora spinosa*	Apiaceae
23	*Eucalyptus citriodora*	Myrtaceae
24	*Eugenia caryophyllata*	Myrtaceae
25	*Foeniculum sativum*	Apiaceae
26	*Foeniculum vulgare*	Apiaceae
27	*Helichrysum arenarium*	Asteraceae
28	*Hypericum perforatum*	Hypericaceae
29	*Illicium verum*	Liliaceae
30	*Juniperus excelsa*	Cupressaceae
31	*Laurus nobilis*	Lauraceae
32	*Lavandula intermedia*	Lamiaceae
33	*Lavandula officinalis*	Lamiaceae
34	*Lippia rugosa*	Verbenaceae
35	*M. albanica*	Lamiaceae
36	*M. thymifolia*	Lamiaceae
37	*Majorana hortensis*	Lamiaceae
38	*Matricaria chamomilla*	Asteraceae
39	*Mentha arvensis*	Lamiaceae
40	*Mentha piperita*	Lamiaceae
41	*Mentha spicata*	Lamiaceae
42	*Micromeria dalmatica*	Lamiaceae
43	*Monarda* spp.	Lamiaceae
44	*Myrrhis odorata*	Apiaceae

(continued)

Table 9.13 (continued)

Sl. no.	Name of the plant	Family
45	*Nepeta rtanjensis*	Lamiaceae
46	*Ocimum basilicum*	Lamiaceae
47	*Ocimum majorana*	Lamiaceae
48	*Ocimum sanctum*	Lamiaceae
49	*Origanum vulgare*	Lamiaceae
50	*Origanum onites*	Lamiaceae
51	*Portenschlagiella ramosissima*	Apiaceae
52	*Rosmarinus officinalis*	Lamiaceae
53	*Seseli annuum*	Apiaceae
54	*Seseli globiferum*	Apiaceae
55	*Seseli montanum* subsp. *tommasinii*	Apiaceae
56	*Seseli rigidum*	Apiaceae
57	*Seseli scardica*	Apiaceae
58	*Salvia brachyodon*	Lamiaceae
59	*Salvia fruticose*	Lamiaceae
60	*Salvia officinalis*	Lamiaceae
61	*Salvia pomifera* subsp. *Calycina*	Lamiaceae
62	*Salvia sclarea*	Lamiaceae
63	*Sardinian desoleana*	Lamiaceae
64	*Sassafras albidum*	Lauraceae
65	*Satureja hortensis*	Lamiaceae
66	*Satureja montana*	Lamiaceae
67	*Satureja thymbra*	Lamiaceae
68	*Seseli montanum* subsp. *tommasinii*	Apiaceae
69	*Syzygium aromaticum*	Myrtaceae
70	*Tagetes patula*	Asteraceae
71	*Thymbra spicata*	Lamiaceae
72	*Thymus pulegioides*	Lamiaceae
73	*Thymus vulgaris*	Lamiaceae
74	*Trachyspermum ammi*	Apiaceae
75	*Zataria multiflora*	Lamiaceae

crossing the cell membrane and entering into the eukaryotic cell. Being lipophilic in their chemical nature, they can easily cross the cell and interrupt in sterol biosynthesis leading to growth retardation and finally cell death. As sterols are the maintenance, compounds of cellular integrity treatment with EO cause fungal cell death. Metabolic processes like respiration, replication, transcription, and translation are inhibited. Membrane permeability is drastically changed as they cause swelling and disruption of protein–lipid–protein membrane. Leakage of useful ions like Ca^{2+} and K^+ causes cell death. Thymol, carvacrol, eugenol, and related phenolic compounds cause H^+ and K^+ leakage and water imbalance and deplete intracellular high-energy molecule (ATP). Essential oils are extracted from almost every parts of a plant, for example, roots, fruits, barks, twigs, leaves, seeds, and flowers, by several extraction procedures that include hydro and steam distillation, cold pressing, and

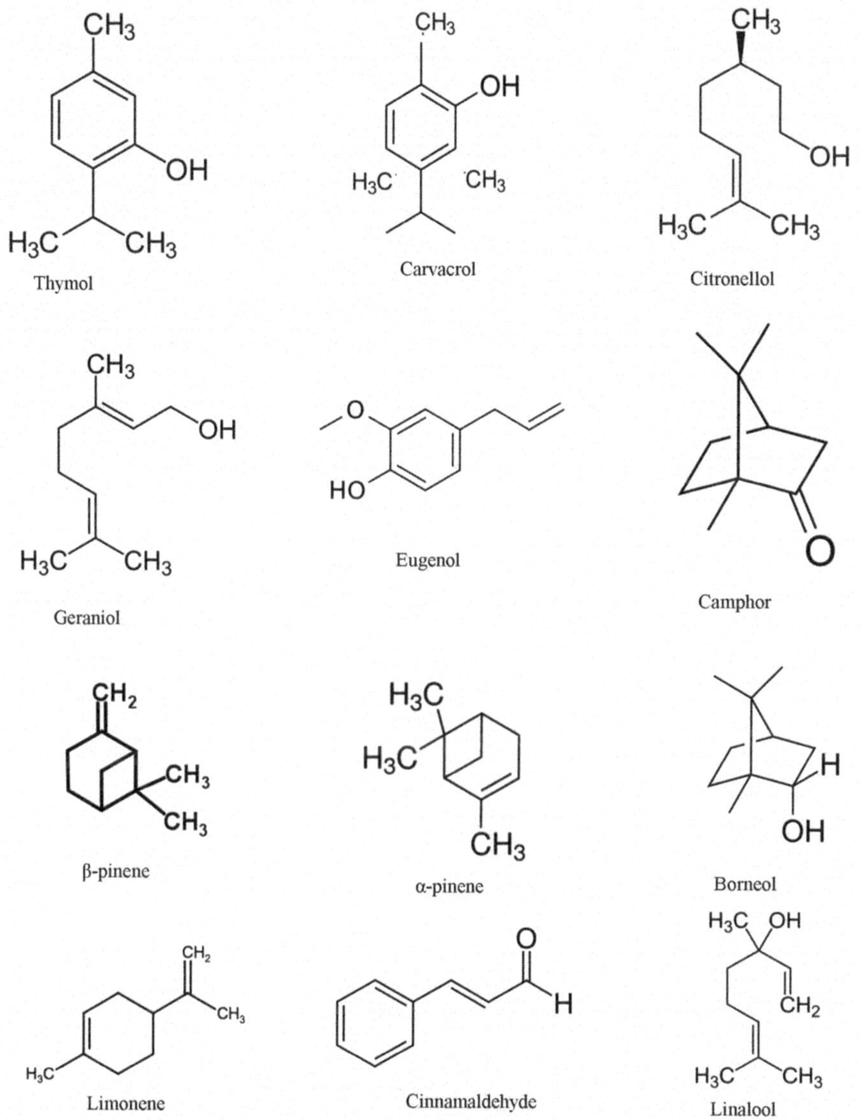

Fig. 9.7 Common ingredients of essential oils of antifungal importance

fermentation. The antifungal efficacy is checked by direct contact of the essential oil components and fungal hypha and poison food method, following micro or broth dilution techniques, or in vivo fumigation assay is also performed in case of field trials.

Essential oils from leaves of *Chenopodium ambrosioides*, a member of Amaranthaceae family, are effective against storage fungi *Aspergillus flavus*, *A. glaucus*, *A. niger*, *A. oryzae*, *Colletotrichum gloeosporioides*, C. musae, *Fusarium oxysporum*, and *Fusarium semitectum* (Jardim et al. 2008). Lemongrass oil from Cymbopogon citratus and Cymbopogon martini are potent inhibitors of *Botrytis cinerea*, *Rhizoctonia solani*, *Aspergillus tamari*, *A. fumigatus*, and *A. conicus* (Tzortzakis and Economakis 2007; Mishra et al. 2015). The members of Lamiaceae family are well known for their pungent odor and are tested for their antifungal activity by agar and broth dilution methods (Roby et al. 2013; Omidbeygi et al. 2007). Essential oils extracted from *Laurus nobilis*, *Syzygium aromaticum*, and *Origanum vulgare* are effective antifungal compounds against two pathogens of rice, *Fusarium culmorum* and *Fusarium verticillioides* (Rosello et al. 2015). Essential oils from Cymbopogon exhibited antifungal activities against rot molds (Soundharrajan et al. 2003). Antifungal activities of peppermint and sweet basil were tested against plant pathogenic fungi *S. sclerotiorum*, *Rhizopus stolonifer*, and *Mucor* sp. (Edris and Farrag 2003). Antifungal activity of β-dolabrin, γ-thujaplicin, and 4-acetyltropolone was tested against *Pythium aphanidermatum* IFO 32440 (Morita et al. 2004). Boyraz and Ozcan (2006) tested the antifungal activity of the essential oils isolated from wild Turkish summer savory (*Satureja hortensis*). Essential oils (carvacrol, thymol, p-cymene) extracted from *Origanum acutidens* are effective against phytopathogens. Growth of *A. humicola*, *Colletotrichum gloeosporioides*, *Rhizoctonia solani*, and *Phytophthora cactorum* was inhibited by the essential oil of *Asarum heterotropoides* var. *mandshuricum* (Dan et al. 2010). Though there are several reports of essential oils being potent anti-phytopathogenic (*Penicillium purpurogenum*, *Rhizopus stolonifer*, *Spondylocladium austral*, *Penicillium digitatum*, *Penicillium luteum*, *Monilinia laxa*, *Curvularia lunata*, etc.) in nature, still there are some problems regarding their maximum use and optimum effectivity. That includes their volatile natures, requirement of close systems, and degradation of EOs by oxidation due to presence of extreme amount of hydrogenated compounds (Kim et al. 2003).

9.9 Conclusion and Future Prospects

We are nourished by Mother Nature. So it is our prime duty to keep up the normal equilibrium of natural parameters. But in a way to seek solutions, some steps taken toward success may have negative impact on our environment. To fight against the fungal pathogens for the ensuring of better crop productivity, use of chemical fungicide is just another example of that fact. But we must emphasize on products from direct natural origin over the chemically synthesized one. Natural products are the best weapon to fight fungal pathogenic diseases on economically important crop species. They are less toxic, stable, and of no side effects when used in crop fields. The crying need of modern era is obtaining pathogen-free crop species in one hand and assurance of environmental sustainability on the other. Fungal and bacterial products are already used in large scales followed by the plants' secondary metabolites. Phytoalexins as internal molecules are the plants' own defense system. The detailed biochemical analysis of the phytoalexins and study of their regulatory mechanisms are opening up new horizons for universal use of phytoalexin inducing elicitors as plant defense enhancers. Mycorrhizae provide the basic line of physical barrier against pathogenic invasion, and reports include their ability to enhance plant growth, thus making the plant nonsusceptible to fungal attack. Endophyte on the other hand can enhance the plants' defense system by direct incorporation and open up popular angles of green immunization or plant vaccination. Researches on these fields are still scanty, but in the near future, they could lead to the ultimate solution of fungal pathogenic crop loss.

References

Abdu-Allah GAM, Abo-Elyousr KAM (2017) Effect of certain plant extracts and fungicides against powdery mildew disease of Grapevines in Upper Egypt. Arch Phytopathol Plant Protect:1–13

Abu-Ghannam N, Rajauria G (2013) Antimicrobial activity of compounds isolated from algae. In: Functional ingredients from algae for foods and Nutraceuticals. Woodhead, New Delhi, pp 287–306

Achenbach H, Stocker M, Constenla MA (1988) Flavonoid and other constituents of *Bauhinia manca*. Phytochemistry 27:1835–1841

Adrian M, Jeandet P (2012) Effects of resveratrol on the ultrastructure of *Botrytis cinerea* conidia and biological significance in plant/pathogen interactions. Fitoterapia 83:1345–1350

Adrian M, Jeandet P, Veneau J, Weston LA, Bessis R (1997) Biological activity of resveratrol, a stilbenic compound from grapevines, against *Botrytis cinerea*, the causal agent for gray mold. J Chem Ecol 23:1689–1702

Afolayan AJ, Meyer JJ (1997) The antimicrobial activity of 3, 5, 7-trihydroxyflavone isolated from the shoots of *Helichrysum aureonitens*. J Ethnopharmacol 57:177–181

Ahsan T, Chen J, Zhao X, Irfan M, Wu Y (2017) Extraction and identification of bioactive compounds (eicosane and dibutyl phthalate) produced by *Streptomyces* strain KX852460 for the biological control of *Rhizoctonia solani* AG-3 strain KX852461 to control target spot disease in tobacco leaf 2–9. AMB Express 7:54

Ahuja I, Kissen R, Bones AM (2012) Phytoalexins in defense against pathogens. Trends Plant Sci 17:73–90

Ajith PS, Lakshmidevi N (2010) Effect of volatile and non-volatile compounds from *Trichoderma* spp. against *Colletotrichum capsici* incitant of anthracnose on bell peppers. Nat Sci 8:265–269

Alam A, Tripathi A, Vats S, Behra KK, Sharma V (2011) In vitro antifungal efficacies of aqueous extract of *Dumortiera hirsuta* (Swaegr.) Nees against sporulation and growth of postharvest phytopathogenic fungi. Arch Bryol 103:1–9

Al-Askar AA, Baka ZA, Rashad YM, Ghoneem KM, Abdulkhair WM, Hafez EE, Shabana YM (2015) Evaluation of *Streptomyces griseorubens* E44G for the biocontrol of *Fusarium oxysporum* f. sp. *lycopersici*: ultrastructural and cytochemical investigations. Ann Microbiol 65(4):1815–1824

Alban R, Guerrero R, Toro M (2013) Interactions between a root-knot nematode (*Meloidogyne exigua*) and arbuscular mycorrhizae in coffee plant development (*Coffea arabica*). Am J Plant Sci 4:19–23

Alcerito T, Barbo FE, Negri G, Santos DYAC, Meda CI, Young MCM, Chávez D, Blatt CTT (2002) Foliar epicuticular wax of *Arrabidaea brachypoda*: flavonoids and antifungal activity. Biochem Syst Ecol 30:677–683

Aldridge DC, Turner WBJ (1970) Metabolites of *Helminthosporium monoceras*: structures of monocerin and related benzopyrans. Chem Soc Sect C 18:2598–2600

Allen EH, Thomas CA (1971) Trans-trans-3, 11-tridecadiene5, 7, 9-triyne-1,2-diol, an antifungal polyacetylene from diseased safflower (*Carthamus tinctorius*). Phytochemistry 10:15791582

Alpha CJ, Campos M, Jacobs-Wagner C, Strobel SA (2015) Mycofumigation by the volatile organic compound-producing fungus *Muscodor albus* induces bacterial cell death through DNA damage. Appl Environ Microbiol 81:1147–1156

Amer MA, Abou-El-Seud (2008) Mycorrhizal fungi and Trichoderma harzianum as biocontrol agents for suppression of Rhizoctonia solani damping off disease of tomato. Commun Agric Appl Biol Sci 73(2):217–232

Amin F, Razdan VK, Mohiddin FA, Bhat KA, Sheikh PA (2010) Effect of volatile metabolites of *Trichoderma* species against seven fungal plant pathogens in vitro. J Phytol 2:34–37

Anitha R, Murugesan K (2005) Production of gliotoxin on natural substrates by *Trichoderma virens*. J Basic Microbiol 45(1):12–19

Aramsirirujiwet Y, Gumlangmak C, Kitpreechavanich V (2016) Studies on antagonistic effect against plant pathogenic fungi from endophytic fungi isolated from *Houttuynia cordata* Thunb. and screening for siderophore and indole-3-acetic acid production. KKU Res J 21(1):55–66

Arruda RS, Mesquini RM, Schwan-Estrada KRF, Nascimento JF (2012) Effect of mushroom extracts in the induction of phytoalexins and in the control of soy oidium in a greenhouse. Biosci J 28:164–172

Atkinson S, Berta G, Hooker JE (1994) Impact of mycorrhizal colonisation on root architecture, root longevity and the formation of growth regulators. In: Gianinazzi S, Schuepp H (eds) Impact of Arbuscular Mycorrhizas on sustainable agriculture and natural ecosystems. ALS, Birkhauser Verlag, Basel, pp 47–60

Atmosukarto I, Castillo U, Hess WM, Sears J, Strobel G (2005) Isolation and characterization of *Muscodor albus* I-41.3s, a volatile antibiotic producing fungus. Plant Sci 169:854–861

Atri A, Singh N, Oberoi H (2019) Influence of seed priming on the development of pearl millet downy mildew (*Sclerospora graminicola*). Indian Phytopathol 72:1–7

Axford LC, Simpson TJ, Willis CL (2004) Synthesis and incorporation of the first Polyketide synthase free intermediate in Monocerin biosynthesis. Angew Chem Int Ed 43:727–730

Azcon-Aguilar C, Barea JM (1996) Mycorrhiza 6:457464

Aziz A, Poinssot B, Daire X, Adrian M, Bezier A, Lambert B, Joubert JM, Pugin A (2003) Laminarin elicits defense responses in grapevine and induces protection against *Botrytis cinerea* and *Plasmopara viticola*. Mol Plant-Microbe Interact 16:1118–1128

Azura NAB, Yusoff M, Tan GYA, Jegadeesh R, Appleton DR, Vikineswary S (2016) *Streptomyces sanglieri* which colonised and enhanced the growth of *Elaeis guineensis* Jacq. Seedlings was antagonistic to *Ganoderma boninense* in in vitro studies. J Ind Microbiol Biotechnol 43(4):485–493

Bagyaraj DJ (2006) Current status of biological control of plant diseases using antagonistic organisms in India. PDBC Pub Bangalore, Bangalore, pp 125–134

Bagyaraj DJ, Chawla G (2012) Status and prospects for enhancing the uptake of antagonistic organisms for nematode Management in India. NBAII Pub Bangalore, Bangalore, pp 74–89

Bahceevli AK, Kurucu S, Kolak U, Topçu G, Adou E (2005) Alkaloids and aromatics of *Cyathobasis fruticulosa* (Bunge) Aellen. J Nat Prod 68:956–958

Banerjee RD (2000) Antimicrobial activities of bryophytes a review. Perspect Indian Bryol:55–74

Banerjee RD, Sen SP (1979) Antibiotic activity of bryophytes. Bryologist 82:141–153

Banerjee D, Strobel GA, Booth E, Geary B, Sears J, Spakowicz D, Busse S (2010a) An endophytic *Myrothecium inundatum* producing volatile organic compounds. Mycosphere 1(3):229–240

Banerjee D, Strobel GA, Geary B, Sears J, Ezra D, Liarzi O, Coombs J (2010b) *Muscodor albus* strain GBA, an endophytic fungus of *Ginkgo biloba* from United States of America, produces volatile antimicrobials. Mycology 1(3):179–186

Baum C, El-Tohamy W, Gruda N (2015) Increasing the productivity and product quality of vegetable crops using arbuscular mycorrhizal fungi: a review. Sci Hortic (Amsterdam) 187:131–141

Ben Salah I, Aghrouss S, Douira A, Aissam S, El Alaoui-Talibi Z, Filali-Maltouf A, El Modafar C (2018) Seaweed polysaccharides as bio-elicitors of natural defenses in olive trees against Verticillium wilt of olive. J Plant Interact 13:248–255

Berdy J (2012) Thoughts and facts about antibiotics: where we are now and where we are heading. J Antibiot 65:385–395
Bodade RG, Borkar PS, Arfeen MS, Khobragade CN (2008) In vitro screening of bryophytes for antimicrobial activity. J Med Plan Theory 7:23–28
Bodker L, Kjoller R, Rosendahl S (1998) Effect of phosphate and the arbuscular mycorrhizal fungus *Glomus intraradices* on disease severity of root rot of peas (*Pisum sativum*) caused by *Aphanomyces euteiches*. Mycorrhiza 8:169–174
Bokel N, Diyasenu MNC, Gunatilaka AAL, Kraus W, Sotheeswaran S (1988) Canaliculatol, an antifungal resveratrol trimer from Stemonoporous canaliculatus. Phytochemistry 27:377–380
Bonde MR, Millar RL, Ingham JL (1973) Induction and identification of sativan and vestitol as two phytoalexins from *Lotus corniculatus*. Phytochemistry 12(12):2957–2959
Brady SF, Wagenaar MM, SinghMP JJE, Clardy J (2000) The cytosporones, new octaketide antibiotics isolated from an endophytic fungus. Org Lett 2:4043–4046
Braga MR (1991) Phytoalexins induction in Rubiaceae. J Chem Ecol 17:1079–1090
Browne LM, Conn KL, Ayer WA, Tewari JP (1991) The camalexins: new phytoalexins produced in the leaves of Camelina sativa (Cruciferae). Tetrahedron 47:3909–3914
Brownlee HE, McEuen AR, Hedger J, Scott IM (1990) Anti-fungal effects of cocoa tannin on the witches' broom pathogen *Crinipellis pernicious*. Physiol Mol Plant Pathol 36:39–48
Budi SW, Tuinen DV, Martinotti G, Gianinazzi S (1999) Applied and environmental microbiology. Am Soc Microbiol 65(11):5148–5150
Bungtongdee N, Sopalun K, Laosripaiboon W, Iamtham S (2019) The chemical composition, antifungal, antioxidant and antimutagenicity properties of bioactive compounds from fungal endophytes associated with Thai orchids. J Phytopathol 167:56–64
Cantrell CL, Schrader KK, Mamonov LK, Sitpaeva GT, Kustova TS, Dunbar C (2005) Isolation and identification of antifungal and antialgal alkaloids from *Haplophyllum sieversii*. J Agric Food Chem 53:7741–7748
Cao L, Qiu Z, You J, Tan H, Zhou S (2005) Isolation and characterization of endophytic Streptomyces antagonists of Fusarium wilt pathogen from surface sterilized banana roots. FEMS Microbiol Lett 247:147–152
Cartwright P (1981) Isolation and characterization of two phytoalexins from rice as momilactones A and B. Phytochemistry 20:535–537
Castillo U, Strobel G, Ford E, Hess W, Porter H, Jensen J et al (2002) Munumbicins, wide- spectrum antibiotics produced by Streptomyces NRRL 30562, endophytic on *Kennedia nigriscans*. Microbiology 148:2675–2685
Castillo UF, Strobel GA, Mullenberg K, Condron MM, Teplow DB, Folgiano V et al (2006) Munumbicins E-4 and E-5: novel broad-spectrum antibiotics from Streptomyces NRRL 3052. FEMS Microbiol Lett 255:296–300
Cavalcanti LS (2005) Aspectos bioquímicos e moleculares da resistência induzida. In: _____. et al. (Eds.). Indução de resistência em plantas a patógenos e insetos. Piracicaba: FEALQ, 81:124
Cavin AL, Hay AE, Marston A, Stoeckli H, Scopelliti R, Diallo D, Hostettmann K (2006a) Bioactive diterpenes from the fruits of *Detarium microcarpum*. J Nat Prod 69:768–773
Channakeshava C, Pankaja NS (2018) In vitro evaluation of fungicides, plant extracts and biocontrol agents against brown leaf spot of paddy. Int J Curr Microbiol App Sci 7(5):127–132
Cavin AL, Hay AE, Marston A, Stoeckli H, Scopelliti R, Diallo D, Hostettmann K (2006b) Bioactive diterpenes from the fruits of *Detarium microcarpum*. J Nat Prod 69:768–773
Chaube HS, Pundhir VS (2005) Crop diseases and their management. PHI Learning Private Limited, Delhi
Chen T, Lu J, Kang B, Lin M, Ding L, Zhang L, Chen G, Chen S, Lin H (2018) Antifungal activity and action mechanism of ginger oleoresin against *Pestalotiopsis microspora* isolated from Chinese olive fruits. Front Microbiol 9:2583
Chen X, Sang X, Li S, Zhang S, Bai L (2010) Studies on a chlorogenic acid-producing endophytic fungi isolated from Eucommia ulmoides Oliver. J Ind Microbiol Biotechnol 37:447–454
Cheng SS, Liu JY, Chang EH, Chang ST (2008) Antifungal activity of cinnamaldehyde and eugenol congeners against wood-rot fungi. Bioresour Technol 99:5145–5149

Cheung RCF, Wong JH, Pan WL, Chan YS, Yin CM, Dan XL, Wang HX, Fang EF, Lam SK, Ngai PHK (2014) Antifungal and antiviral products of marine organisms. Appl Microbiol Biotechnol 98:3475–3494

Chludil HD, Muniain CC, Seldes AM, Maier MS (2002) Cytotoxic and antifungal triterpene glycosides from the Patagonian Sea cucumber *Hemoiedema spectabilis*. J Nat Prod 65:860–865

Cho JY, Moon JH, Seon KY, Park KH (1998) Antimicrobial activity of 4-Hydroxybenzoic acid and trans 4-hydroxycinnamic acid isolated and identified from rice hull. Biosci Biotechnol Biochem 62:2273–2276

Chowdhary K, Kaushik N (2018) Diversity and antifungal activity of fungal endophytes of *Asparagus racemosus* Willd. Agric Res 8:27–35

Claydon N, Grove JF, Pople MJ (1979) Insecticidal secondary metabolic products from the entomogenous fungus Fusarium larvarum. In Vertebr Pathol 33:364–367

Cojocaru M, Droby S, Glotter E, Goldman A, Gottlieb HE, Jacoby B, Prusky D (1986) 5-(12-Heptadecenyl)-resorcinol, the major component of the antifungal activity in the peel of mango fruit. Phytochemistry 25:1093–1095

Cole MD, Bridge PD, Dellar JE, Fellows LE, Cornish MC, Anderson JC (1991) Antifungal activity of neo-clerodane diterpenoids from Scutellaria. Phytochemistry 30:1125–1127

Conn KL, Browne LM, Tewari JP, Ayer WA (1994) J Plant Biochem Biotechnol 3:125

Copping LG (2004) The manual of biocontrol agents, 3rd edn. BCPC Publications, Alton, p 702

Corcuff R, Mercier J, Tweddell R, Arul J (2011) Effect of water activity on the production of volatile organic compounds by *Muscodor albus* and their effect on three pathogens in stored potato. Fungal Biol 115:220–227

Cordelier S, Ruffray P, Fritg B, Kauffmann S (2003) Biological and molecular comparison between localized and systemic acquired resistance induced in tobacco by *Phytophthora megasperma* glycoprotein elicitin. Plant Mol Biol 51:109–118

Costa FG, Zucchi TD, de Melo IS (2013) Biological control of phytopathogenic fungi by endophytic actinomycetes isolated from maize (*Zea mays* L). Braz Arch Biol Technol 56(6):948–955

Cowan MM (1999) Plant products as antimicrobial agents. Clin Microbiol Rev 12:564–582

Coxon DT (1980) Identification of three hydroxyflavan phytoalexins from daffodil bulbs. Phytochemistry 19:889891

Coxon DT (1982) Phytoalexins from other plant families. In: Bailey JA, Mansfield JW (eds) Phytoalexins. Blackie, Glasgow/London, pp 106–132

Crawford DL, Lynch JM, Whipps JM, Ousley MA (1993) Isolation and characterization of actinomycete antagonists of a fungal root pathogen. Appl Environ Microbiol 59(11):3889–3905

Cuenca-Estrella M, Mellado E, Diaz-Guerra TM, Monzon A, Rodriguez-Tudela JL (2000) Susceptibility of fluconazole-resistant clinical isolates of *Candida* spp. to echinocandin LY303366, itraconazole and amphotericin B. J Antimicrob Chemother 46:475–477

Curvêlo CRS, Rodrigues FA, Silva LC, Nascimento KJT, Berger PG (2013) Biochemical defense mechanisms in cotton plants against Ramularia leaf spot mediated by silicon. Bragantia 72:41–51

Dabur R, Chhillar AK, Yadav V, Kamal PK, Gupta J, Sharma GL (2005) In vitro antifungal activity of 2-(3,4-dimethyl-2,5-dihydro-1H-pyrrol-2-yl)-1-methylethyl pentanoate, a dihydro – pyrrole derivative. J Med Microbiol 54:549–552

Dahiya JS, Rimmer SR (1988) Phytoalexin accumulation in tissues of *Brassica napus* inoculated with *Leptosphaeria maculans*. Phytochemistry 27:3105–3107

Daisy BH, Strobel GA, Castillo U, Ezra D, Sears J, Weaver DK, Justin B (2002) Naphthalene, an insect repellent, is produced by *Muscodor vitigenus*, a novel endophytic fungus. Runyon Microbiol 148:3737–3741

Dan Y, Liu HY, Gao WW, Chen SL (2010) Activities of essential oils from *Asarum heterotropoides* var. mandshuricum against five phytopathogens. Crop Prot 29:295–299

Dávila MD, Gallegos MG, Castillo HFD, Fuente CYM, Olivas FA, Siller CM (2016) Antagonist actinomycetes metabolites against plant pathogens fungi of agricultural importance. J Sci Res Stud 3(11):224–230

Deice Raasch-Fernandes L, Bonaldo SM, de Jesus Rodrigues D, Magela Vieira-Junior G, Regina Freitas Schwan-Estrada K, Rocco da Silva C (2019) Induction of phytoalexins and proteins related to pathogenesis in plants treated with extracts of cutaneous secretions of southern Amazonian Bufonidae amphibians. PLoS One 14(1): e0211020

Dayan FE, Cantrell CL, Duke SO (2009) Natural products in crop protection. Bioorg Med Chem 17:4022–4034

de Bertoldi C, de Leo M, Braca A, Ercoli L (2009) Bioassay-guided isolation of allelochemicals from Avena sativa L.: allelopathic potential of flavone C-glycosides. Chemoecology 19:169–176

De Corato U, Salimbeni R, De Pretis A, Avella N, Patruno G (2017) Antifungal activity of crude extracts from brown and red seaweeds by a supercritical carbon dioxide technique against fruit postharvest fungal diseases. Postharvest Biol Technol 131:16–30

de Vasconcellos RLF, Cardoso EJBN (2009) Rhizospheric streptomycetes as potential biocontrol agents of *Fusarium* and *Armillaria* pine rot and as PGPR for *Pinus taeda*. BioControl 54:807

DeBoer C, Meulman PA, Wnuk RJ, Peterson DH (1970) Geldanamycin, a new antibiotic. J Antibiot (Tokyo) 23:442–447

Defago G, Kern H (1988) Induction of Fusarium solani mutants insensitive to tomatine, their pathogenicity and aggressiveness to tomato fruits and pea plants. Physiol Plant Pathol 22:29–37

Delaunois B, Cordelier S, Conreux A, Clément C, Jeandet P (2009) Molecular engineering of resveratrol in plants. Plant Biotechnol J 7:2–12

Deng Y, Nicholson RA (2005) Antifungal properties of surangin B, a coumarin from *Mammea longifolia*. Planta Med 71:364–365

Deora GS, Guhil N (2015) Phytochemical analysis and antifungal activity of moss *Bryum cellulare* against some phytopathological fungi. Int J Pharm Sci Res 6(2):688–691

Deora GS, Guhil N (2016) Studies on antifungal potential of *Bryum cellulare* against spore germination of fungus *Curvularia lunata*. Int J Pharm Sci Res 7(1):353–357

Deora GS, Jain NS (2008) In vitro antifungal activity of *Plagiochasma appendiculatum* against *Alternaria solani*. J Pure Applied Microbiol 2:569–572

Deora GS, Suhalka D (2017) Evaluation of bryophyte for green fungicides as alternative treatment to control plant pathogen. Int J Pharmacog Phytochem Res 9(10): 1373–1379

Deverall BJ (1982) Introduction. In: Bailey JA, Mansfield JW (eds) Phytoalexins. Blackie, Glasgow/London, pp 19–20

Devys M, Barbier M (1990) Oxidative ring contraction of the phytoalexin cyclobrassinin: a way to brassilexin. J Chem Soc Perkin Trans I:2856

Devys M, Barbier M, Loiselet I, Rouxel T, Sarniguet A, Kollmann A, Bousquet J (1988) Brassilexin, a novel sulphur-containing phytoalexin from *Brassica juncea* L., (cruciferae). Tetrahedron Lett 29:6447

Dewi TK, Dwi A, Antonius S (2017) Secondary metabolites production by actinomycetes and their antifungal activity. ICBS conference proceedings, international conference on biological science KnE. Life Sci 3:256–264

Dey A, De JN (2011) Antifungal bryophytes: a possible role against human pathogens and in plant protection. Res J Bot 6(4):129–140

Dhatwalia VK, Sati OP, Tripathi MK, Kumar A (2009) Isolation, characterization and antimicrobial activity at diverse dilution of wheat puroindoline protein. World J Agric Sci 5:297–300

Dhingra OD, Mizubuti ESG, Napoleao IT, Jham G (2001) Free fatty acid accumulation and quality loss of stored soybean seeds invaded by *Aspergillus ruber*. Seed Sci Technol 29:193–203

Dixon RA, Lamb CJ (1999) Molecular communication in interactions between plants and microbial pathogens. Annu Rev Plant Physiol Plant Mol Biol 41:229367

Dochhil H, Dkhar MS, Barman D (2013) Seed germination enhancing activity of endophytic Streptomyces isolated from indigenous ethno-medicinal plant *Centella asiatica*. Int J Pharm Bio Sci 4(1):256–262

Dos Anjos ÉCT, Cavalcante UMT, Gonçalves DMC, Pedrosa EMR, Santos VF, Maia LC (2010) Interactions between an arbuscular mycorrhizal fungus (*Scutellospora heterogama*) and the root-knot nematode (*Meloidogyne incognita*) on sweet passion fruit (*Passiflora alata*). Braz Arch Biol Technol 53:801–809

Draz IS, Elkhwaga AA, Elzaawely AA, El-Zahaby HM, Anter Ismail AW (2019) Application of plant extracts as inducers to challenge leaf rust of wheat. Egypt J Biol Pest Control 29:6

Du R, Liu J, Sun P, Li H, Wang J (2017) Inhibitory effect and mechanism of Tagetes erecta L. fungicide on Fusarium oxysporum f. sp. niveum. Sci Rep 7(14442):1–13

Dukare AS, Prasanna R, Dubey SC, Nain L, Chaudhary V, Singh R, Saxena AK (2011) Evaluating novel microbe amended composts as biocontrol agents in tomato. Crop Prot 30:436–442

Duke (2004) Biological activity summary for cocoa (*Theobroma cacao* L.). J Med Food 3:115–119

Durango D, Pulgarin N, Echeverri F, Escobar G, Quiñones W (2013) Effect of salicylic acid and structurally related compounds in the accumulation of phytoalexins in cotyledons of common bean (*Phaseolus vulgaris* L.) cultivars. Molecules 18:10609–10628

Echeverri F, Torres F, Quiñones W, Escobar G, Archbold R (2012) Phenylphenalenone phytoalexins, will they be a new type of fungicide? Phytochem Rev 11:1–12

Edris AE, Farrag ES (2003) Antifungal activity of peppermint and sweet basil essential oils and their major aroma constituents on some plant pathogenic fungi from the vapor phase. Nahrung 47:117–121

El-Gendy MMA, EL-Bondkly AMA (2010) Production and genetic improvement of a novel antimycotic agent, saadamycin, against dermatophytes and other clinical fungi from endophytic Streptomyces sp. Hedaya48. J Ind Microbiol Biotechnol 37:831–841

El-Shatoury S, El-Kraly O, El-Kazzaz W, Dewedar A (2009) Antimicrobial activities of Actinomycetes inhabiting *Achillea Fragrantissima* (Family: Compositae). Egypt J Nat Toxins 6(2):1–15

El-Tarabily KA (2003) An endophytic chitinase-producing isolate of *Actinoplanes missouriensis*, with potential for biological control of root rot of lupine caused by *Plectosporium tabacinum*. Aust J Bot 51:257–266

El-Tarabily KA, Hardy GESJ, Sivasithamparam K (2010) Performance of three endophytic actinomycetes in relation to plant growth promotion and biological control of *Pythium aphanidermatum*, a pathogen of cucumber under commercial field production conditions in the United Arab Emirates. Eur J Plant Pathol 128:527–539

El-Tarabily KA, Sivasithamparam K (2006) Nonstreptomycete actinomycetes as biocontrol agents of soil-borne fungal plant pathogens and as plant growth promoters. Soil Biol Biochem 38:1505–1520

Endo K, Kanno E, Oshima Y (1990) Structures of antifungal diarylheptenones, gingerenones A, B, C and isogingerenone B, isolated from the rhizomes of *Zingiber officinale*. Phytochemistry 29:797

Ezra D, Castillo UF, Strobel GA, Hess WM, Porter H, Jensen JB, Condron MAM, Teplow DB, Sears J, Maranta M, Hunter M, Weber B, Yaver D (2004a) Coronamycins, peptide antibiotics produced by a verticillate *Streptomyces* sp. (MSU-2110) endophytic on *Monstera* sp. Microbiology 150:785–793

Sánchez-Ortiz BL, Sánchez-Fernández RE, Duarte G, Lappe-Oliveras P, Macías-Rubalcava ML (2016) Antifungal, anti-oomycete and phytotoxic effects of volatile organic compounds from the endophytic fungus *Xylaria sp.* strain PB3f3 isolated from *Haematoxylum brasiletto*. J Appl Microbiol 120(5):1313–1325

Ezra D, Hess WM, Strobel GA (2004b) New endophytic isolates of *Muscodor albus*, a volatile-antibiotic-producing fungus. Microbiology 150:4023–4031

Ezra D, Strobel GA (2003) Effect of substrate on the bioactivity of volatile antimicrobials produced by *Muscodor albus*. Plant Sci 165:1229–1238

Fernández RES, Diaz D, Duarte G, Oliveras PS, Sánchez S, Macías-Rubalcava ML (2016) Antifungal volatile organic compounds from the Endophyte *Nodulisporium* sp. strain GS4d2II1a: a qualitative change in the intraspecific and interspecific interactions with *Pythium aphanidermatum*. Microb Ecol 71(2):347–364

Feliziani E, Landi L, Romanazzi G (2015) Preharvest treatments with chitosan and other alternatives to conventional fungicides to control postharvest decay of strawberry. Carbohydr Polym 132:111–117

Franich RA, Gadgil PD, Shain L (1983) Fungistatic effects of *Pinus radiata* needle epicuticular fatty and resin acids on *Dothistroma pini*. Physiol Plant Pathol 23:183–195

Frankmölle WP, Larsen LK, Caplan FR, Patterson GML, Knubel G, Levine IA, Moore RE (1992) Blue-green alga Anabaena laxa. I isolation and biological properties. J Antibiot (Tokyo) 45:1451–1457

Friend J (2016) Plant Phenolics, lignification. Prog Phytochem 7:197

Fritz M, Jakobsen I, Lyngkjaer MF, Thordal Christensen H, Kühnemann P (2006) Arbuscular mycorrhiza reduces susceptibility of tomato to *Alternaria solani*. Mycorrhiza 16:413–419

Fu J, Zhou Y, Li H-F, Ye Y-H, Guo J-H (2011) Antifungal metabolites from *Phomopsis* sp. By254, an endophytic fungus in *Gossypium hirsutum*. Afr J Microbiol Res 5:1231–1236

Fukuta M et al (2007) Comparative efficacies in vitro of antibacterial, fungicidal, antioxidant, and herbicidal activities of momilactones a and B. J Plant Interact 2:245251

Gahotri D, Chaturvedi P (2011) Antifungal and antibacterial potential of methanol and chloroform extracts of *Marchantia polymorpha* L. Arch Phytopathol Plant Protect 44:736–731

Galeotti F, Barile E, Curir P, Dolci M, Lanzotti V (2008) Flavonoids from carnation (*Dianthus caryophyllus*) and their antifungal activity. Phytochem Lett 1:44–48

Gamalero E, Pivato B, Bona E, Copetta A, Avidano L, Lingua G (2010) Interactions between a fluorescent pseudomonad, an arbuscular mycorrhizal fungus and a hypo virulent isolate of *Rhizoctonia solani* affect plant growth and root architecture of tomato plants. Plant Biosyst 144:582–591

Gangwar M, Dogra S, Gupta UP, Kharwar RN (2014) Diversity and biopotential of endophytic actinomycetes from three medicinal plants in India. Afr J Microbiol Res 8(2):184–191

Gangwar M, Kaur N, Saini P, Kalia A (2015) The diversity, plant growth promoting and antimicrobial activities of endophytic actinomycetes isolated from *Emblica officinalis* Gaertn. Int J Adv Res 3(4):1062–1071

Garcıa-Mier L, RG G-G'l, VM M'n-O'n, Verduzco-Cuellar BR, TorresPacheco I (2013) Agriculture and bioactives: achieving both crop yield and phytochemicals. Int J Mol Sci 14:4203–4222

Gayathri P, Muralikrishnan V (2013) Isolation and characterization of Endophytic Actinomycetes from mangrove plant for antimicrobial activity. Int J Curr Microbiol App Sci 2(11):78–89

Geigert J (1973) Two phytoalexins from sugar beet (*Beta vulgaris*) leaves. Tetrahedron 29:2703–2706

Genovese G, Leitner S, Minicante SA, Lass-Florl C (2013) The Mediterranean red alga *Asparagopsis taxiformis* has antifungal activity against Aspergillus species. Mycoses 56:516–519

Gholami M, Khakvar R, Niknam G (2014) Introduction of some new endophytic bacteria from Bacillus and Streptomyces genera as successful biocontrol agents against *Sclerotium rolfsii*. Arch Phytopathol Plant Protect 47(1):122–130

Goa KI, Barradell LB (1995) Fluconazole: an update of its pharmacodynamic and pharmacokinetic properties and therapeutic use in major superficial and systemic mycoses in immunocompromised patients. Drugs 50:658–690

Grayer RJ, Kokubun T (2001) Plant-fungal interactions: the search for phytoalexins and other antifungal compounds from higher plants. Phytochesmistry 56:253–263

Greger H, Hofer O, Kiihlig H, Wurz G (1992) Sulfur containing cinnamides with antifungal activity from *Glycosmis cyanocarpa*. Tetrahedron 48:1209–1218

Gross D, Porzel A, Schmidt J (1994) Phytoalexin emit Indolstruktur aus kohlrabi (*Brassica oleracea* var. *gongylodes*). Z Naturforsch 49c:281–285

Grosskinsky DK, Naseem M, Abdelmoshem UA, Plickert N, Engelke T, Griebel T, Zeier J, Novak O, Strand M, Pfeifhofer H (2011) Cytokinins mediate resistance against pseudomonas syringae in tobacco through increased antimicrobial phytoalexin synthesis independent of salicylic acid signaling. Plant Physiol 2011(157):815–830

Grove JF, Pople MJ (1979) Metabolic products of *Fusarium larvarum* fuckel. The fusarentins and the absolute configuration of monocerin. Chem Soc Perkin Trans 1 (1979):2048–2051

Guimarães SS, Mazaro SM, Freddo ÁR, Wagner Júnior A (2015) Potential of horsetail (*Equisetum* sp.) preparations in the synthesis of defense metabolites in soy (*Glycine max* L.) cotyledons and the effect on the growth of Rhizoctonia solani Kuhn, in vitro. Rev Bras Plantas Med 17:143–149

Güimil S, Chang HS, Zhu T, Sesma A, Osbourn A, Roux C, Ioannidis V, Oakeley EJ, Docquier M, Descombes P, Briggs SP, Paszkowski U (2005) Comparative transcriptomics of rice reveals an ancient pattern of response to microbial colonization. Proc Natl Acad Sci 102:8066–8070

Guo L, Wu JZ, Han T, Cao T, Rahman K, Qin LP (2008) Chemical composition, antifungal and antitumor properties of ether extracts of *Scapania verrucosa* Heeg. and its endophytic fungus *Chaetomium fusiformis*. Molecules 13:2114–2125

Gurgel LA, Sidrim JJC, Martins DT, Filho CV, Rao VS (2005) In vitro antifungal activity of dragon's blood from *Croton urucurana* against dermatophytes. J Ethnopharmacol 97:409–412

Gutjahr C, Paszkowski U (2013) Multiple control levels of root system remodelling in arbuscular mycorrhizal symbiosis. Plant Sci 4:204

Gwa VI, Nwankiti AO, Ekefan EJ (2018) Antifungal effect of five aqueous plant extracts on Mycelial growth of *Penicillium Expansum* isolated from rotted yam tubers in storage. Acta Sci Agric 2(6)

Hage-Ahmed K, Moyses A, Voglgruber A, Hadacek F, Steinkellner S (2013) Alterations in root exudation of intercropped tomato mediated by the arbuscular mycorrhizal fungus *Glomus mosseae* and the soil borne pathogen *Fusarium oxysporum* f.sp. lycopersici. J Phytopathol 161:763–773

Hahn MG, Darvill AG, Albersheim P (2008) Host-pathogen interactions: XIX. The endogenous elicitor, a fragment of a plant cell wall polysaccharide that elicits phytoalexin accumulation in soybeans. Plant Physiol 68:1161–1169

Hammer EC, Pallon J, Wallander H, Olsson PA (2011) Tit for tat? A mycorrhizal fungus accumulates phosphorus under low plant carbon availability. FEMS Microbiol Ecol 76:236–244

Harding VK, Heale JB (1981) The accumulation of inhibitory compounds in the induced resistance response of carrot root slices to *Botrytis cinerea*. Physiol Plant Pathol 18:7–15

Har-Nun N, Meyer AM (1990) Cucurbitacins protect cucumber tissue against infection by *Botrytis cinerea*. Phytochemistry 29:787–791

Harper JK, Arif AM, Ford EJ, Strobel GA, Porco JA, Tomer DP, KL O¢N, Grant DM (2003) Pestacin: a 1,3-dihydro isobenzofuran from *Pestalotiopsis microspora* possessing antioxidant and antimycotic activities. Tetrahedron 59:2471–2476

Harris JE, Dennis C (1977) The effect of post-infectional potato tuber metabolites and surfactants on zoospores of Oomycetes. Physiol Plant Pathol 11:163–169

Hartmann G, Niehaus F (1974) The isolation of xanthoxylin from the bark of Phytophthora and Hendersonula-infected Citrus lemon and its fungitoxic effect. Phytopathology 81:97–113

Hasegawa M, Mitsuhara I, Seo S, Okada K, Yamane H, Iwai T, Ohashi Y (2014) Analysis on blast fungus-responsive characters of a flavonoid phytoalexin sakuranetin; accumulation in infected rice leaves, antifungal activity and detoxification by fungus. Molecules 19:11404–11418

Hawksworth DL (1991) The fungal dimension of biodiversity: magnitude, significance, and conservation. Mycol Res 95:641–655

Hawksworth DL (2001) The magnitude of fungal diversity: the 1±5 million species estimate revisited. Mycol Res 105:1422–1431

Hayakawa M, Nonomura H (1987) Efficacy of artificial humic acid is a selective nutrient in HV agar used for the isolation of actinomycetes. J Ferment Technol 65:609–616

He XZ, Dixon RA (2000) Genetic manipulation of isoflavone 7-O-methyltransferase enhances biosynthesis of 4′-O-methylated isoflavonoid phytoalexins and disease resistance in alfalfa. Plant Cell 12:1689–1702

Herger G, Klingauf F, Mangold D, Pommer EH, Scheter M (1988) Die Wirkung von Auszigen aus dem Sachalin- Staudenknoterich *Reynoutria sachalinensis* (F. Schmidt) Nakai gegen Plizkrankheiten, insbesondere Echte Mehltauplize. Nachrchtenblatt der Deutsche Pflanzenschutzdienst 40:56–60

Homma Y, Kato Z, Hirayama F, Konno K, Shirahama H, Suzui T (1989) Production of antibiotics by *Pseudomonas cepacia* as an agent for biological control of soilborne plant pathogens. Soil Biol Biochem 21:723–728

Hoof LV, Berghe DAV, Vlietinck AJ (2008) Screening of poplar trees for antibacterial, antifungal and antiviral activity. Biol Plant 22:265–273

Howell CR, Stipanovic RD (1980) Suppression of *Pythium ultimum* induced damping -off of cotton seedlings by *Pseudomonas fluorescens* and its antibiotic, pyoluterin. Phytopathology 70:712–715

Huang LD, Backhouse D (2004) Effects of *Fusarium* species on defence mechanisms in sorghum seedlings. N Z Plant Protection 57:121–124

Huang R, Li GQ, Zhang J, Yang L, Che HJ, Jiang DH, Huang HC (2011) Control of post harvest botrytis fruit rot of strawberry by volatile organic compounds of *Candida intermedia*. Phytopathology 101:859–869

Huang WY, Cai YZ, Hyde KD, Corke H, Sun M (2008) Biodiversity of endophytic fungi associated with 29 traditional Chinese medicinal plants. Fungal Divers 33:61–75

Huffaker A et al (2011) Novel acidic sesquiterpenoids constitute a dominant class of pathogen-induced phytoalexins in maize. Plant Physiol 56:2082–2097

Ibraheem IB, Hamed SM, Abd elrhman AA, Mohamed Farag F, Abdel-Raouf N (2017) Antimicrobial activities of some brown macroalgae against some soil borne plant pathogens and in vivo management of *Solanum melongena* root diseases. Aust J Basic Appl Sci 11:157–168

Ibrahima M, Kaushik N, Sowemimoa A, Chhipab H, Koekemoerc T, Venterc M, Odukoyaa OA (2017) Antifungal and antiproliferative activities of endophytic fungi isolated from the leaves of *Markhamia tomentosa*. Pharm Biol 55(1):590–595

Igarashi Y (2004) Screening of novel bioactive compounds from plant-associated actinomycetes. Actinomycetology 18:63–66

Igarashi Y, Iida T, Sasaki T, Saito N, Yoshida R, Furumai T (2002) Isolation of actinomycetes from live plants and evaluation of anti phytopathogenic activity of their metabolites. Actinomycetology 16:9–13

Inderiati S, Franco CMM (2008) Isolation and Identification of Endophytic Actinomycetes and their Antifungal Activity. J Biotech Res Trop Reg 1. (Spl. Ed.)

Ingham JL (1982) Phytoalexins from the Leguminosae. In: Bailey JA, Mansfield JW (eds) Phytoalexins. Blackie, Glasgow/London, pp 21–80

Ingham JL, Tahara S, Harborne JB (1983) Fungitoxic isoflavones from *Lupinus albus* and other Lupinus species. Z Naturforsch 38c:194–200

Isono K, Suzuki S (1979) The polyoxins: pyrimidine nucleoside peptide antibiotics inhibiting fungal cell wall biosynthesis. Heterocycles 13:133

Karneda Y, Asano N, Yamaguchi T, Matsui K (1987) Validoxylamines as trehalase inhibitors. Anrihiot (Tokyo) 40:563–565

Islam TM, Hashidoko Y, Deora A, Ito T, Tahara S (2005) Suppression of damping-off disease in host plants by the rhizoplane bacterium *Lysobacter* sp. strain SB-K88 is linked to plant colonization and antibiosis against soilborne peronosporomycetes. Appl Environ Microbiol 71:3786–3796

Ismail A, Salah KBH, Ahmed M, Mastouri M, Bouraoui A, Aouni M (2014) Antibacterial and antifungal activities of brown alga *Zonaria tournefortii* (JV Lamouroux). Allelopath J 34:143–153

Ivantiskaya LP, Singal SM, Bibikova MV, Vostrov SN (1978) Direct isolation of Micromonospora on selective media with gentamicin. Antibiot 23:690–692

Jambunathan R, Butler LG, Bandsyopadkyay R, Mughogho K (1986) Polyphenol concentrations in grain, leaf and callus tissues of mold-susceptible and mold-resistant sorghum cultivars. J Agric Food Chem 34:425–429

Jardim CM, Jham GN, Dhingra OD, Freire MM (2008) Composition and antifungal activity of the essential oil of the Brazilian *Chenopodium ambrosioides* L. J Chem Ecol 34(9):1213–1218

Jaulneau V, Lafitte C, Jacquet C, Fournier S, Salamagne S, Briand X, Esquerré-Tugayé M-T, Dumas B (2010) Ulvan, a sulfated polysaccharide from green algae, activates plant immunity through the jasmonic acid signaling pathway. J Biomed Biotechnol 2010:525291

Jeandet P, Clément C, Courot E, Cordelier S (2013) Modulation of phytoalexin biosynthesis in engineered plants for disease resistance. Int J Mol Sci 14:14136–14170

Jeandet P, Delaunois B, Aziz A, Donnez D, Vasserot Y, Cordelier S, Courot E (2012) Metabolic engineering of yeast and plants for the production of the biologically active hydroxystilbene, resveratrol. J Biomed Biotechnol 2012:1–14

Jeandet P, Delaunois B, Conreux A, Donnez D, Nuzzo V, Cordelier S, Clément C, Courot E (2010) Biosynthesis, metabolism, molecular engineering and biological functions of stilbene phytoalexins in plants. Biofactors 36:331–341

Jeandet P, Hébrard C, Deville MA, Cordelier S, Dorey S, Aziz A, Crouzet J (2014) Deciphering the role of phytoalexins in plant-microorganism interactions and human health. Molecules 19(11):18033–18056

Jimenez LD, Ayer WA, Tewari JP (1997) Phytoalexins produced in the leaves of *Capsella bursa-pastoris* (shepherd's purse). Phytoprotection 78:99–103

Johnson C, Brannon DR, Kuc J (1973) Xanthotoxin: a phytoalexin of *Pastinaca sativa* root. Phytochemistry 12:2961–2962

Jung SC, Martinez-Medina A, Lopez-Raez JA, Pozo MJ (2012) Mycorrhiza-induced resistance and priming of plant defenses. J Chem Ecol 38:651–664

Kafur A, Khan AB (2011) Isolation of endophytic actinomycetes from *Catharanthus roseus* (L.) G. Don leaves and their antimicrobial activity. Iranian. J Biotechnol 9(4):301–306

Karagiannidis N, Bletsos F, Stavropoulos N (2002) Effect of Verticillium wilt (*Verticillium dahliae* Kleb.) and mycorrhiza (*Glomus mosseae*) on root colonization, growth and nutrient uptake in tomato and eggplant seedlings. Sci Hortic 94:145–156

Keen NT, Littlefield LJ (1975) The possible association of phytoalexins with resistant gene expression in flax to *Melampsora lini*. Physiol Plant Pathol 14:275280

Khallil AM, Daghman IM, Fady AA (2015) Antifungal potential in crude extracts of five selected brown seaweeds collected from the western Libya coast. J Microbiol Mod Tech 1:103

Khan MS, Ahmad I (2011) In vitro antifungal, anti-elastase and anti-keratinase activity of essential oils of Cinnamomum-, Syzygium- and Cymbopogon-species against *Aspergillus fumigatus* and *Trichophyton rubrum*. Phytomedicine 19:48–55

Kim SI, Roh JY, Kim DH, Lee HS, Ahn YJ (2003) Insecticidal activities of aromatic plant extracts and essential oils against *Sitophilus Oryzae* and *Callosobruchus chinensis*. J Stored Prod Res 39(3):293–303

Kim YM, Lee CH, Kim HG, Lee HS (2004) Anthraquinones isolated from *Cassia tora* (Leguminosae) seed show an antifungal property against phytopathogenic fungi. J Agric Food Chem 52:6096–6100

Kimura M, Yamaguchi I (1996) Recent development in the use of Blasticidin S, a microbial fungicide, as a useful reagent in molecular biology. Pestic Biochem Physiol 56:243

Klarzynski O, Plesse B, Joubert JM, Yvin JC, Kopp M, Kloareg B, Fritig B (2000) Linear b-1,3 glucans are elicitors of defense responses in tobacco-France. Plant Physiol 124:1027–1037

Kloepper JW, Ryu CM, Zhang S (2004) Induce systemic resistance and promotion of plant growth by *Bacillus* spp. Phytopathology 94:1259–1266

Kobayashi K, Nishino C, Tomita H, Fukushima M (1987) Antifungal activity of pisiferic acid derivatives against the rice blast fungus. Phytochemistry 26:3175–3179

Koga J et al (1997) Functional moiety for the antifungal activity of phytocassane E, a diterpene phytoalexin from rice. Phytochemistry 44:249–253

Kokubun T, Harborne JB (1995) Phytoalexin induction in the sapwood of plants of the Maloideae (Rosaceae): biphenyls or dibenzofurans. Phytochemistry 40:1649–1654

Koumoutsi A, Chen XH, Henne A, Liesegang H, Gabriele H, Franke P, Vater J, Borris R (2004) Structural and functional characterization of gene clusters directing non-ribosomal synthesis of bioactive lipopeptides in *Bacillus amyloliquefaciens* strain FZB42. J Bacteriol 186:1084–1096

Kraiczy P, Haase U, Gencic S, Flindt S, Anke T, Brandt U, von Jagow G (1996) Eur J Biochem 235:54

Krishna Kishore G, Pande S, Harish S (2007) Evaluation of essential oils and their components for broad-spectrum antifungal activity and control of late leaf spot and crown rot diseases in peanut. Plant Dis 91(4):375–379

Krzaczkowski L, Wright M, Gairin JE (2008) Bryophytes, a potent source of drugs for tomorrow's medicine? Med Sci 24:947–953

Krzyzaniak Y, Trouvelot S, Negrel J, Cluzet S, Valls J, Richard T, Bougaud A, Jacquens L, Klinguer A, Chiltz A, Adrian M, Héloir M-C (2018) A plant extract acts both as a resistance inducer and an oomycide against grapevine downy mildew. Front Plant Sci 9:1085

Küster E (1959) Outline of a comparative study of criteria used in characterization of the actinomycetes. Int Bull Bacteriol Nomencl Taxon 9:97–104

Küster E, Williams ST (1964) Selection of media for isolation of Streptomycetes. Nature 202:928–929

Kuc J (1982) Phytoalexins from the Solanaceae. In: Bailey JA, Mansfield JW (eds) Phytoalexins. Blackie, Glasgow/London, pp 81–105

Kudalkar P, Strobel G, Ul-Hassan SR, Geary B, Sears J (2012) *Muscodor sutura*, a novel endophytic fungus with volatile antibiotic activities. Mycoscience 53:319–325

Kumar S, Kaushik N (2013) Endophytic fungi isolated from oil-seed crop *Jatropha curcas* produces oil and exhibit antifungal activity. PLoS One 8(2):e56202

Kumar S, Kaushik N, Proksch P (2013) Identification of antifungal principle in the solvent extract of an endophytic fungus Chaetomium globosum from *Withania somnifera*. Springerplus 2(1):37

Kumar S, Kaushik N, Edrada-Ebel R, Ebel R, Proksch P (2011) Isolation, characterization, and bioactivity of endophytic fungi of *Tylophora indica*. World J Microbiol Biotechnol 27:571–577

Mends MT, Yu E, Strobel GA, Riyaz-Ul-Hassan S, Booth E, Geary B, Sears J, Taatjes CA, Hadi MZ (2012) An endophytic *Nodulisporium* sp. producing volatile organic compounds having bioactivity and fuel potential. J Pet Environ Biotechnol 03:117

Kumar V, Karunaratne V, Sanath MR, Meegalle K, Macleod JK (1990) Two fungicidal phenylethanones from *Euodia lunu-ankenda* root bark. Phytochemistry 29:243–245

Labbe C, Faini F, Villagram C, Coll J, Rycroft DS (2005) Antifungal and insect antifeedant 2-phenylethanol esters from the liverwort *Balantiopsis cancellata* from Chile. J Agric Food Chem 53:247–249

Lacey LA, Horton DR, Jones DC, Headrick HL, Neven LG (2009) Efficacy of the biofumigant fungus *Muscodor albus* (Ascomycota: Xylariales) for control of codling moth (Lepidoptera: Tortricidae) in simulated storage conditions. J Econ Entomol 102(1):43–49

Lacey LA, Neven LG (2006) The potential of the fungus, *Muscodor albus*, as a microbial control agent of potato tuber moth (Lepidoptera: Gelechiidae) in stored potatoes. J Invertebr Pathol 9:1195–1198

Lago JH, Ramos CS, Casanova DC, Morandim AA, Bergamo DC, Cavalheiro AJ, Bolzani VS, Furlan M, Guimaraes EF, Young MC, Kato MJ (2004) Benzoic acid derivatives from piper species and their fungitoxic activity against *Cladosporium cladosporioides* and *C. sphaerospermum*. J Nat Prod 67:783

Langcake P, Pryce RJ (1976) The production of resveratrol by *Vitis vinifera* and other members of the Vitaceae as a response to infection or injury. Physiol Plant Pathol 9:77–86

Larsen J, Bodker L (2003) Interactions between pea root-inhabiting fungi examined using signature fatty acids. New Phytol 149:487–493

Leclere V, Bechet M, Adam A, Guez JS, Wathelet B, Ongena M, Thonart P, Gancel F, Chollet-Imbert M, Jacques P (2005) Mycosubtilin overproduction by *Bacillus subtilis* BBG100 enhances the organism's antagonistic and biocontrol activities. Appl Environ Microbiol 71:4577–4584

Lee CW et al (1999) Momilactones a and B in rice straw harvested at different growth stages. Biosci Biotechnol Biochem 63:1318–1320

Lee HB, Lee HK, Ahn YJ (2003) Antibacterial activity of oriental medicinal plant extracts toward *Helicobacter pylori*. Agric Chem Biotechnol 46:97–99

Lee SO, Kim HY, Choi GJ, Lee HB, Jang KS, Choi YH (2009) Mycofumigation with *Oxyporus latemarginatus* EF069 for control of postharvest apple decay and Rhizoctonia root rot on moth orchid. J Appl Microbiol 106:1213–1219

Li H, Qing C, Zhang Y, Zhao Z (2005a) Screening for endophytic fungi with antitumour and antifungal activities from Chinese medicinal plants. World J Microbiol Biotechnol 21:1515–1519

Li JY, Strobel G, Harper J, Lobkovsky E, Clardy J (2000) Cryptocin, a potent tetramic acid antimycotic from the endophytic fungus *Cryptosporiopsis* cf. *quercina*. Org Lett 2:767–770
Li S, Zhang Z, Cain A, Wang B, Long M, Taylor J (2005b) Antifungal activity of camptothecin, trifolin, and hyperoside isolated from *Camptotheca acuminata*. J Agric Food Chem 53:32–37
Liangbin H, Hongbo L, Junliang S, Zeng J (2012) Effect of laminarin on *Aspergillus Flavus* growth and aflatoxin production. Adv Mater Res 342:1168–1171
Liarzi O, Bar E, Lewinsohn E, Ezra D (2016) Use of the endophytic fungus *Daldinia cf. concentrica* and its volatiles as bio-control agents. PLoS One 11(12):0168242
Linderman RG (1994) In: Mycorrhizae and plan health. ASA Special Pub. No. 54, Madison
Lioussanne L, Beauregard M, Hamel C et al (2009) Advances in Mycorrhizal science and technology. NRC Research Press, Ottawa
Liu H, Wang J, Zhao J, Lu S, Wang J, Jiang W, Ma Z, Zhou L (2009a) Isoquinoline alkaloids from *Macleaya cordata* active against plant microbial pathogens. Nat Prod Commun 4:1557–1560
Liu L, Liu S, Chen X, Guo L, Che Y (2009b) Pestalofones A-E, bioactive cyclohexanone derivatives from the plant endophytic fungus *Pestalotiopsis fici*. Bioorg Med Chem 17:606–613
Liu M, Wang G, Xiao L, Xu A, Liu X, Xu P, Lin X (2014) Bis(2,3-dibromo-4,5-dihydroxybenzyl) ether, a marine algae derived bromophenol, inhibits the growth of *Botrytis cinerea* and interacts with DNA molecules. Mar Drugs 12:3838–3851
Lopes G, Andrade PB, Valentão P (2015) Screening of a marine algal extract for antifungal activities. Methods Mol Biol 1308:411–420
Lu C, Shen Y (2003) A new macrolide antibiotic with antitumor activity produced by *Streptomyces* sp. CS, a commensal microbe of *Maytenus hookeri*. J Antibiot 56(4):415–418
Lu C, Shen Y (2007) A novel ansamycin, naphthomycin k from *Streptomyces* sp. J Antibiot 60(10):649–653
Lu H, Xou WX, Meng JC, Hu J, Tan RX (2000) New bioactive metabolites produced by *Colletotrichum* sp., an endophytic fungus in *Artemisia annua*. Plant Sci 151:67–73
Ludwig W, Klenk HP (2005) Overview: a phylogenetic backbone and taxonomic framework for prokaryotic systematics. In: Brenner DJ, Krieg NR, Staley JT, Garrity GM (eds) Bergey's manual of systematic bacteriology, The Proteobacteria, part A, Introductory essays, vol 2, 2nd edn. Springer, New York, pp 49–65
Mabrouk SS, El-Shayeb NMA, El-Refai AH, Sallam LAR, Hamdy AA (1985) Inhibitory activities of some marine algae on aflatoxin accumulation. Appl Microbiol Biotechnol 22:152–155
Macias RML, Hernandez BBE, Jimenez EM, Gonzalez MC, Glenn AE, Hanlin RT, Hernandez OS, Saucedo GA, Muria GJM, Anaya AL (2008) Naphthoquinone spiroketal with allelochemical activity from the newly discovered endophytic fungus Edenia gomezpompae. Phytochemistry 69(5):1185–1196
Usman MG, Ahmadu T, Adamu JN, Aisha M (2018) Plant disease control: understanding the roles of toxins and phytoalexins in host-pathogen interaction. PJSRR 4(1):54–66
Mahmud H, Hossain I, Ahmad MU (2018) Plant extracts, BAU-biofungicide and fungicides in controlling some important diseases of Rice cv. BRRI Dhan40. Am J Phytomed Clin Ther 6(2):7
Manjunath M, Prasanna R, Nain L, Dureja P, Singh R, Kumar A, Jaggi S, Kaushik BD (2010) Biocontrol potential of cyanobacterial metabolites against damping off disease caused by *Pythium aphanidermatum* in solanaceous vegetables. Arch Phytopathol Plant Protect 43:666–677
Mann R, Mattner S, Allen D, Porter I, Edwards J (2008) Bioprospecting for endophytes from Australian flora with mycofumigation potential. In: 3rd international biofumigation symposium, CSIRO Discovery Centre, Canberra, Australia 80
Manojlovic NT, Solujic S, Sukdolak S, Milosev M (2005) Antifungal activity of *Rubia tinctorum*, *Rhamnus frangula* and *Caloplaca cerina*. Fitoterapia 76:244–246
Mansfield JW (1999) Antimicrobial compounds and resistance: the role of phytoalexins and antianticipins. In: Slusarenko AJ, RSS F, LC VL (eds) Mechanisms of resistance to plant diseases. Kluwer, Amsterdam
Marston A, Gafner F, Dossagi SF, Hostettmann K (1988) Fungicidal and molluscicidal saponins from *Dolichos kilimandscharicus*. Phytochemistry 27:1325–1326

Marston A, Hamburger M, Sordat-Disercns I, Msonthi JD, Hostettmann K (1993) Xanthones from *Polygala nyikensis*. Phytochemistry 33:809–812
Mathias-Mundy E, McCorkle C (1995) Ethnoveterinary medicine and development: a review of the literature. In: Warren DM, Surrerwer L, Broshenka D (eds) The cultural dimension of indigenous knowledge systems. Intermediate Technology Publications, London
Matiello J, Bonaldo SM (2013) Elicitor activity of phytoalexins in soy and sorghum by extracts and tinctures of medicinal plant species. Rev Bras Plantas Med 15:541–550
Matiello J, Raasch-Fernandes LD, Berber GCM, Trento RM, Bonaldo SM (2016) Synthesis of phytoalexins in soy and sorghum by extracts and tinctures from three forest species. Revista em Agronego ´cio e Meio Ambiente 9:617–633
Matuso A, Yuki S, Nakayama M (1986) Structures of ent-herbertane sesquiterpenoids displaying antifungal properties from the liverwort *Herberta adunca*. J Chem Soc Perkin Trans 1:701–710
Katoch M, Pull S (2017) Endophytic fungi associated with *Monarda citriodora,* an aromatic and medicinal plant and their biocontrol potential. Pharm Biol 55(1):1528–1535
Mekuria T, Steiner U, Hindorf H, Frahm JP, Dehne HW (2005) Bioactivity of bryophyte extracts against *Botrytis cinerea*, *Alternaria solani* and *Phytophthora infestans*. J Appl Bot Food Qual 79:89–93
Mercier J, Jiménez JI (2004) Control of fungal decay of apples and peaches by the biofumigant fungus *Muscodor albus*. Postharvest Biol Technol 31:1–8
Mercier L, Lafitte C, Borderies G, Briand X, Esquerré-Tugayé MT, Fournier J (2001) The algal polysaccharide carrageenans can act as an elicitor of plant defense. New Phytol 149:43–51
Mery BA, Joubert JM, Hoareau A (2013) Laboratoires Goëmar. Parc Technopolitain atalante, Saint-malo Cedex
Meshram V, Gupta M, Saxena S (2015) *Muscodor ghoomensis* and *Muscodor indica*: new endophytic species based on morphological features, molecular and volatile organic analysis from Northeast India. Sydowia 67:133–146
Meshram V, Kapoor N, Chopra G, Saxena (2017) *Muscodor camphora*, a new record from *Cinnamomum camphora*. Mycosphere 8(4):568–582
Meshram V, Kapoor N, Saxena S (2013) *Muscodor kashayum* sp. nov. – a new volatile antimicrobial producing endophytic fungus. Mycology 4(4):196–204
Meshram V, Saxena S, Kapoor N (2014) *Muscodor strobelii*, a new endophytic species from South India. Mycotaxon 128:93–104
Meyer JR, Linderman RG (1986) Response of subterranean clover to dual inoculation with vesicular-arbuscular mycorrhizal fungi and a plant growth-promoting bacterium, *Pseudomonas putida*. Soil Biol Biochem 18:185–190
Mikkelsen MD et al (2003) Modulation oh CYP79 genes and glucosilate profiles in Arabidopsis by defense pathways. Plant Physiol 131:298–308
Miles DH, Chittawong V, Hedin PA, Kokpol U (1993) Potential agrochemicals from leaves of *Wedelia biflora*. Phytochemistry 32:1427–1429
Miles OH, Rosa de Medeiros JM, Chittawong Y, Hedin PA, Swithenbunk C, Lidert Z (1991) 30-formyl-20, 40,60-trihydroxydihydrochalcone from *Psidium acutangulum*. Phytochemistry 30:1131–1132
Miller JD (1995) Fungi and mycotoxins in grain: implications for stored product research. J Stored Prod Res 31:1–16
Mini Priya R (2012) Endophytic actinomycetes from Indian medicinal plants as antagonists to some phytopathogenic fungi. Sci Rep 1(4):259
Mishra PK, Kedia A, Dubey NK (2015) Chemically characterized *Cymbopogon martinii* (Roxb.) Wats. essential oil for shelf life enhancer of herbal raw materials based on antifungal, antiaflatoxigenic, antioxidant activity and favorable safety profile. Plant Biosyst 150:1313–1322
Mitchell AM, Strobel GA, Moore E, Robison R, Sears J (2010) Volatile antimicrobials from *Muscodor crispans*, a novel endophytic fungus. Microbiology 156:270–277
Moezelaar R, Braam C, Zomer J, Gorris LGM, Smid EJ (1999) Volatile plant metabolites for postharvest crop protection. In: Lyr H, Russell PE, Dehne HW, Sisler HD (eds) Modern fungicides and antifungal compounds II, 1st edn. Intercept Limited, Andover, pp 453–467

Monde K, Sasaki K, Shirata A, Takasugi M (1990a) 4-Methoxybrassinin, a sulphur-containing phytoalexin from Brassica oleracea. Phytochemistry 29:1499–1500

Monde K, Sasaki K, Shirata A, Takasugi M (1991) Brassicanal C and two dioxindoles from cabbage. Phytochemistry 30:2915–2917

Monde K, Katsui N, Shirata A, Takasugi M (1990b) Brassicanal A and B, novel sulfur-containing phytoalexins from the Chinese cabbage *Brassica campestris* L. ssp *pekinensis*. Chem Lett:209–210

Monde K, Tamura K, Takasugi M, Kobayashi K, Somei M (1994) Dehydro-4-methoxycyclobrassinin, a sulfur-containing phytoalexin isolated from turnip *Brassica campestris* L. ssp. *rapa*. Heterocycles 38:263–267

Moon SS, Lu Chen J, Moore RE, Patterson GML (1992) Calophycin, a fungicidal cyclic decapeptide from the terrestrial blue-green alga *Calothrix fusca*. J Organomet Chem 57:1097–1103

Morita Y, Matsumura E, Okabe T, Fukui F, Ohe T, Ishida N, Inamori Y (2004) Biological activity of β-dolabrin, γ- thujaplicin, and 4-acetyltropolone, hinokitiol-related compounds. Biol Pharm Bull 27:1666–1669

Moyne AL, Shelby R, Cleveland TE, Tuzun S (2001) Bacillomycin D: an iturin with antifungal activity against *Aspergillus flavus*. J Appl Microbiol 90:622–629

Mueller-Riebau F, Berger B, Yegen O (1995) Chemical composition and fungitoxic properties to phytopathogenic fungi of essential oils of selected aromatic plants growing wild in Turkey. J Agric Food Chem 43:2262–2266

Muirhead IF, Deverall BJ (1984) Evaluation of 3, 4-dihydroxybenzaldehyde, dopamine and its oxidation products as inhibitors of *Colletotrichum musae* (Berk. and Curt.) arx in green banana fruits. Aust J Bot 32:375–382

Mukherjee KP, Saha K, Saha PB, Pal M, Das J (1996) Antifungal activities of the leaf extract of *Cassia tora* Linn. (Fam. Leguminosae). Phytoter Res 10:521–522

Müller KO, Börger H (1940) Experimentelle Untersuchungen über die Phytophthora Resistenz der Kartoffel. Arb Biol Reichsanst Land Forstwirtsch 23:189–231

Nagata T, Tsushida T, Hamaya E, Enoki N (1985) Camellidins, antifungal saponins isolated from *Camellia japonica*. Agric Biol Chem 49(4):1181–1186

Naoumkina M, Farag MA, Sumner LW, Tang YH, Liu CJ, Dixon RA (2007) Different mechanisms for phytoalexin induction by pathogen and wound signals in Medicago truncatula. Proc Natl Acad Sci 104:17909–17915

Ng TB, Ye XJ, Wong JH, Fang EF, Chan YS, Pan W, Ye XY, Sze SCW, Zhang KY, Liu F, Wang HX (2011) Glyceollin, a soybean phytoalexin with medicinal properties. Appl Microbiol Biotechnol 90(1):59–68

Nisimura T, Meguro A, Hasegawa S, Nakagawa Y, Shimizu M, Kunho H (2002) An Endophytic Actinomycete, *Streptomyces* sp. AOK-30, isolated from mountain laurel and its antifungal activity. J Gen Plant Pathol 68:390–397

Ogundare AO, Adetuyi FC, Akinyosoye FA (2006) Antimicrobial activities of *Vernonia tenoreana*. Afr J Biotechnol 5(18):1663–1668

Olanrewaju OS, Babalola OO (2019) Streptomyces: implications and interactions in plant growth promotion. Appl Microbiol Biotechnol 103:1179–1188

Oliveira CM, Regasini LO, Silva GH, Pfenning LH, Young MCM, Berlinck RGS, Bolzani VS, Araujo AR (2011) Dihydro isocoumarins produced by *Xylaria* sp. and *Penicillium* sp., endophytic fungi associated with *Piper aduncum* and *Alibertia macrophylla*. Phytochem Lett 4:93–96

Oliveira JSB, Maia AJ, Schwan-Estrada KRF, Bonato CM, Carneiro SMTPG, Picoli MHS (2014) Activation of biochemical defense mechanisms in bean plants for homeopathic preparations. Afr J Agric Res 9:971–981

Om Y, Yamaguchi I, Misato T (1984) Inhibition of protein biosynthesis by mildiomycin, an anti-mildew substance. Nihon Noyaku Gakkaishi J Pestic Sci 9:317

Omidbeygi M, Barzegar M, Hamidi Z, Naghdibadi H (2007) Antifungal activity of thyme, summer savory and clove essential oils against *Aspergillus flavus* in liquid medium and tomato paste. Food Control 18(12):1518–1523

Orole OO, Adejumo TO (2009) Activity of fungal endophytes against four maize wilt pathogens. Afr J Microbiol Res 3(12):969–973

Osman MEH, El-Sheekh MM, Metwally MA, Ismail AA, Ismail MM (2011) Efficacy of some agricultural wastes in controlling root rot of *Glycine max* L. induced by *Rhizoctonia solani*. Asian J Plant Pathol 5:16–27

Merriman PR, Price RD, Kollmorgen AJF, Piggott T, Ridge EH (1974) Effect of seed inoculation with *Bacillus subtilis* and *Streptomyces griseus* on the growth of cereals and carrots. Aust J Agric Res 25:219–226

Pacher T, Bacher M, Hofer O, Greger H (2001) Stress induced carbazole phytoalexins in Glycosmis species. Phytochemistry 58:129–135

Padmakumar K, Ayyakkannu K (1997) Seasonal variation of antibacterial and antifungal activities of the extracts of marine algae from southern coasts of India. Bot Mar 40:507–515

Park JH, Choi GJ, Lee HB, Kim KM, Jung HS, Lee SW, Jang KS, Cho KY, Kim JC (2005) Griseofulvin from *Xylaria* sp. Strain F0010, and endophytic fungus of *Abies holophylla* and its antifungal activity against plant pathogenic fungi. J Microbiol Biotechnol 15:112–117

Parniske M (2008) Arbuscular mycorrhiza: the mother of plant root endosymbiosis. Nat Rev Microbiol 6:763–775

Park MS, Ahn J, Choi GJ, Choi YH, Jang KS, Kim JC (2010) Potential of the volatile producing fungus *Nodulisporium* sp. CF016 for the control of postharvest diseases of apple. Plant Pathol J 26:253–259

Passari AK, Mishra VK, Saikia R, Gupta VK, Singh BP (2015) Isolation, abundance and phylogenetic affiliation of endophytic actinomycetes associated with medicinal plants and screening for their in vitro antimicrobial biosynthetic potential. Front Microbiol 6:273

Passari AK, Mishra VK, Gupta VK, Saikia R, Singh BP (2016) Distribution and identification of Endophytic Streptomyces species from *Schima wallichii* as potential biocontrol agents against fungal plant pathogens. Pol J Microbiol 65(3):319–329

Paszkowski U (2006) Mutualism and parasitism: the yin and yang of plant symbioses. Curr Op Plant Biol 9:364–370

Paulert R, Talamini V, Cassolato JEF, Duarte MER, Noseda MD, Smania A, Stadnik MJ (2009) Effects of sulfated polysaccharide and alcoholic extracts from green seaweed *Ulva fasciata* on anthracnose severity and growth of common bean (*Phaseolus vulgaris* L.). J Plant Dis Protect 116:263–270

Paulitz TC, Belanger RR (2001) Biological control in greenhouse systems. Annu Rev Phytopathol 39:103–133

Paxton JD (1980) A new working definition of the term "phytoalexin". Plant Dis 64:734

Pedras MSC, Khan AQ (1996) Biotransformation of the Brassica phytoalexin brassicanal A by blackleg fungus. J Agric Food Chem 44:3403–3407

Pedras MSC, Khan AQ, Taylor JL (1997) Phytoalexins from Brassicas: overcoming plants' defenses. ACS Symp Ser 658:155–166

Pedras MSC, Sorensen JL (1998) Phytoalexin accumulation and antifungal compounds from the crucifer wasabi. Phytochemistry 49:1959–1965

Pedras MS, Abdoli A (2017) Pathogen inactivation of cruciferous phytoalexins: detoxification reactions, enzymes and inhibitors. RSC Adv 7:23633–23646

Pejin B, Sabovljević A, Soković M, Glamočlija J, Ćirić A, Vujičić M, Sabovljević M (2012) Antimicrobial activity of *Rhodobryum ontariense*. Hemijska industrija, Research note, in press, ISSN 0367-598X

Pelaez F, Cabello A, Platas G, Diez MT, del Val AG, Basilio A, Martan I, Vicente F, Bills GF, Giacobbe RA, Schwartz RE, Onishi JC, Meinz MS, Abruzzo GK, Flattery AM, Kong L, Kurtz MB (2000) The discovery of enfumafungin, a novel antifungal compound produced by an endophytic *Hormonema* species biological activity and taxonomy of the producing organisms. Syst Appl Microbiol 23:333–343

Pezet R, Pont V (1990) Ultrastructural observations of pterostilbene fungitoxicity in dormant conidia of *Botrytis cinerea* Pers. J Phytopathol 129:29–30

Pittet A (1998) Natural occurrence of mycotoxins in foods and feeds – an update review. Revue de Medecine Veterinire 49:479–492

Pont V, Pezet R (1990) Relation between the chemical structure and biological activity of hydroxystilbenes against *Botrytis cinerea*. J Phytopathol 130:1–8

Potterat O, Stoeckli-Evans H, Msonthi JD, Hostettmann K (1987) Two new antifungal naphthoxirene derivatives and their glucosides from Sesamum angolense WELW. Helv Chim Acta 70:1551–1557

Poussio GB, Abro MA, Hajano BA, Khaskheli MI, Rajput AI, Memon SA (2018) Potential of plant extracts and fungicides for managing *Fusarium oxysporum* f. Sp *lycopersici*. Pak J Phytopathol 30(01):75–81

Prusky D, Keen NT, Eaks J (1983) Further evidence for the involvement of a pre-formed antifungal compound in the latency of *Colletotrichum gloeosporioides* on unripe avocado fruits. Physiol Mol Plant Pathol 22:189–198

Purkayastha RP (2017) Progress in phytoalexin research during the past 50 years. In: Handbook of phytoalexin metabolism and action. Routledge, Boca Raton, pp 1–39

Rashad YM, Aseel DG, Hafez EE (2018) Antifungal potential and defense gene induction in maize against Rhizoctonia root rot by seed extract of *Ammi visnaga* (L.) Lam. Phytopathol Mediterr 57(1):73–88

Ratnayake Bandara BM, Wimalasiri WR (1988) Diterpene alcohols from *Croton lacciferus*. Phytochemistry 27:225–226

Ratnayake Bandara BM, Hewage CM, Karun-aratne V, Wannigama P, Adikaram NKB (1992) An antifungal chromene from *Eupatorium riparium*. Phytochemistry 31:1983–1985

Reddy BN, Raghavender CR (2007) Outbreaks of aflatoxicoses in India. Afr J Food Agric Nutr Dev 7(5):1–15

Reuveni M, Tuzun S, Cole JS, Siegel R, Nesmith WC, Kuc J (1987) Removal of duvatrienediols from the surface of tobacco leaves increases their susceptibility to blue mold. Physiol Mol Plant Pathol 30:441

Reymond P, Farmer EE (1998) Jasmonate and salicylate as global signals for defense gene expression. Curr Opin Plant Biol 1:404–411

Righini H, Roberti R, Baraldi E (2018) Use of algae in strawberry management. J Appl Phycol 30:3551–3564

Ristić MD, Duletić-Laušević S, Knežević-Vukčević J, Marin PD, Simić D, Vukojević J, Janaćković P, Vajs V (2000) Antimicrobial activity of essential oils and ethanol natural products from plants and Fungi as fungicides 227 extract of *Phlomis fruticosa* L. (Lamiaceae). Phytoter Res 14:267–271

Rizk MA (2006) Growth activities of the sugar beet pathogens *Sclerotium rolfsii* Sacc. *Rhizoctonia solani* Kühn. and *Fusarium verticillioides* Sacc. Under cyanobacterial filtrates stress. Plant Pathogol J 5:212–215

Roberti R, Galletti S, Burzi PL, Righini H, Cetrullo S, Perez C (2015) Induction of defense responses in zucchini (*Cucurbita pepo*) by *Anabaena* sp. water extract. Biol Control 82:61–68

Roberti R, Righini H, Reyes CP, Roberti R, Righini H, Reyes CP (2016) Activity of seaweed and cyanobacteria water extracts against *Podosphaera xanthii* on zucchini. Ital J Mycol 45

Robeson DJ, Strobel GA (1982) Agric Biol Chem 46:2681–2683

Roby MHH, Sarhana MA, Selima KA, Khalela KI (2013) Antioxidant and antimicrobial activities of essential oil and extracts of fennel (*Foeniculum vulgare* L.) and chamomile (*Matricaria chamomilla* L.). Ind Crop Prod 44:437–445

Román M, Fernández I, Wyatt T, Sahrawy M, Heil M, Pozo MJ (2011) Elicitation of foliar resistance mechanisms transiently impairs root association with arbuscular mycorrhizal fungi. J Ecol 99:36–45

Rosello CJ, Sempere FF, Sanz BI, Chiralt BMA, Santamarina SMP (2015) Antifungal activity and potential use of essential oils against *Fusarium culmorum* and *Fusarium verticillioides*. J Essent Oil Bear Plants 18(2):359–367

Rungin S, Indanand C, Suttiviriya P, Kruasuwan W, Jaemsaeng R, Thamchaipenet A (2012) Plant growth enhancing effects by a siderophore producing endophytic streptomycete isolated from a Thai jasmine rice plant (*Oryza sativa* L. cv. KDML105). Antonie Van Leeuwenhoek 102:463–472

Sabovljevic A, Sokovic M, Glamoclija J, Citric A, Vujicic M, Pejin B, Sabovljevic M (2011) Bioactivities of extracts from some axenically farmed and naturally grown bryophytes. J Med Plant Res 5:656–671

Sabovljevic A, Sokovic M, Sabovljevic M, Grubisic D (2006) Antimicrobial activity of *Bryum argenteum*. Fitoterapia 77:144–145

Sadrati N, Daoud H, Zerroug A, Dahamna S, Bouharati S (2013) Screening of antimicrobial and antioxidant secondary metabolites from endophytic fungi isolated from wheat (*Triticum durum*). J Plant Protect Res 53(2):129–136

Saha S, Walia S, Kumar J, Parmar BS (2010) Structure biological activity relationships in triterpenic saponins: the relative activity of protobassic acid and its derivatives against plant pathogenic fungi. Pest Manag Sci 66:825–831

Sandra AI, Wright CH, Zumoff LS, Steven VB (2001) *Pantoea agglomerans* strain EH318 produces two antibiotics that inhibit *Erwinia amylovora* in vitro. Appl Environ Microbiol 67:282–292

Sanzani S, Schena L, Ippolito A (2014) Effectiveness of phenolic compounds against citrus green mould. Molecules 19:12500–12508

Sanzani SM, de Girolamo A, Schena L, Solfrizzo M, Ippolito A, Visconti A (2009) Control of *Penicillium expansum* and patulin accumulation on apples by quercetin and umbelliferone. Eur Food Res Technol 228:381–389

Savaroglu F, Fylyk Iscen C, Oztopeu-Vatan P, Kabadere S, Ilhan S, Uyar R (2011) Determination of antimicrobial and antiproliferative activities of the aquatic moss *Fontinalis antipyretica* Hedw. Turk J Biol 35:361–369

Saxena S, Meshram V, Kapoor N (2015) *Muscodor tigerii* sp. nov.-volatile antibiotic producing endophytic fungus from the Northeastern Himalayas. Ann Microbiol 65(1):47–57

Saxena S, Meshram V, Kapoor N (2014) *Muscodor darjeelingensis*, a new endophytic fungus of *Cinnamomum camphora* collected from northeastern Himalayas. Sydowia 66(1):55–67

Scher JM, Speakman JB, ZApp J, Becker H (2004) Bioactivity guided isolation of antifungal compounds from the liverwort *Bazzania trilobata* (L.) S. F Gray. Phytochemistry 65:2583–2588

Schmelz EA, Huffaker A, Sims JW, Christensen SA, Lu X, Okada K, Peters RJ (2014) Biosynthesis, elicitation and roles of monocot terpenoid phytoalexins. Plant J 79:659–678

Schulz B, Sucker J, Aust HJ, Krohen K, Ludewig K, Jones PG, Doring D (1995) Biologically active secondary metabolites of endophytic *Pezicula* sp. Mycol Res 99:1007–1015

Scott FE, Simpson TJ, Trimble LA, Vederas JCJ (1984) Chem Soc Chem Commun 12:756–758

Shain L, Miller JB (1982) Pinocembrin: an antifungal compound secreted by leaf glands of eastern cottonwood. Phytopathology 72:877–880

Shan W, Zhou Y, Liu H, Yu X (2018) Endophytic actinomycetes from tea plants (*Camellia sinensis*): isolation, abundance, antimicrobial, and plant-growth-promoting activities. Hindawi BioMed Research 2018:1–12

Shanahan P, O'Sullivan DJ, Simpson P, Glennon JD, O'Gara F (1992) Isolation of 2,4-Diacetylphloroglucinol from a fluorescent pseudomonad and investigation of physiological parameters influencing its production. Appl Environ Microbiol 58:353–358

Sharma HSS, Fleming C, Selby C, Rao JR, Martin T (2014) Plant bio-stimulants: a review on the processing of macroalgae and use of extracts for crop management to reduce abiotic and biotic stresses. J Appl Phycol 26:465–490

Sharp JK, Valent B, Albersheim P (1984) Purification and partial characterization of a b-glucan fragment that elicits phytoalexin accumulation in soybean. J Biol Chem 259:11312–11320

Shen X-Y, Cheng Y-L, Cai C-J, Fan L, Gao J et al (2014) Diversity and antimicrobial activity of Culturable Endophytic Fungi isolated from Moso bamboo seeds. PLoS One 9(4):e95838

Shimizu M (2011) Endophytic actinomycetes: biocontrol agents and growth promoters. In: Maheshwari DK (ed) Bacteria in agrobiology: plant growth responses. Elsevier Academic Press, San Diego, pp 201–220

Shimizu M, Nakagawa Y, Sato Y, Furumai T, Igarashi Y, Onaka H, Yoshida R, Kunoh H (2000) Studies on endophytic actinomycetes (I) Streptomyces sp. isolated from rhododendron and its antifungal activity. J Gean Plant Pathol 66:360–366

Shirzadian S, Afshari Azad H, Khalghani J (2009) Introduction study of antifungal activities of bryophyte extracts. Iran J Plant Pest Dis 77:1–22

Shlezinger N, Minz A, Gur Y, Hatam I, Dagdas YF, Talbot NJ, Sharon A (2011) Anti-apoptotic machinery protects the necrotrophic fungus Botrytis cinerea from host-induced apoptotic-like cell death during plant infection. PLoS Pathog 7:1002185

Shutsrirung A, Chromkaew Y, Pathom-Aree W, Choonluchanon S, Boonkerd N (2014) Diversity of endophytic actinomycetes in mandarin grown in northern Thailand, their phytohormone production potential and plant growth promoting activity. Soil Sci Plant Nutr 59(3):322–330

Silva-Hughesa AF, Wedgeb DE, Cantrellb CL, Carvalhoa CR, Panb Z, Moraesc RM, Madoxxe VL, Rosaa LH (2015) Diversity and antifungal activity of the endophytic fungi associated with the native medicinal cactus *Opuntia humifusa* (Cactaceae) from the United States. Microbiol Res 175:67–77

Singh AK, Pandey MB, Singh S, Singh AK, Singh UP (2008) Antifungal activity of securinine against some plant pathogenic fungi. Mycobiology 36:99–101

Singh M, Rawat AK, Govindarajan R (2007) Antimicrobial activity of some Indian mosses. Fitoterapia 78:156–158

Singh R, Kalra A, Parameswaran TN, Srinivas KVNS et al (2012) Plant Pathol 61:700–708

Singh SK, Strobel GA, Knighton B, Geary B, Sears J, Ezra D (2011) An Endophytic *Phomopsis* sp. possessing bioactivity and fuel potential with its volatile organic compounds. Microb Ecol 61(4):729–739

Siri-udom S, Suwannarach N, Lumyong S (2016) Existence of *Muscodor vitigenus*, *M. equiseti* and *M. heveae* sp. nov. in leaves of the rubber tree (*Hevea brasiliensis* Müll. Arg.), and their biocontrol potential. Ann Microbiol 66:437–448

Sitton D, West CA (1975) Casbene: an antifungal diterpene produced in cell-free extracts of *Ricinus communis* seedlings. Phytochemistry 14:1921–1925

Skaltsa H, Lazari D, Panagouleas C, Georgiadov E, Garcia B, Sokovic M (2000) Sesquiterpene lactones from *Centaurea thessala* and *Centaurea attica*: antifungal activity. Phytochemistry 55:903–908

Slusarenko AJ, Fraser RS, Van Loon LC (2012) Mechanisms of resistance to plant diseases. Springer, Dordrecht

Smith FA, Smith SE (2011) What is the significance of the arbuscular mycorrhizal colonisation of many economically important crop plants? Plant Soil 348:63–79

Smith GS (1988) The role of phosphorous nutrition in interactions of vesicular arbuscular mycorrhizal fungi with soil borne nematodes and fungi. Phytopathology 78:371–374

Smith KP, Havey MJ, Handelsman J (1993) Suppression of cottony leak of cucumber with *Bacillus cereus* strain UW85. Plant Dis 77:139–142

Smith SE, Gianinazzi Pearson V, Koide R et al (1994) Management of Mycorrhiza in agriculture, horticulture and forestry. Kluwer, Dordecht, pp 103–113

Song YC, Li H, Ye YH, Shan CY, Yang YM, Tan RX (2004) Endophytic naphthopyrone metabolites are co-inhibitors of xanthine oxidase, SW1116 cell and some microbial growths. FEMS Microbiol Lett 241:67–72

Sopalun K, Strobel GA, Wm H, Worapong J (2003) A record of *Muscodor albus*, an endophyte from *Myristica fragrans* in Thailand. Mycotaxon 88:239–247

Soundharrajan RS, Belusamy R, Ramasamy R, Selladurai M, Srinivasan N (2003) Antifungal activity of some essential oils. J Agric Food Chem 51:7596–7599

Spendley PJ, Bird PM, Ride JP, Leworthy DP (1982) Two novel antifungal alka-2,4-dienals from *Triticum aestivum*. Phytochemistry 21:2403–2404

Springob K, Kutchan TM (2009) Introduction to the different classes of natural products. In: Osbourn AE, Lanzotti V (eds) Plant-derived natural products synthesis, function, and application. Springer, New York, pp 3–50

Srividya S, Thapa A, Bhat DV, Golmei K, Dey N (2012) *Streptomyces* sp. 9p as effective biocontrol against chilli soilborne fungal phytopathogens. Eur J Exp Biol 2(1):163–173
Stadnik MJ, Freitas MB (2014) Algal polysaccharides as source of plant resistance inducers. Trop Plant Pathol 39:111–118
St-Arnaud M, Vujanovic V (2007) Mycorrhizae in crop production. The Haworth Press, Inc, New York, pp 67–122
Stinson M, Ezra D, Hess WM, Sears J, Strobel G (2003) An endophytic *Gliocladium* sp. of *Eucryphia cordifolia* producing selective volatile antimicrobial compounds. Plant Sci 165:913–922
Stirk WA, Reinecke DL, van Staden J (2007) Seasonal variation in antifungal, antibacterial and acetyl cholinesterase activity in seven south African seaweeds. J Appl Phycol 19:271–276
Stoessl A (1980) Phytoalexins – a biogenetic perspective. J Phytopathol 99(3):251–272
Storck M, Sacristan MD (1995) The role of phytoalexins in the seedling resistance to Leptosphaeria maculans in some crucifers. Z Naturforsch 50c:15–20
Strobel GA, Spang S, Kluck K, Hess WM, Sears J, Livinghouse T (2008a) Synergism among volatile organic compounds resulting in increased antibiosis in *Oidium* sp. FEMS Microbiol Lett 283:140–145
Strobel G, Singh SK, Riyaz-Ul-Hassan S, Mitchell AM, Geary B, Sears J (2011) An endophytic/pathogenic *Phoma* sp. from creosote bush producing biologically active volatile compounds having fuel potential. FEMS Microbiol Lett 320:87–94
Ul-Hassan R, Syed SG, Geary B, Sears J (2013) An Endophytic *Nodulisporium* sp. from Central America producing volatile organic compounds with both biological and fuel potential. J Microbiol Biotechnol 23(1):29–35
Strobel GA, Knighton B, Kluck K, Ren Y, Livinghouse T, Griffin M, Spakowicz D, Sears J (2008b) The production of mycodiesel hydrocarbons and their derivatives by the endophytic fungus *Gliocladium roseum* (NRRL 50072). Microbiology 154:3319–3328
Strobel GA, Kluck K, Hess WM, Sears J, Ezra D, Vargas PN (2007) *Muscodor albus* E-6, an endophyte of *Guazuma ulmifolia* making volatile antibiotics: isolation, characterization and experimental establishment in the host plant. Microbiology 153:2613–2620
Subhisha S, Subhramoniam A (2005) Antifungal activities of a steroid from *Pallavicinia lyellii*, a liverwort. Indian J Pharm 37:304–308
Sunilkumar G, Campbell LM, Pukhaber L, Stipanovic RD, Rathore KS (2006) Engineering cottonseed for use in human nutrition by tissue-specific reduction of toxic gossypol. Proc Natl Acad Sci 103:18054–18059
Suwannarach N, Bussaban B, Hyde KD, Lumyong S (2010) *Muscodor cinnamomi*, a new endophytic species from *Cinnamomum bejolghota*. Mycotaxon 15(114):15–23
Suwannarach N, Bussaban B, Nuangmek W, Pithakpol W, Jirawattanakul B, Matsui K, Lumyong S (2015) Evaluation of *Muscodor suthepensis* strain CMU-Cib462 as a postharvest biofumigant for tangerine fruit rot caused by *Penicillium digitatum*. J Sci Food Agric 96(1):339–345
Suwannarach N, Kumla J, Bussaban B, Hyde KD, Matsui K, Lumyong S (2013) Molecular and morphological evidence support four new species in the genus Muscodor from northern Thailand. Ann Microbiol 63:1341–1351
Suwannarach N, Kumla J, Bussaban B, Lumyong S (2012) Biocontrol of *Rhizoctonia solani* AG-2, the causal agent of damping-off by *Muscodor cinnamomi* CMU-Cib 461. World J Microbiol Biotechnol 28:3171–3177
Taechowisan T, Lu C, Shen Y, Lumyong S (2007) Antitumor activity of 4-arylcoumarins from endophytic *Streptomyces aureofaciens* CMUAc130. J Cancer Res Ther 3(2):86–91
Taechowisan T, Wanbanjob A, Tuntiwachwuttikul P, Taylor WC (2006) Identification of *Streptomyces* sp. Tc022, an endophyte in *Alpinia galanga*, and the isolation of actinomycin D. Ann Microbiol 56:113–117
Takasugi M (1979) Structures of moracins E, F, G and H, new phytoalexins from diseased mulberry. Tetrahedron Lett 28:4675–4678

Takasugi M, Katsui N, Shirata A (1986) Isolation of three novel sulphur-containing phytoalexins from the Chinese cabbage *Brassica campestris* L. ssp. *pekinensis* (Cruciferae). J Chem Soc Chem Commun 14:1077–1078

Takasugi M, Monde K, Katsui N, Shirata A (1988) Novel sulfur-containing phytoalexins from the Chinese cabbage *Brassica campestris* L. ssp. *pekinensis* (Cruciferae). Bull Chem Soc Jpn 61:285–289

Tanaka N, Yamaguchi H, Umezawa H (1966) Mechanism of kasugamycin action on polypeptide synthesis. J Biochem 60:429

Tan HM, Cao LX, He ZF, Su GJ, Lin B, Zhou SN (2006) Isolation of endophytic actinomycetes from different cultivars of tomato and their activities against *Ralstonia solanacearum* in vitro. World J Microbiol Biotechnol 22:1275–1280

Taware R, Abnave P, Patil D, Rajamohananan PR, Raja R, Soundararajan G, Kundu GC, Ahmad A (2014) Isolation, purification and characterization of Trichothecinol-A produced by endophytic fungus *Trichothecium* sp. and its antifungal, anticancer and antimetastatic activities. Sustainable Chem Processes 2:8

te Welscher YM, ten Napel HH, Balague MM, Souza CM, Riezman H, de Kruijff B, Breukink EJ (2008) Biol Chem 283:6393

Telaxka FJ, Jaski JM, Baldin D, Scariot E, Grosselli M, Moura GS (2014) Antibacterial activity and induction of phytoalexins in bean plants by homeopathic preparations of essential oil of *Eucalyptus globulus*. Congresso Paranaense de Agroecologia 9:1–5

Thomashow LS, Bonsall RF, Weller DM (2002) Antibiotic production by soil and rhizosphere microbes *in situ*. In: Manual of environmental microbiology, 2nd edn. ASM Press, Washington, DC, pp 638–647

Tian XL, Cao LX, Tan HM, Zeng QG, Jia YY, Han WQ, Zhou SH (2004) Study on the communities of endophytic fungi and endophytic actinomycetes from rice and their antipathogenic activities in vitro. World J Microbiol Biotechnol 20:303–309

Ting ASY, Mah SW, TEE CS (2010) Identification of volatile metabolites from fungal endophytes with biocontrol potential towards *Fusarium Oxysporum* F. sp. *cubense* race 4. Am J Agric Biol Sci 5(2):177–182

Toljander JF, Lindahl BD, Paul LR, Elfstr MA, Finlay RD (2007) Influence of arbuscular mycorrhizal mycelial exudates on soil bacterial growth and community structure. FEMS Microbiol Ecol 61:295–304

Toral L, Rodriguez M, Bejar V, Sampedro I (2018) Antifungal activity of Lipopeptides from *Bacillus* XT1 CECT 8661 against *Botrytis cinerea*. Front Microbiol 9:1315

Tomsheck AR, Strobel GA, Booth E, Geary B, Spakowicz D, Knighton B, Floerchinger C, Sears J, Liarzi O, Ezra D (2010) *Hypoxylon sp.*, an Endophyte of *Persea indica*, producing 1,8-cineole and other bioactive volatiles with fuel potential. Microb Ecol 60:903–914

Toyomasu T et al (2008) Diterpene phytoalexins are biosynthesized in and exuded from the roots of rice seedlings. Biosci Biotechnol Biochem 72(2):562–567

Tsuji J, Zook M, Jackson EP, Cage DA, Hammerschmidt R, Somerville S (1992) Plant Physiol 98:1304

Tzortzakis NG, Economakis CD (2007) Antifungal activity of lemongrass (*Cymbopogon citratus* L.) essential oil against key postharvest pathogens. Innov Food Sci Emerg Technol 8:253–258

Vanden B, DRomer HF, Improvisi I, Lozano-Chiu M, Rex JH, Sanglard D (1998) Antifungal drug resistance in pathogenic fungi. Med Mycol 36:119–128

Ulloa-Benítez Á, Medina-Romero YM, Sánchez-Fernández RE, Lappe-Oliveras P, Roque-Flores G, Duarte Lisci G, Herrera Suárez T, Macías-Rubalcava ML (2016) Phytotoxic and antimicrobial activity of volatile and semi-volatile organic compounds from the endophyte *Hypoxylon anthochroum* strain Blaci isolated from *Bursera lancifolia* (Burseraceae). J Appl Microbiol 121(2):380–400

Romero YMM, Flores GR, Rubalcava MLM (2017) Volatile organic compounds from endophytic fungi as innovative postharvest control of *Fusarium oxysporum* in cherry tomato fruits. Appl Microbiol Biotechnol 101:8209–8222

VanEtten HD, Bateman DF (1971) Studies on the mode of action of the phytoalexin phaseolin. Phytopathology 61:1363–1372
VanWees SC, Chang HS, Zhu T, Glazebrook J (2003) Characterization of the early response of Arabidopsis to *Alternaria brassicicola* infection using expression profiling. Plant Physiol 132:606–617
Veljic M, Durie A, Sokovic M, Ciric A, Glamochja J, Marin PD (2009) Antimicrobial activity of methanol extracts of *Fontinalis Antipyretica, Hypnum Cupressiforme* and *Ctenidium molluscum*. Arch Biol Sci 61:225–229
Vera J, Castro J, Gonzalez A, Moenne A (2011) Sea weed polysaccharides and derived oligosaccharides stimulate defense responses and protection against pathogens in plants. Mar Drugs 9:2514–2525
Verma VC, Surendra KG, Kumar A, Mishra A, Kharwar RN, Gange AC (2009) Endophytic Actinomycetes from *Azadirachta indica* a. Juss.: Isolation, Diversity, and anti-microbial activity. Microb Ecol 57(4):749–756
Vierheilig H, Steinkellner S, Khaosaad T (2008) The biocontrol effect of mycorrhization on soil borne fungal pathogens and the autoregulation of the AM symbiosis: one mechanism, two effects? In: Varma A (ed) Mycorrhiza. Springer, Berlin, pp 307–320
Vigo C, Norman JR, Hooker JE (2000) Biocontrol of the pathogen *Phytophthora parasitica* by arbuscular mycorrhizal fungi is a consequence of effects on infection loci. Plant Pathol 49:509–514
Viturro CF, Dela-Fuente JR, Maier MS (2004) 5-Methylcoumaranones from *Mutisia friesiana* and their bioactivity. J Nat Prod 67:778
Volpin H, Elkind Y, Okon Y, Kapulnik Y (1994) A vesicular Arbuscular Mycorrhizal fungus (*Glomus intraradix*) induces a Defense response in alfalfa roots. Plant Physiol 104:683–689
Volpin H, Phillips DA, Okon Y, Kapulnik Y (1995) Suppression of an Isoflavonoid Phytoalexin Defense response in Mycorrhizal alfalfa roots. Plant Physiol 108(4):1449–1454
Vos CM, Yang Y, De Coninck B, Cammue BPA (2014) Fungal (−like) biocontrol organisms in tomato disease control. Biol Control 74:65–81
Walters DR, Ratsep J, Havis ND (2013) Controlling crop diseases using induced resistance: challenges for the future. J Exp Bot 64:1263–1280
Wang SY, Chen PF, Chang ST (2005) Antifungal activities of essential oils and their constituents from indigenous cinnamon (*Cinnamomum osmophloeum*) leaves against wood decay fungi. Bioresour Technol 96:813–818
Wang X, Manrang Z, Gao J, Pu T, Bilal M, Wang Y, Zhang X (2018) Antifungal activity screening of soil actinobacteria isolated from Inner Mongolia, China. Biol Control 127:78–84
Wang Y, Wang G, Wang L, Xu X, Xia J, Huang X, Wu Y, Zhang C (2010) Isolation and identification of an endophytic fungus of *Polygonatum cyrtonema* and its antifungal metabolites. Wei Sheng Wu Xue Bao 50:1036–1043
Ward EWB et al (1975) Loroglossol: an orchid phytoalexin. Phytopathology 65:632–633
Wasimfiroz RM, Hosagoudar GN, Narayanaswamy H, Patil B (2018) Evaluation of fungicides, bio-agents and plant extracts against *Pyricularia oryzae*. Int J Chem Stud 6(2):1875–1879
Wei ZM, Laby RJ, Zurnoff CH, Bauer DW, He SY, Collmer A, Beer SV (1992) Science 257:85
Wharton P, Nicholson R (2000) Temporal synthesis and radiolabelling of the sorghum 3-deoxyanthocyanidin phytoalexins and the anthocyanin, cyanidin 3-dimalonyl glucoside. New Phytopatol 145:457–469
Wilhite SE, Lunsden RD, Strancy DC (2001) Peptide synthetase gene *in Trichoderma virens*. Appl Environ Microbiol 67:5055–5062
Williams ST, Davies FL (1965) Use of antibiotics for selective isolation and enumeration of actinomycetes in soil. J Gen Microbiol 38:251–262
Wippich C, Wink M (1985) Biological properties of alkaloids. Influence of quinolizidine alkaloids and gramine on the germination and development of Powderly mildew, *Erysiphe graminis* f. sp. hordei. Experientia 41:1477–1479
Wolters B (1964) Die Verbreitung antifungaler Eigenschaften bei Moosen. Planta 62:88–96

Woodward S, Pearce RN (1988) The role of stilbenes in resistance of Sitka spruce (*Picea sitchensis* (bong) Carr) to entry of fungal pathogens. Physiol Mol Plant Pathol 33:127

Worapong J, Strobel G, Daisy B, Castillo UF, Baird G, Hess WM (2002) *Muscodor roseus* anam. Sp. nov., an endophyte from *Grevillea pteridifolia*. Mycotaxon 81:463–475

Worapong J, Strobel GA, Ford EJ, Li JY, Baird G, Hess WM (2001) *Muscodor albus* anam. Sp. Nov., an endophyte from *Cinnamomum zeylanicum*. Mycotaxon 79:67–79

Xie CF, Lou HX (2008) Chemical constituents from the Chinese bryophytes and their reversal of fungal resistance. Curr Org Chem 12:619–628

Xu Q, Wang J, Huang Y, Zheng Z, Song S, Zhang Y, Su W (2004) Metabolites from mangrove endophytic fungus Dothiorella sp. Acta Oceanol Sin 23:541–547

Yang MH, Wang JS, Luo JG, Wang XB, Kong LY (2009) Tetran or triterpenoids from *Chisocheton paniculatus*. J Nat Prod 72:2014–2018

Yao H, Tian S (2005) Effects of pre-and post-harvest application of salicylic acid or methyl jasmonate on inducing disease resistance of sweet cherry fruit in storage. Postharvest Biol Technol 35:253–262

Yao YQ, Lan F, Qiao YM, Wei JG, Huang RS, Li LB (2017) Endophytic fungi harbored in the root of *Sophora tonkinensis* Gapnep: diversity and biocontrol potential against phytopathogens. Microbiol Open 6:e437

Yemele-Bouberte M, Krohn K, Hussain H, Dongo E, Schulz B, Hu Q (2006) Tithoniamarin and tithoniamide: a structurally unique isocoumarin dimer and a new ceramide from *Tithonia diversifolia*. Nat Prod Res 20:842–849

Yoganandam GP, Gowri R, Biswas D (2009) Evaluation of *Wedelia Biflora* (Linn) D.C for anthelmintic and antimicrobial activity. J Pharm Res 2:375–377

Yoon MY, Cha B, Kim JC (2013) Recent trends in studies on botanical fungicides in agriculture. Plant Pathol J 29(1):1–9

Yoon MY, Choi GJ, Choi YH, Jang KS, Park MS, Cha B, Kim JC (2010) Effect of polyacetylenic acids from *Prunella vulgaris* on various plant pathogens. Lett Appl Microbiol 51:511–517

Yoon MY, Choi NH, Min BS, Choi GJ, Choi YH, Jang KS, Han SS, Cha B, Kim JC (2011) Potent in vivo antifungal activity against powdery mildews of pregnane glycosides from the roots of *Cynanchum wilfordii*. J Agric Food Chem 59:12210–12216

Young MCM, Braga MR, Dietrich SMC, Gottlieb HE, Trevisan LMV, Bolzani VS (1992) Fungitoxic non-glycosidic iridoids from *Alibertia macrophylla*. Phytochemistry 31:3433

Yu J, Wu Y, He Z, Li M, Zhu K, Gao B (2018) Diversity and antifungal activity of endophytic fungi associated with *Camellia oleifera*. Mycobiology 46(2):85–91

Yuan Y, Feng H, Wang L, Li Z, Shi Y, Zhao L et al (2017) Potential of Endophytic Fungi isolated from cotton roots for biological control against Verticillium wilt disease. PLoS One 12(1):e0170557

Zaker M (2016) Natural plant products as eco-friendly fungicides for plant diseases control-a review. Agriculturists 14(1):134–141

Zernova OV, Lygin AV, Pawlowski ML, Hill CB, Hartman GL, Widholm JM, Lozovaya VV (2014) Regulation of plant immunity through modulation of phytoalexin synthesis. Molecules 19:7480–7496

Zhang J, Wang JD, Liu CX, Yuan JH, Wang XJ, Xiang WS (2014) A new prenylated indole derivative from endophytic actinobacteria *Streptomyces* sp. neau-D50. Nat Prod Res 28(7):431–437

Zhang JF, Li YB, Li CL, Jiang JQ (2006) Studies on chemical constituents in root tuber of *Cynanchum auriculatum*. Zhongguo Zhong Yao Za Zhi 31:814–816

Zhang CL, Wang GP, Mao LJ, Komon-Zelazowska M, Yuan ZL, lin FC, Drzhinina IS, Kubicek CP (2010) *Muscodor fengyangensis* sp. nov. from Southeast China: morphology, physiology and production of volatile compounds. Fungal Biol 114:797–808

Zhang W, Krohn K, Draeger S, Schulz B (2008) Bioactive isocoumarins isolated from the endophytic fungus *Microdochium bolleyi*. J Nat Prod 71:1078–1081

Zhao G, Zhao J, Peng L, Zou L, Wang J, Zhong L, Xiang D (2012) Effects of yeast polysaccharide on growth and flavonoid accumulation in *Fagopyrum tataricum* sprout cultures. Molecules 17:11335–11345

Zhao K, Li J, Shen M, Chen Q, Liu M, Ao X, Liao D, Gu Y, Xu K, Ma M, Yu X, Xiang Q, Chen J, Zhang X, Penttinen P (2018) Actinobacteria associated with chinaberry tree are diverse and show antimicrobial activity. Sci Rep 8:11103

Zhao K, Penttinen P, Guan T, Xiao J, Chen Q, Xu J et al (2011) The diversity and anti-microbial activity of endophytic actinomycetes isolated from medicinal plants in Panxi plateau China. Curr Microbiol 62:182–190

Zheng Y, Sheng J, Zhao R, Zhang J, Lv S, Liu L, Shen L (2011) Preharvest l -arginine treatment induced postharvest disease resistance to *Botrytis cinerea* in tomato fruits. J Agric Food Chem 59:6543–6549

Zhou CX, Liu JY, Ye WC, Liu CH, Tang RX (2003) Neoverataline a and B, two antifungal alkaloids with a novel carbon skeleton from *Veratrum taliense*. Tetrahedron 59:5743–5747

Zin NM, Sarmin NIM, Ghadin N, Basri DF, Sidik NM, Hess WM, Strobel GA (2007) Bioactive endophytic streptomycetes from the Malay peninsula. FEMS Microbiol Lett 274:83–88

Zook M, Leege L, Jackson D, Hammerschmidt R (1998) Camalexin accumulation in Arabis lyrata. Phytochemistry 49:2287–2289

Zou WX, Meng JC, Lu H, Chen GX, Shi GX, Zhang TY, Tan RX (2000) Metabolites of *Colletotrichum gloeosporioides*, an endophytic fungus in *Artemisia mongolica*. J Nat Prod 63:1529–1530

Natural Products as Eco-Friendly Bactericides for Plant Growth and Development

10

Shruthy Ramesh and Preetha Radhakrishnan

Abstract

Nowadays, the eco-friendly disease control methods in plants are gaining widespread attention. To control plant diseases, there are different methods, namely, chemical/physical/biological methods or a combination of these. The usage of chemicals is hazardous for humans as well as the ecosystem. At the same time, biological control is having the ability to control the ecosystem in its own natural balanced rate and thereby improving the growth and development of plants. This chapter discusses about crop management and biological control of pests and disease management in agriculture with an emphasis on the advantages/disadvantages. This chapter also deals with microbes as biological controls for the growth and development of the agriculture, fungal biological control agents, biomolecules as biological control agents, biological control by using plant extracts, current scenario, and future aspects in biological control in the agriculture in detail.

Keywords

Biological control · Natural products · Crop management · Bactericides

S. Ramesh · P. Radhakrishnan (✉)
Department of Food Process Engineering, School of Bioengineering, SRM Institute of Science and Technology, Kattankulathur, Tamilnadu, India
e-mail: preethar@srmist.edu.in

J. Singh, A. N. Yadav (eds.), *Natural Bioactive Products in Sustainable Agriculture*, https://doi.org/10.1007/978-981-15-3024-1_10

10.1 Introduction

The protection, growth, and development of agriculture by a natural and eco-friendly way are an essential part in planning and policy of every country. The usage of chemicals is hazardous for humans as well as the ecosystem (Ayansina and Oso 2006). The growth and development of the agriculture sector depend on both the prevention of the diseases and pests. Agricultural practice of monoculture including wheat, rice, maze, cotton, and cone in the same plantation field for repeated process will result in exhaustion of the soil variety, soil nutrients, and beneficial microbes and also reduce the availability of ground water. Due to the deterioration of soil quality, agricultural crops are prone to diseases and compel the farmers to use the pesticides and synthetic fertilizers. The usage of the chemical fertilizers is a serious threat to the beneficial microflora (Shaikh and Gachande 2015). Instead of using chemicals, there are varieties of biological controls or natural/eco-friendly products capable to prevent the disease and thereby promote the growth of the agriculture and improved crop productivity and quality.

10.2 Overview of Crop Management

The growth of agricultural plants depends on the environmental factors such as soil nutrients, natural water source, and presence of beneficial soil microbes. The environmental factors are mainly classified into chemical and physical factors. The physical factors refer to temperature, light, water, and so on. Among chemical factors the role of growth regulators and pesticides is very prominent (Okwute 2012). The chemical factors have common property and have a big impact even at a low dosage (Hirai and Kimura 1979). The chemical factors are categorized into two based on their practical uses:

(i) Growth-regulating compounds
(ii) Pesticidal substances

The growth regulating compounds are acceptable to use intentionally for the growth and development of crops. However, the pesticidal substances used to control weeds, pathogenic organisms, and insects will cause serious side effects and are categorized as accidental contaminants (Lena and Rao 1997). The mechanisms of the chemical control with examples are given in Fig. 10.1.

The main side effects of the usage of the chemical controls are (Xian 1989) given below:

(i) The usage of the chemical control can cause disease-resistant weeds, pests, and plant pathogens.

(ii) Persistence of many pesticides (e.g., DDT) for a long time in the environment leads to bioaccumulation, which leads to several health issues such as nerve disorder, reproductive disorder and cancer in humans as well as other living organisms.

(iii) Evolution of new and resistant pests is also possible, for example, the usage of DDT to control the insects in lemon trees causes the outbreak of insects capable of sucking the juice by attaching plant.

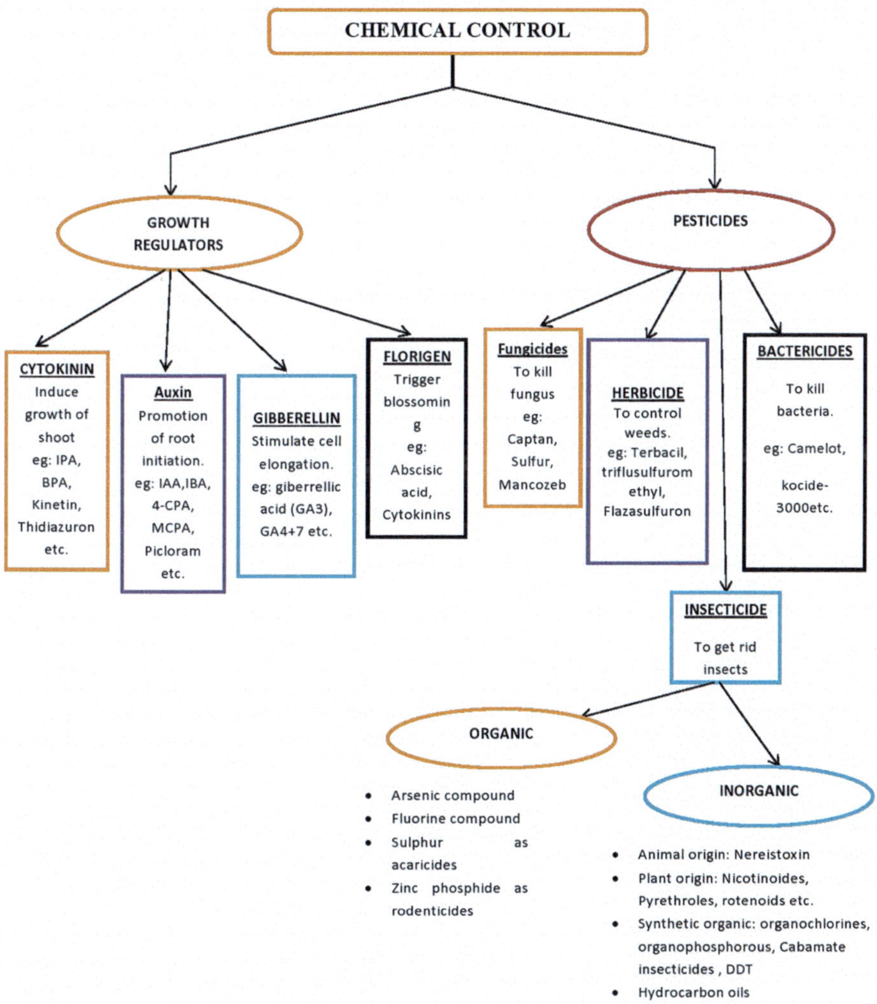

Fig. 10.1 Classification and examples of chemical control. (Okwute 2012)

10.3 Biological Control of Pests and Disease Management in Agriculture: Its Advantages and Disadvantages

Biological control is the process of usage of naturally occurring or genetically modified organism or gene products for the reduction of disease and pests. The biological control are mainly classified into three methods, which are as follows (Baker 1985):

(i) Classical approach (pest population regulation)
(ii) Exclusionary system (microbial barrier for protection)
(iii) Self-defense system (immunization and resistance)

The major biological control agents used in agriculture are avirulent strains of pathogens, natural enemies/antagonists, and immunized plants that defend it. The applications of the biological controls for pests are given in Table 10.1. The integrated crop and pest management for plant protection obey eight principles and are given below (Baker and Cook 1974):

(a) Understanding of the agroecosystem production limits
(b) Proper crop rotation
(c) Retaining/maintaining of the soil organic matter
(d) Usage of a hygienic planting material
(e) Selection of pest-resistant and well-adapted plant or seed for cropping
(f) Reduction of the nutritional and environmental stress
(g) Amplification of the usage of beneficial microbes
(h) Minimal usage of pesticides

Table 10.1 Application of the biological control for pests. (Thakur and Sohal 2013)

Types of controls	Biological control methods	Natural enemies/ antagonists	Mode of action in plants
Traditional method	Parasites	*Lactobacillus*	Induced resistance
Traditional method	Mixed cultivation	*P. syringae*	Host plant resistance
Traditional method	Usage of trap plant	*Sorghum bicolor* L. *Nezara viridula* L.	Volatile compound–based resistance
Traditional method	Sowing of crops in a dense manner	*Geocoris* spp. *Cryptolaemus montrouzieri*	Parasitoids
Genetically engineered plants	Modified growth habits for enhanced resistance against pathogens	*Pseudomonas fluorescens* *Bacillus thuringiensis*	Genetically modified vector

Table 10.2 Advantages and disadvantages of biological control agents. (Sharma et al. 2009)

Advantages	Disadvantages
Protection throughout the crop growth and development	It is not having the complete curative effect
There is no toxic effect to plants	The effectiveness against the plant disease is very slow
Less costly and cheaper compared with other methods	Act against only to specific disease
Specific to plant disease	Handling of biological agents requires skilled persons to prevent contamination
Spread easily in the soil	They are having short shelf life
Increase the crop yield	Biological controls are having less effectiveness than fungicides
Safe for environment as well as humans and animals	Environmental condition should be followed for the better efficiency of the biological control agents
Helps in growth and development of plants by supporting beneficial microbes present in soil	
No residual problems	
Ease of manufacturing and handling;	
it can be easily mixed with fertilizers and manures	

Application of biological control agents has both advantages and disadvantages and are given in Table 10.2.

The mechanism involved in biological control based on interspecies antagonism is of three types: direct antagonism, mixed path antagonism, and indirect antagonism (Sharma et al. 2009). Direct antagonism refers to the usage of biological control methods which act directly or in a high-degree manner against selective pathogens (Nega 2014). The mechanism of direct antagonism is a kind of hyperparasitism or predation. *Lysobactor enzymogenes* for example against fungal and bacterial diseases of Xanthomonadaceae family. Similarly, lytic or nonlytic Mycoviruses for controlling plant-pathogenic fungi such as *Macrophomina phaseolina, Colletotrichum truncatum, Rhizoctonia solani* and *Diaporthe longicolla* (Sharma et al. 2009). At the same time, indirect antagonism refers to the activities that do not affect directly the targeted pathogens. The main mechanisms involved in indirect antagonism are competition and enhancement of resistance in host. The main mechanism involved in competition includes scavenging of nutrients using siderophore, consumption of leachates, and competition for niche. The mechanism also includes induction of host resistance at molecular level, phytohormone-mediated induction, and resistance associated with direct contact of pathogens (Raaijmakers and Weller 2001).

In the case of mixed path antagonism, the mechanisms involved are based on lytic enzymes, physical/chemical inference, antibiotics, and unregulated waste products (Fig. 10.2). The pathogen-mediated diseases of plants can be treated by using glucanase, chitinases, and proteases. The pathogens of plants can also be

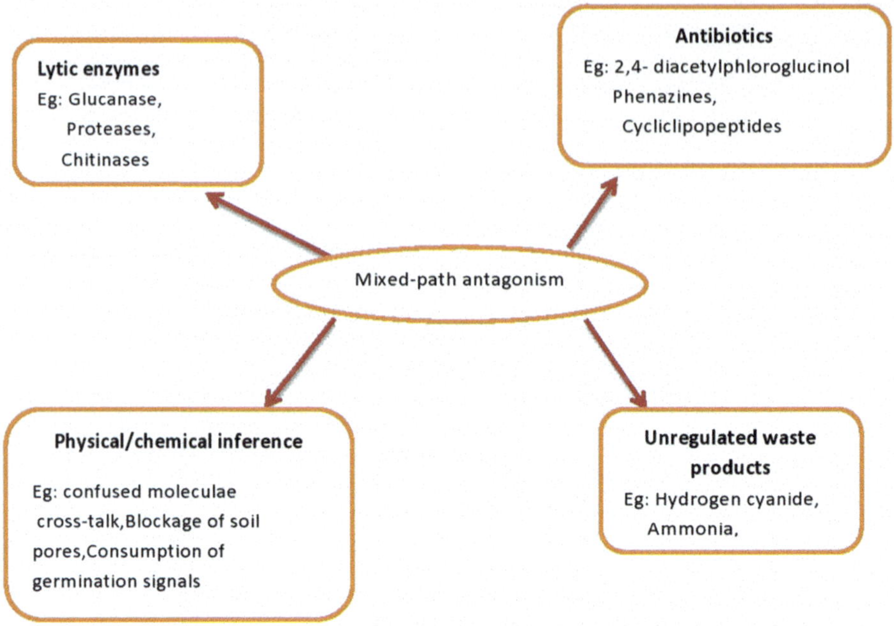

Fig. 10.2 Mechanism involved in mixed-path antagonism. (Sharma et al. 2009, Preetha et al. 2010)

treated by using antibiotics such as cyclopeptides and 2,4-diacetylphloroglucinol phenazines (Sharma et al. 2009).

10.4 Microbes as Biological Controls for the Growth and Development of the Agriculture

There are many microbes which will help in the growth and development of the plants. The introduction of the beneficial bacteria such as *Rhizosphere*, *Bacillus*, *Pseudomonas*, *Arthrobacter* and *Serratia* has been proven that can be able to control the fungal disease (Saleem and Kandasamy 2002). For example, *Pseudomonas fluorescens* produces antimicrobial polyketides against fungal pathogens (*Pythium species*) of cucumber species (Girlanda et al. 2001). The *Bacillus* species is having more notable capabilities to control soilborne fungal pathogen, stimulation of plant growth, and control wilt disease due to *Rhizoctonia solani* in brinjal and also having the vitro and in vitro antagonism property against fungal pathogens such as *Fusarium semitectum* and *Fusarium Solani* (Dekaboruah and Dileep Kumar 2003). Saleem and Kandasamy (2002) reported that the *Bacillus* species is having the capability to control the *Curvularia lunata* causing leaf spot disease and grain mold in dryland crops such as sorghum. Similarly, the wilt of radish can also be controlled by the help of *Pseudomonas strains*, thereby improving the growth of the plants.

Bacillus thuringiensis (BT) is the most popular bacterial pesticide which covers 90% of the total share of bioinsecticide as well as biopesticide market (Baum et al. 1999). The mechanism involved in BT is the presence of an insecticidal crystal protein which has toxic effect to the insect species (Milner 1994). For example, beetle larvae destroy the leaves of the agricultural crops which can be destroyed by BT protein. Immediately after the ingestion of the BT crystal proteins from the treated leaves, the insects will stop feeding. After a few minutes, crystal protein (endotoxin) hydrolyzes in the insect's gut and results in blood poisoning. Within the time gap of 1–3 days, the death of the insects will take place (Milner et al. 2003). The main drawback of the BT is the repeated exposure leading to the emergence of the resistant pests due to repeated exposure.

There are many virus-based biological control agents which can affect more than 1100 species of mites and insects. *Baculovirus* are the rod-shaped viruses having DNA and are coated with protein-based inclusion body (Zvereva and Pooggin 2012). The mechanism of killing insect involves rapid multiplication in epithelial cells and dissolving the protein in insects midgut and the release of the viral particles from the inclusion body (Deshpande 2000). Most of the virus-based biological control is host-specific, and they do not have the ability to attack different species of pests. Compared to bacterial control agents, viral biological control agents are slow and are not stable under ultraviolet rays from the sun. The advantages of the usage of the virus-based biological controls are that they do not cause any threat to the wildlife as well as humans. The productions of the virus-based biological control are very costly; therefore, commercialization has been limited (Possiede et al. 2009). Some commonly used varieties of virus-based biological agents in agriculture are *Nuclear Polyhedrosis Virus* (NPV) and *Granulosis virus*. NPV has the ability to affect corn earthworm, cabbage loppers, alfalfa looper, cabbageworm, cotton leaf worm, tobacco, cotton bollworm, army worm, gypsy moth, pine saw fly, and almond moth.

The *Granulosis virus* is used to control cabbage looper, fall webworm, and mosquitoes. *Granulosis virus* was isolated from caterpillar species. The devastation of the mites/insects by the *Granulosis virus* results in white color, and then the insect stops feeding the plant/crop, and finally the larvae body contents get liquefied. Finally, the cuticle ruptures and thereby releases viral particles. This process will take place within 8 days and collapse the entire host population (Hoffmann and Frodsham 1993).

10.5 Fungal Biological Control Agents

Fungi are also used in agriculture as biological control agents. The devastation of the fungi to the insects/pests occurs through the cuticle and which is attributed with the help of hydrolytic enzymes such as lipases and chitinase (Patel et al. 2014). The destruction of the agricultural pests/insects by fungal biological control agent involves mainly two mechanisms such as mycoparasitism and entomopathogenic. Mycoparasitism is the interrelationship between the fungus parasite and host fungus. And entomopathogenic fungi are the fungi that act as parasites which are capable of killing the insects and pests. *Trichoderma* and the *Gliocladium* are the two well-reported fungi capable of controlling fungi pathogens in plants by mycoparasition (Yoon et al. 2013). They are having the ability to produce toxic metabolites against the host fungi pathogens. The *Gliocladium* produce chemicals which bind with the host fungus spores/mycelium and destruct the spores/mycelium (Shuping 2016). In the case of *Trichoderma* which attaches the host spores by hook/coiling and penetrates into host by secreting cell wall lytic enzymes (Schafer 1994).

10.6 Biomolecules as Biological Control Agents

Nowadays, there is a great demand for biopesticides due to better awareness about higher environmental risks of synthetic pesticides. Biomolecules such as oligochitosan and chitosan are well-documented biological control agents, due to their biodegradability, biocompatibility, and the nontoxic nature. Chitosan is the second most abundant biopolymer which controls pathogenic microorganisms by preventing growth/sporulation and reducing spore viability/germination (Vander 1998). Chitosan also disrupts host cells and induces defense responses in different host plants by inhibiting and/or inducing different kinds of biochemical interactions during the pathogen attack in plants (Vander 1998). Chitosan has been used for different post- and preharvest disease managements. Moreover, chitosan can enhance biodiversity in the rhizosphere (Gupta et al. 1992). Chitosan has an important role in protecting the plant from various diseases, and the mechanism of the chitosan in reducing the plant disease is given in Fig. 10.3 (Schickler and Chet 1997).

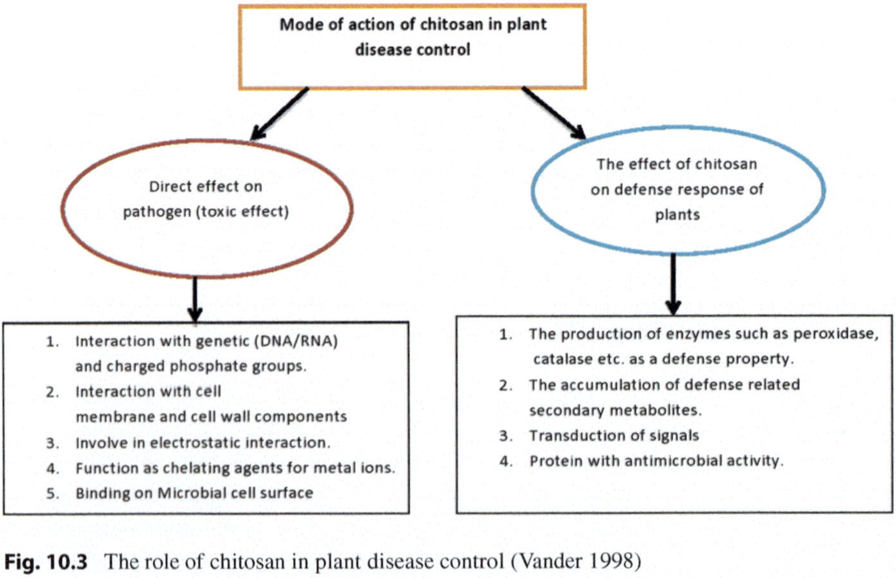

Fig. 10.3 The role of chitosan in plant disease control (Vander 1998)

However, environmental impact of chitosan is not evaluated for sustainable agricultural practice.

Chitosan controls pathogenic microbes by reducing viability, growth, sporulation, and germination of spores. Chitosan also disrupts pathogen by enhancing defense responses in host plant during the plant–pathogen interaction (St. Leger et al. 1991). As per the reports, the plant can also acquire the improved control to the stressful growth condition after the successful application of the chitosan (Mathivanan et al. 1998). The application of the beneficial microorganisms along with the chitosan is an efficient biological control method in integrated pest management tool in agriculture (Mendonsa et al. 1996).

10.7 Biological Control by Using Plant Extracts

The antifungal activity of the plant extracts has an impact on the growth and protection of plant. Some plants have the ability to protect themselves from the phytopathogens (Martinez 2012). The plant extracts from neem (phytoanticipins) and *Capsicum frutescens* (phytoalexins) have antimicrobial activity (Ribera and Zuniga 2012). The plants that have the aforementioned properties are able to prevent the growth of the fungi pathogens (Masoko and Eloff 2005). Since plant extracts contain different concentrations of various antifungal compounds, the chances of developing resistance in pathogens are very rare. For example, the natural extract/compound obtained from *Citrus paradisi* contains 7-geranoxy coumarin having the in vivo and in vitro antifungal property against *Penicillium digitatum* and *Penicillium italicum* (Chitarra et al. 2004). In another study, the shelf life of the orange was extended up to 6 days due to the usage of the natural plant extract as biological control (Tripathi and Dubey 2004). Other antifungal compounds from plant extracts for biological control of agricultural pathogens are extensively discussed in Table 10.3 (Mahlo et al. 2010).

10.8 Current Scenario and Future Aspects in Biological Control in the Agriculture

Nowadays, the food security and well-functioning ecosystem are beaming questionable due to overuse of chemical control agents. At this context, many researchers are concentrating on biological methods for individual pest control or for protection of a particular crop (Charles et al. 2016). To retain sustainability, agricultural biodiversity plays an important role. The efficient usage of the biological control contributes economic benefits, and its role in control of naturally occurring parasitoids, predators, and pathogen has been studied by (Power 2010). The usage of the biological control system makes the ecosystem a self-balancing system since the existing natural enemies of agricultural pathogens can protect agricultural crops from pests in a natural way (Landis et al. 2000).

Table 10.3 Inhibiting property of various plant extract as fungicide

Natural plant extracts	Organism affecting (pathogen)	References
Citrus paradisi	*P. italicum*	Tripathi and Dubey (2004)
	P. digitatum	Mahlo and Eloff (2014)
Alchornea cordifolia, *Annona muricata*, *Allium sativum*	*Botryodiplodia theobromaa*	Gurib-Fakim (2006)
Kaempferol from *Acacia nilotica*	*P. italicum*	Ribera and Zuniga (2012)
Lawsone from *henna*	*Fusarium oxysporum f.* sp. *Melonis*	Ribera and Zuniga (2012)
Naphthoquinones from *Eleutherine*	*Cladosporium sphaerospermum*	
Sesquiterpene and *zerumbone*	*R. solani*	Okwute (2012)
Bucida buceras	*Trichoderma harzianum*	Mahlo et al. (2010)
	P. janthinellum	
	Fusarium oxysporum	
	Penicillium expansum	
Breonadia salicina	*Penicillium* species	Mdee et al. (2009)
Campuloclinium macrocephalum leaf extract	*Colletotrichum gloeosporioides*	Mdee et al. (2009)

The importance in the self-balance of the ecosystem and natural food chain/food web was studied by Kean et al. (2003) for sustainable agriculture. He has also suggested maintenance of habitat quality by increasing natural resources, application of selective chemicals, and temporary usage of pesticides to achieve sustainable environment for balanced ecosystem.

There are many studies such as CRISPR (clustered regularly interspaced short palindromic repeats) and BINGO (Breeding Invertebrates for Next Generation BioControl) as an initiative to improve the production of biological control agents for improved plant growth and development. CRISPR was introduced by Yoshizumi Ishino and his colleagues (Osaka University) in 1987, and BINGO was introduced by Marie Sklodowska Curie Innovative Training Network for the improvement of the performance and production of biological control. In both the initiatives, it seeks augmentative biocontrol by genomic techniques for the enhanced and improved production/performance of the biological control agents (Pannebakker and Beukeboom 2016). The biological control agents are mainly used to bring down the pest population in an economical way (Barratt 2011). However sometime, the usage of biological control itself could make parasitoid attack (Barlow et al. 2004).

Recent development in the field of agriculture, ecology, and taxonomy contributes a better understanding of function and implementation of biological control methods and also provided the information regarding the natural way of disease control such as application of bactericides for better production in the agriculture sector.

10.9 Conclusion

The chapter discussed the possibilities of the usage of biological control methods for the growth and development of agriculture. The usage of microbes such as bacteria, fungi, and virus for biological control reduces the usage of the chemical control agents and thereby improves the sustainability of ecosystem in natural way. Biomolecules such as chitosan and oligochitosan are also capable of promoting the biological control of the pathogenic microbes in agriculture. The antifungal activities of the plant extracts against agricultural pathogens which affect the growth of the plants are also gaining attention nowadays. Moreover, biological control methods are more economical, and hence it is affordable for farmers and also help to achieve sustainability in agriculture sector.

References

Patel N, Desai P, Patel N, Jha A, Gautam HK (2014) Agronanotechnology for plant fungal disease management: a review. Int J Curr Microbiol App Sci 3(10):71–84

Ayansina ADV, Oso BA (2006) Effect of two commonly used herbicides on soil microflora at two different concentrations. African J Biotech 5:129–132

Baker KF, Cook RJ (1974) Biological control of plant pathogens. W. H. Freeman and Co, San Francisco, 433 pp. (Book, reprinted in 1982, Amer. Phytopathol. Soc., St. Paul, Minnesota)

Baker R (1985) Biological control of plant pathogens: definitions. In: Hoy MA, Herzog DC (eds) Biological Control in Agricultural IPM Systems. Academic Press, New York, pp 25–39. 600 pp

Barlow ND, Barratt BIP, Ferguson CM, Barron MC (2004) Using models to estimate parasitoid impacts on non-target host abundance. Environ Entomol 33(4):941–948

Barratt BIP (2011) Assessing safety of biological control introductions. CAB Rev:Perspect Agric vet Sci Nutr Nat Resour 6(042):1–12

Baum JA, Johnson TB, Carlton BC (1999) *Bacillus thuringiensis* natural and recombinant bioinsecticide products. In: Hall FR, Mean JJ (eds) Methods in biotechnology-biopesticides: use and delivery, vol 5. Humana Press, Inc., Totowa, pp 189–209

Charles JG, Gardner-Gee R, Hunt S, MacDonald F, Davis V (2016) Preemptive assessment of the biosafety of a classical biocontrol agent: responses of Trissolcus japonicus to New Zealand's native Pentatomidae prior to the arrival of BMSB. Paper presented at the international congress of entomology, Orlando, FL

Chitarra GS, Abee T, Rombouts FM, Posthumus MA, Dijksterhuis J (2004) Germination of *Penicillium paneum* conidia is regulated by 1-octen-3-ol, a volatile self-inhibitor. Appl Environ Microbiol 70(5):2823–2829

Dekaboruah HP, Dileep Kumar BS (2003) *Rhizoctonia* wilt suppression of brinjal (*Solanum melongena* L) and plant growth activity by *Bacillus* BS2. Ind J Exp Biol 41(6):627–631

Deshpande MV (2000) Mycopesticide production: contribution of fungal morphologies in the biopesticide formulations. Nat Bot Soc 55:11–17

Girlanda M, Perotto S, Moenne-Loccoz Y, Bergero R, Lazarri A, Defago G, Bonfante P, Luppi AM (2001) Impact of biocontrol *Pseudomonas ûuorescens* CHA0 and a genetically modiûed derivative on the diversity of culturable fungi in the cucumber rhizosphere. Appl Environ Microbiol 67:1851–1864

Gupta SC, Leathers TD, El-Sayed GN, Ignoffo CM (1992) Insect cuticle-degrading enzymes from the entomogenous fungus *Beauveria bassiana*. Exp Mycol 16:132–137

Hirai T, Kimura K (1979) Resistance to blast disease of rice plants grown in the non-manured and chemicals non-applied paddy field for long period. Mem Fac Agric Kinki University 12:189–194

Hoffmann MP, Frodsham AC (1993) Natural Enemies of Vegetable Insect Pests. Cooperative Extension, Cornell University, Ithaca 63

Kean JM, Wratten SD, Tylianakis JM, Barlow ND (2003) The population consequences of natural enemy enhancement, and implications for conservation biological control. Ecol Lett 6:604–612

Landis DA, Wratten SD, Gurr GM (2000) Habitat management to conserve natural enemies of arthropods pests in agriculture. Annu Rev Entomol 45:175–201

Lena QM, Rao GN (1997) Chemical fractionation of cadmium, copper, nickel, and zinc in contaminated soils. J Environ Qual 26:259–264

Mahlo SM, Eloff JN (2014) Acetone leaf extracts of *Breonadia salicina* (Rubiaceae) and ursolic acid protects oranges against infection by *Penicillium* species. S Afr J Bot 93:48–53

Mahlo SM, McGaw LJ, Eloff JN (2010) Some tree leaf extracts have good activity against plant fungal pathogens. Crop Prot 29:1529–1533

Martinez JA (2012) Natural fungicides obtained from plants, fungicides for plant and animal diseases, Dr. Dharumadurai Dhanasekaran (Ed.), ISBN. 978–953. 307–804-5

Masoko P, Eloff JN (2005) The diversity of antifungal compounds of six south African *Terminalia* species (Combretaceae) determined by bioautography. Afr J Biotechnol 4:1425–1431

Mathivanan N, Kabilan V, Murugesan K (1998) Purification, characterization and anti-fungal activity from *Fusarium chlamydosporum*, a mycoparasite to groundnut rust, *Puccinia arachidis*. Can J Microbiol 44:646–651

Mdee LK, Masoko P, Eloff JN (2009) The activity of extracts of seven common invasive plant species on fungal phytopathogens. S Afr J Bot 75:375–379

Mendonsa ES, Vartak PH, Rao JU, Deshpande MV (1996) An enzyme from *Myrothecium verrucaria* that degrades insect cuticles for biocontrol of *Aedes aegypti* mosquito. Biotechnol Lett 18:373–376

Milner AM (1994) Colonization and succession of invertebrate communities in a new stream in Glacier Bay National Park, Alaska. Fresh Bio 32:387–400

Milner R, Samson P, Morton R (2003) Persistence of conidia of *Metarhizium anisopliae* in sugarcane ûelds. Biocontrol Sci Tech 13(5):507–516

Nega A (2014) Review on concepts in biological control of plant pathogens. J Biol Agric and Healthc 4(27):33–35

Okwute SK (2012) Plants as potential sources of pesticidal agents: a review, pesticides – advances in chemical and botanical pesticides. In: Soundararajan RP Dr, editor. InTech https://doi.org/10.5772/46225

Pannebakker B, Beukeboom LW (2016) BINGO: Breeding invertebrates for next generation biocontrol. Paper presented at the XXV international congress of entomology, Orlando, Florida, pp 25–30

Possiede YM, Gabardo J, Kava-Cordeiro V, Galli-Terasawa LV, Azevedo JL, Glienke C (2009) Fungicide resistance and genetic variability in plant pathogenic strains of *Guignardia citricarpa*. Braz J Microbiol 40:308–313

Power AG (2010) Ecosystem services and agriculture: tradeoffs and synergies. Philos Trans R soc B Biol Sci 365(1554):2959–2971

Preetha R, Seena J, Prathapan S, Vijayan KK, Jayaprakash NS, Philip R, Singh ISB (2010) An inhibitory compound produced by *Pseudomonas* with effectiveness on *Vibrio harveyi*. Aquac Res 41:1452–1461

Raaijmakers JM, Weller DM (2001) Exploiting genotypic diversity of 2,4-diacetylphloroglucinol-producing *Pseudomonas spp.*: characterization of superior root-colonizing *P. fluorescens* strain Q8r1-96. Appl Environ Microbiol 67(6):2545–2554

Ribera AE, Zuniga G (2012) Induced plant secondary metabolites for phytopathogenic fungi control: a review. J Soil Sci Plant Nutr 12(4):893–911

Saleem B, Kandasamy (2002) Antagonism of *Bacillus* species (strain BC 21) towards *Curvu larialunata*. Curr Sci 82(12):1456–1457

Schafer W (1994) Molecular mechanisms of fungal pathogenicity to plants. Annu Rev Phytopathol 32:461–477

Schickler H, Chet I (1997) Heterologous chitinase gene expression to improve plant defense against phytopathogenic fungi. J Industrial Microbiol and biotechnol 19:196–201

Shaikh NF, Gachande BD (2015) Effect of organic bio-booster and inorganic inputs on rhizosphere Mycoflora population and species diversity of wheat. Int Jour of Sci and Res 4:295–302

Sharma RR, Singh D, Singh R (2009) Biological control of postharvest diseases on fruits and vegetables by microbial antagonists: a review. Biol Control 50:205–221

Shuping DSS (2016) Development of an antifungal product from *Melianthus comosus* (Melianthaceae) that can be used to control plant fungal pathogens. PhD thesis, University of Pretoria, Pretoria, South Africa

St. Leger RJ, Cooper RM, Charnley AK (1991) Characterization of chitinase and chitobiase produced by the entomopathogenic fungus *Metarhizium anisopliae*. J Invertebr Pathol 58:415–426

Tripathi P, Dubey NK (2004) Exploitation of natural products as alternative strategy to control post-harvest fungal rotting of fruits and vegetables. Postharvest Biol Technol 32:235–245

Vander P (1998) Comparison of the ability of partially acetylated chitosans and chito oligosaccharides to elicit resistance reactions in wheat leaves. Plant Physiol 118:1353–1359

Xian X (1989) Effect of chemical forms of cadmium, zinc, and lead in polluted soils on their uptake by cabbage plants. Plant Soil 113:257–264

Yoon M-Y, Cha B, Kim JC (2013) Recent trends in studies on botanical fungicides in agriculture. Plant Pathol J 29(1):1–9

Zvereva AS, Pooggin MM (2012) Silencing and innate immunity in plant defense against viral and non-viral pathogens. Virus 4:2578–2597

Natural Biological Products from Plants as Rodenticides

11

Jatinder Singh and Anis Mirza

Abstract

Infestation by rodent is considered as one of the main pest problems since it not only affects health but also causes serious damages to agricultural fields and households including transportation business. Numerous methods and approaches to control rodent infestation are being tried, such as environmental, cultural, mechanical, and chemical methods including their combination. It is also possible that in biological control methods the use of reproduction inhibitors or rodent predators or diseases may ultimately be developed for rodent control purposes. Moreover, different kinds of chemicals that are used to control rodents are harmful to mammals particularly human beings. Such control also adds cost to the approach. Due to probable toxicity of the chemical compound, various alternatives like natural extracts should be considered. Hence, the use of plant natural extracts as a rat repellent practice may be an improved alternative approach. Various natural extracts are experimented on different rat species in a behavioral mode of study. These kinds of natural products are easily available in present society and have no adverse effect on environment and mammals. The rodenticidal effects of many plant extracts solved by various chemical substances could be deliberated upon under various conditions, and that has given very satisfactorily outcome. It is very possible to implement these results in problematic areas. We can, therefore, accomplish that such natural extracts can deter the rodents. Still, some more studies need to be carried out to see whether these extracts are of practical use.

Keywords

Rodents · Rodenticides · Biological products · Pesticides · Plant extracts

J. Singh · A. Mirza (✉)
Department of Horticulture, School of Agriculture, Lovely Professional University, Phagwara, Punjab, India
e-mail: anis.19474@lpu.co.in

J. Singh, A. N. Yadav (eds.), *Natural Bioactive Products in Sustainable Agriculture*, https://doi.org/10.1007/978-981-15-3024-1_11

11.1 Introduction

Rodents were killed by rodenticides which are considered as pesticides. Rodents are not rats and mice but also include squirrels, woodchucks, chipmunks, porcupines, nutria, and beavers. Rodents are found in nature and play a very important role and may require control. They can damage crops, violate housing codes, transmit disease, and also cause ecological damage. Using rodenticides for controlling rodents is a communal approach; however, most of rodenticides are toxic to human being. For example, on a ship, these rodents (rats) can cause much damage to food and cargo. Moreover, droppings of these rodents contain microorganism that causes various diseases (WHO 1988). It is estimated that the total cost of devastation by rats in the United States may be as high as $19 billion per annum (Pimentel et al. 2005), while in India, information accessible on economic losses including the damage due to rodents in various crop sectors, forestry and horticulture, poultry farms, and urban and rural residences and storage services exhibited that enduring damage ranging from 2% to 15% continued throughout the country, and sometimes up to 100% loss of the field crop was also observed in severe damage (Parshad 1999). Many approaches are practiced to control this infestation such as biological, mechanical, environmental, cultural, and various chemical methods. Various techniques are beneficial under different situations; like in a ship, mechanical method (trapping) is a decent method to keep down the rat number and is more effective and practical than other techniques (WHO 1988), whereas the use of chemicals to check rodent's population is the better methodology in South Asia country (Parshad 1999). Various rodenticides can be categorized into fast-acting and slow-acting rodenticides (WHO 1988). Anticoagulants type of rodenticides like diphenadione and warfarin must be swallowed for several succeeding days before they act effectively. At the same time, bromethalin and zinc phosphide are also acknowledged as acute rodenticide types and often destroy particular rodents with a single dose. Nevertheless, most rodenticides are noxious to human beings and their effectiveness depends upon the assortment of a suitable compound as bait. Moreover, the method adopted and application time are very important and need to be considered before application. One major problem of using bait is that if the rats have taken preliminary nonfatal dose, it can depress the rats from taking further bait known as bait shyness. Chemicals used are not only restricted to rodenticides, but some are repellents like thiram, copper oxychloride, cycloheximide, tributyltin, and beta-nitrostyrene that had been effectively established under laboratory conditions (Tigner 1966; Parshad 1999). Foremost factors that restrict the use of chemical repellent substances may possibly be being hazardous in management and causing food contamination, when applied on packing or individual boxes containing food items for humans (Tigner 1966). Because of possible toxicity, various alternatives like natural extracts should be deliberated. Consequently, using natural plant extracts as a rat repellent substance may be a better substitute. Numerous plant extracts can be investigated whether they could prevent a rat infestation when testing in the behavioral model. Observed substances were peppermint oil, bergamot oil, geranium oil, and wintergreen oil. They were applied either as single material or in combination with each other. The rats exhibited the maximum activity during the commencement of the night time test. It may therefore be concluded that these natural plant extracts can deter the rats as decided by rat's actions under circular open field conditions. Still, more investigations are required to evaluate whether these natural extracts are practical in actual environment. In order to assess the effectiveness of rat repellents, there are a number of laboratory assessment methods like food acceptance experiment, graded strip test, and barrier test (Weeks 1959). In food recognition test, the repellent substance is mixed with food to be provided, and the efficiency is based on the quantity of food eaten by a rat in observed period of time. Unfortunately, in this investigation, some chemical substances were able to make the food not acceptable but were unsuccessful to deter the rats. During barrier test, hungry rodents must be trained to pass through repellent-coated paper obstruction, and time required to cross the paper may be recorded. In graded strip evaluation, a rat has to pass through the paper (ribbon-like strip) coated with diverse concentration of repellent chemicals to get to food which is placed under each paper piece, and then the repellent action is considered based on numbers of obtained food items such as peanuts. For the other two tests, the rats must be trained and hungry; furthermore, the instruments must be specifically made and evaluated with carefulness. The open field condition is a behavioral model generally used for study of anxiety or locomotor activity. During this experiment, the rats are permitted to travel freely in the apparatus for a period of time, and then time spent in each segment of the machine or the number of lines crossed by rodents is evaluated according to test type. Circular open field situation is also considered for measurement of locomotor action of rats. During this study, the circular open field situation was adapted in order to perceive 24-h action of the rats exposed to different repellents. The aim of this study was, therefore, to perceive the natural performance of rats in the circular-type open field when exposed to different natural repellents or if these repellents were as effective as assessed in species of rodents.

Rodenticides are pesticides that may damage all kinds of rodents. These are formulated as baits with some attractive ingredients like peanut butter, molasses, etc. These types of baits may provide short-time control of rodent. Human beings, their pets, and wildlife are analogous to rodents, so they may also be harmed by such rodenticides. Although rodents play vital roles in nature, they may require some kind of regulation. Rodents especially rats and mice can harm the crops, spread various diseases, and also cause ecological damage.

Rodents are also mammals like human beings, dogs, and cats. Rodenticides have the similar kind of effect when eaten by any mammal. Their effect on birds is very important and significant. Usually various rodenticides are formulated as poisonous baits, which are deliberated to attract animals. Additives may include peanut butter, molasses, or fish oil. Especially baits that are used in agriculture or natural areas may contain vegetables, grains, fruits, or ground meat. Being flavored, they may be attractive to children and pets, so keep them out of their reach, or use them in area which is difficult to approach. So keeping the above in view, focus has definitely shifted from the prevailing use of one chemical to new concept called integrated pest management (IPM). In other words, we can say that the focus is on biological control, and other natural resources have increased with reduced dependency on chemicals (Schmutterer 1981). Since the ancient times, different plant extracts, having pesticidal characteristics, have been used by humans that continues to the modern time (Fellows 1979). Use of such pesticides is more common in developing countries, where plants are grown frequently and are cheaper than the man-made chemical pesticides (El-Gengaihi et al. 1997). Various plants have been studied and evaluated for pesticidal properties, and many of them have been reported to be active. These plants were phytochemically examined to decide their chemical compositions. The cardenolides, steroidal glycosides, etc. are deliberated as the most important of all the naturally happening products. During ancient times, extract from these plants was used for arrow poisoning or as drugs (Hussein 1991).

11.2 Damage

It is reported by Nowak (1999) that rodents is composed of 1400 species and it is the major and largest group in taxonomic study containing creatures. Their territories are common but of different types. Some rodent kinds are adaptable and comparatively small but a prolific, mysterious one. They have constantly increasing incisors which require constant corroding by nibbling activity. It is supposed that they have productive potential, though there is substantial inconsistency between different species regarding the age during first reproduction phase, the number of litters per year, and size of litters. Various rodent species have economic, social, ecological, and scientific uses. They reutilize various minerals and nutrients, distribute crop especially grain seeds and also spores, aerate fields, and influence plant activities. Rodents provide meat and fur-like items. Some rodent kinds are used in medical science for experiment purpose. Furthermore, they are used as prey base for many predatory species. Worldwide, less than 5% rodent species are considered as serious pests. Prakash (1988) and Witmer and Singleton (2012) published a list of their genera and species. Many health and economic complications can result from interaction of rodent and human beings. It has been reported by Marsh (1988) and Witmer and Singleton (2012) that rodents can damage standing agricultural crops and stored grains, orchards and forests, property (structures, cables), rangelands, and natural flora and faunal. It was assessed that in Asia alone, the extent of grains damaged by rodents can provide necessary nourishment to 200 million individuals for 365 days (Singleton et al. 2003). Singleton et al. (1999) and Witmer and Singleton (2012) described that to control rodent species which causes the damage, various managing strategies are employed with different aspects of biotic–abiotic–-cultural with level of the damage and involved rodent population. Such kind of impairment activities can be insane when the population of rodents occur thoroughly (Singleton et al. 2010). Generally, the commensal rodent species include Norway rat (*R. norvegicus*), black rat, Polynesian rat (*Rattus exulans*), and house mouse. Rodents live in adjacent vicinity to human beings and are influenced by the favorable conditions. Rodents (rats) have widely spread throughout different areas in the world and lead to lot of damage to warehoused foods, etc. Furthermore,

Angel et al. (2009) and Witmer and Pitt (2012) described that they have also been particularly harmful to limited ecologies when introduced, mostly accidently to dissimilar islands. Rodenticides have been completely depended upon time factor, regardless of the several methods accessible to reduce rodent population or damage caused by them (Witmer and Eisemann 2007; Witmer et al. 2007). Nevertheless, there has been emerging concern about the weaknesses posed by most of rodenticides, particularly about the toxicity and perseverance in tissues in case of anticoagulant type of rodenticides (Pelz 2007; Eisemann et al. 2010; Rattner et al. 2012, 2014a, 2014b; Proulx 2014; Nogeire et al. 2015; Pitt et al. 2015). The purpose of this chapter is to detail various alternative techniques to anticoagulant rodenticides for the monitoring of rodents' populations and damage caused due to their activities. Both toxic and nontoxic methods are documented, and the value of an integrated pest management (IPM) technique or rodent management approach based on ecological system is also deliberated. There are small numbers of rodent species which are creating problems, and they require some kind of particular management. According to Singleton et al. (2007), only 4–5% of rodent types pose a noteworthy risk to humans' globally. Different species may be included in commensal rodents, such as rice field rats, black rats, Norway rats, and house mice (*Mus musculus*), as well as field rodents including the lesser bandicoot rat (*Bandicota bengalensis*) that can cause much damage and multimammate mice (*Mastomys* spp.) (Buckle and Smith 2015). Under some situations, reasonable damage only occurs when enormously overplentiful populations build up during population occurrences (Delattre et al. 1992; Singleton et al. 2007) and in the tropics (Doungboupha et al. 2003; Leirs et al. 1997). Adversative effects of rodents are many and include impairment of preharvest crop and postharvest damage to stored grains including infrastructure. Singleton et al. (1999) described that preharvest rodent damage is predominantly common in Africa and Asia. Such kind of preharvest losses in Asia led to reduction in yield of 5–20% in rice crop (Singleton et al. 2003). It is equivalent to an annual loss of 77 million tons of food or so (John 2014), and above all, this is sufficient food to feed 200 million people for duration of a year (Singleton et al. 2003). Epidemic of bamboo rats in Asian continent has more dramatic influences as competition for food between rodents and people and livestock may lead to widespread famine (Singleton et al. 2010). When stored products are destroyed or contaminated by commensal rodents, postharvest kind of losses occur. Furthermore, these rodents contribute to health complications for human beings and diseases to livestock including companion animals like zoonotic diseases. They have various kinds of pathogens which include viruses like hepatitis E virus, hantavirus, and tick-borne encephalitis virus; bacteria (e.g., Borrelia, Rickettsia, and Leptospira); and parasites like toxoplasmosis and various infections like echinococcus infection and giardia infection (Meerburg et al. 2009). The cost accompanying the spread of various rodent-borne diseases is supposed to be analogous to the losses due to rodents in crop production (Bordes et al. 2015). Annual cost of preharvest and postharvest rodent damage, losses of stored goods, and expenditure for disease inhibition and management is likely to surpass US$23 billion (Jacob and Buckle 2018).

Due to the ill effects of rodents on crop production and human health, rodent control is serious in some cropping systems and under urban conditions (Buckle and Smith 2015). Though large-scale strategies of rodent controlling in an agricultural environment include practices that reduce habitat temporally or are unsuitable for rodents (e.g., plowing) and adaptation of biocontrol (e.g., promoting predation), they rely largely on the application of rodenticides to decrease damage to crops. For the controlling of commensal rodents in rural and urban areas, anticoagulant rodenticides often are the weapon mostly used, though the increased resistant populations pose a massive challenge to future approaches. Main difficulty in using poison such as rodenticide is the danger to nontarget species. It can be either by direct consumption of such poisonous bait or indirectly via the ingestion of poisoned prey or meat. This kind of report regarding secondary poisoning which is posed by anticoagulants is frequent worldwide (Eason et al. 2002; Geduhn et al. 2015) that is very important from conservation and biodiversity point of view. The problems related with the use of such poison to manage rodent populations and efforts to increase effectiveness of rodent control have concluded in a call for ecologically based rodent management (Singleton et al. 1999) that involves the use of a toolbox of control methods that are socially, economically, and ecologically suitable.

11.3 Aspects for Development of Natural Pesticides

11.3.1 Novelty

In current years, the prominence on plant protection has shifted from the prevailing chemical pesticides to the integrated pest management; the emphasis is on biological control techniques and other natural kind of resources with reduced dependence on chemical pesticides (Schmutterer 1981). Many research workers recommended that the elementary research must be focused toward the discovery of safe types of pest control approach in order to guarantee high production and conservation of various agriculture products (Schmutterer 1981; Saleh et al. 1986; Qureshi et al. 1991; Afifi et al. 1992; Oji et al. 1994). Different plant extracts have been used as natural pesticides by human beings since before the time of the prehistoric Romans, and this practice continues to the present with several plant species with pesticidal properties (Fellows 1979). Using toxic plants in this context is especially predominant in the developing countries, where plants are grown locally and are cheaper than synthetic pesticides of chemical nature (El-Gengaihi et al. 1997; Gabr et al. 2004). Rodent pests are governed by chemical compounds that cause threats to the health and pollution of the environment and ultimately led to the toxic effect to nontarget species. These kinds of problems provided a motivation to get natural poisonous materials that could be used against rodenticides. Some plant materials like extract of oshar leaves and abamectin (Vertimic biocide) exhibited to be promising in rodent control (Gabr et al. 2004). The secondary metabolites of some plants are good basis for compounds which have wide range of biological events/actions. This kind of variety is largely the outcome of coevolution of plant species with altogether including numerous types of microorganisms and animals. Therefore, secondary compounds from several plants are assumed to have guarantee along with biological activity in plant protection against particular competitor or herbivore pathogen. In this phase, information regarding species, habit, nature, etc. of the harmful pest may provide valuable recommendations in forecasting that what can help to regulate these pests. Such kind of approach has resulted in the innovation of numerous marketable pyrethroid, natural pesticides, etc. Assortment of such compounds and chemical details of plant extracts having vigorous biological actions can be a primary step to compare or synthesize a new artificial compound. Nevertheless, the affirmation of biological action and more development in procedures of contamination removal including structural documentation are ever-changing the odds in courtesy of normal materials.

Bearing this in mind, the possibility of secondary natural plant products is being used in plant–pest relations; during interaction of plant–pest, the possibility of plant secondary metabolite is of main and primary significance; the technique of casually isolating, classifying, and bioassaying these constituents may also be an imperative technique of pesticide discovery. These extractions from the specific plants will show activity against attacking pathogen and help the plant to govern adverse effects due to pest which is responsible for the attack. In spite of these numerous plant secondary metabolites present in natural materials, screening is done properly for such pesticidal activity.

The discovery procedure for these kinds of natural pesticides is more complex than for synthetic one. Conservatively, new chemical has been developed (pesticides/insecticides) by synthesis method, bioassay, and assessment. If the end product is suitably encouraging one, assessable structure–action–relationship-based synthesis of equivalents is used to advance necessary pesticidal characters. The finding of such expected products is intricate in nature due to many issues.

First of all, the decontamination method is performed. It is supposed to be variable for most of the plants, as there is no hard and fast rule for such plants. Moreover, from secondary metabolites, different compounds are assessed in comparatively smaller volumes compared to the quantities of synthesized substances for broadcast pesticide activity. Consequently, bioassays necessitating very less quantities of material will be beneficial in screening of the natural compounds. In the allelochemical field, various methods are available for assaying less quantities of compounds for biological and pesticidal activities that are accessible. In discovery method, structural identification is a foremost and prime requirement, while in certain cases, it is a very problematic process. Ultimately, synthesis along with equivalents must be reflected which is somewhat more complicated. Irrespective of such difficulties, recent contributory analysis and upgraded approaches are lessening the price, levels of complicity, and requisite time in such approaches.

11.3.2 Development

There are limited pesticide compounds that are established as decidedly effective in assessment and are constantly brought to bazaar. Several issues must be deliberated in the pronouncement to improve and market such pesticide. A primary deliberation is the patentability of such compounds. A patent exploration must be done for all these natural compounds as with any synthetic one. Earlier publication of a compound with pesticidal properties could pose patent problems. Paralleled to synthetic materials, there is a lot of available knowledge on the biological activity of these natural products. Due to this purpose, patenting synthetic equivalents without reference of the natural source might be harmless than patenting the natural produce under some circumstances. Toxicological and environmental traits of all compounds must be deliberated. Simply because of this reason that a compound is a natural product and does not confirm that it is harmless. The utmost toxic mammalian poisons are also known as natural materials, and most of these are produced by plants. Introduction of different intensities of toxic natural substances into the environment that would never be established in nature could cause adversative influences. But there is solid evidence that natural products generally have a petite half-life in the environment in comparison to synthetic ones. Actually, the comparatively brief environmental perseverance of such natural compounds can be a problem, because many chemicals must have some kind of lasting activity in order to show some effectiveness. As with pyrethroid compounds, chemical alteration can increase persistence process. After favorable biological action is discovered, extraction of more amounts of the compound for further wide-ranging bioassays can be deliberated. Likewise, equivalents of such compound should be synthesized by modification of the same. Structural management of such compound could result in development of useful action, transformed environmental influences, toxicological properties, or finding such products that can be frugally produced. This may be the case with various natural-occurring compounds that have been used as a source for many commercial compounds, that is, for pesticides (e.g., pyrethroids). Before a conclusion is drawn to synthesize a natural pesticide compound for commercial application, profitable resources including means of manufacture must be assessed. Though this is a decisive question in deliberation of the advancement of any pesticide, it may be much complicated and thought of in case of natural products. Factually, preparations of simple natural product combinations have been used as pesticide chemicals. However, the possible complications in clearing a complicated combination of several biologically active compounds for use may be restrictive in today's governing environment. Therefore, the query that will possibly be deliberated is whether a pure compound will be formed by purification, biosynthesis process, etc., or by traditional chemical method. Beforehand, bearing in mind any other matters, there are two major benefits to the pesticide business from synthesis of such compounds. One of them is heavy investment in personnel and facilities for this kind of methodology. Shifting this tactic may be problematic for employees proficient in disciplines geared to use it. Secondly, copyrights for chemical synthesis often guard the investment that particular company makes in advancement of a synthetic pesticide.

Natural compounds are very much complicated that their synthesis would be excessive, even so, much economically synthesized equivalents with acceptable or better biological action may tip the steadiness toward industrial synthesis. If not, biosynthesis must be considered. There are increasing number of biosynthetic choices.

The easiest technique is to prepare the extract of the compound from field-grown common plants. To advance production, the variety of that particular type that produces the highest amount of that compound must be identified and cultivated under such conditions that will improve their biosynthetic potential. Hereditarily manipulation of the producing plants by classical or biotechnological approaches could also enhance the production of some secondary plant products. For example, low concentrations of ether herbicides, namely diphenyl, can cause enormous upsurges in phytoalexins in a crop variety. Another substitute technique is to culture the compound in cell or tissue culture process. Cell lines that yield higher levels of the required compounds can be quickly chosen. Still, genetic constancy of these characters has been a hurdle in cell or tissue culture for the creation of secondary plant products. Cells that yield and collect huge quantities of probably autotoxic plant secondary metabolites are apparently at metabolic hindrance and are thus designated contrary to under several cell or tissue culture situations. An approach, such as restrained cell column that constantly eradicates such secondary plant products, can increase production by decreasing feedback inhibition of synthesis, possibly increasing generic stability and reducing autotoxicity. Further culture procedures that improve production can also be applied. For example, providing economical synthesized metabolic predecessors can significantly increase biosynthesis of several secondary products. Similarly, various PGR, metabolic blockers, and elicitors can be used to improve the production rate. Genetic engineering and biotechnology may permit for the production of secondary plant products by gene transfer technique to various microorganisms and production by fermentation-like process. This approach is striking because of the prevailing fermentation know-how for production of such plant secondary products. Still, this may be too problematic for complicated secondary nature of products in which many genes govern the alteration of some complex intermediates to the anticipated produce. Under some circumstances, genetic engineering process might also be used to supplement the genetic material for production of plant-based pesticides from one particular plant species to another species to safeguard it from pests. Still, such transgenic management of the complicated metabolism of a higher plant might be enormously problematic. Another simple substitute might be contaminated plant-colonizing microorganisms with the anticipated genetic machinery to yield the natural pesticide.

11.3.3 Plant Extracts

Presently, research has concentrated on how to choose ecological-responsive sterilants but without effecting effluence level and to comprehend control of pest rodents. Tran and Hinds (2012) shortened 13 plant species with antifertility properties on female rodents from plants along with planned application routes and recognized many plant extracts with good probable response. In China, various sterilants based on plant sources that have been tried on rodents include gossypol, *Tripterygium wilfordii*, *Camellia oleifera*, *Ruta graveolens*, colchicine, semen ricini (*Ricinus communis*), radix trichosanthis (*Trichosanthes kirilowii*), neem (*Melia azedarach*) and acrogenous turmeric rhizome (*Curcuma aeruginosa*). Still, all of these compounds had been used as baits for animal experimentation under lab conditions or sprayed for animal scavenging, which had a positive effect; the practical applications have been acceptable (Croxatto 2000). And that the researches on various aspects like pathology, physiology, contraceptive, and pharmacology influences of the shikonin are not adequate (Zhang et al. 2004; Coulson et al. 2008; Conn and Crowley 1991). Shikonin is an extract of *Arnebia euchroma* plant of family Boraginaceae and is commonly dispersed in Xinjiang, Inner Mongolia, and Gansu provinces (Herbert et al. 2006). It has been seen in "Shennong's Herbal" that radices lithospermi has antipyretic, anti-inflammatory, and antifertility influences. Earlier study has focused on anti-HIV, antitumor, and anti-inflammatory influences (Miller et al. 1998; Zhang et al. 2004), while there is occasional, tentative indication of its antifertility properties and the procedure through which this consequence happens. During this experiment, three dissimilar doses of shikonin were used for fertility control in rats (mice), and its probable mechanism of action was considered. It is expected that plant-source-based sterilant, which are environment friendly and without pollution, can be used for sustainable regulation of rodents.

11.3.3.1 *Thevetia peruviana*

An albino strain of non-fasted Norway rats *was* treated with *Thevetia peruviana*, by being fed a bait containing ground grains from the plant or managed aqueous extracts of the kernels by intubation. No abnormal behavioral or mortality or physiological variations were recorded for control group or those applied with crude aqueous extract. Rodents treated with bait comprising fatal doses of ***Thevetia peruviana*** demonstrated signs of poison within 1 h and died after treatment. Poisoning signs involved tail erection and pilomotor, diarrhea, ataxia, paroxysmal tachycardia, and diuresis limb paralysis. Poisoned rats were easily terrified, became aggressive by touching, and showed coprophagic tendencies.

11.3.3.2 Argel

Leaves of argel were dried and grounded to a fine powder. Extraction was done according to the Freedman et al. (1979) method. Leaf powder was saturated in solvent (ethanol 95% or water) for a period of 3 days in brown-painted bottles. These bottles were shaken. Extract was then filtered over anhydrous Na_2So_4 and evaporated till waterlessness under reduced pressure conditions. The extracts were then balanced and kept in cold conditions till use. Then argel powder was mixed carefully with crushed maize grains. The recorded results exhibited that baits of argel (ethanol extract) at various levels were very effective against *R. norvegicus* and result in killing. Attained results also showed that the palatability of male rats was 35.9%, 32.1%, and 33.8%. Temporarily, the palatability and death percentages for female rats for the former's concentrations were 38.7%, 34.9%, and 38.4% and

25%, 25% and 75%, correspondingly. On the other hand, various tried concentrations of argel aquatic extract had diverse effects on rat mortality. The average time mandatory for death was 18.0, 14.5, and 16.5 days for females and 16.0, 12.7, and 8.25 days for males, correspondingly, trailed by argel powder leaves which were 16.0, 15.0, and 12.7 days for females and 14.5, 12.5, and 8.0 days for males, separately, while the mean time required for death in case of aquatic argel extract for the same doses were 13.0, 19.7, and 9.7 days for males and 16.0, 21.0, and 18.0 days for females, correspondingly.

11.3.3.3 Neem Tree (*Azadirachta indica* L.)

Extracts of many plant parts (leaves and bark) have been linked with productiveness, mainly in Indian culture. The plants like neem have many effects in human avoiding troublesome implantation, spermatogenesis, and abortion. In rats, oral application of neem oil (seed) for 18 days disturbed the estrous rotation and encouraged significant alteration in uterine system and results in 70–100% (almost) in rat infertility (females) that soon mated after ending of application (Dhaliwal et al. 1998). These experiments also meaningfully decreased numbers of follicle in all steps of advancement (Dhaliwal et al. 1999). Irregularity in estrous cycles, with extended diestrous stages, was also recorded in female rats when ethanol extracts of the neem plant (flowers) were given orally for a period of 3 weeks (Gbotolorun et al. 2008). Hexane and methanol extracts of neem seeds, after oral application to female rats for a period of 18 days, meaningfully lessened the follicle number at different steps of follicular growth (Roop et al. 2005), and higher doses led to delay in creation of first litters by 52–67 days, correspondingly (Morovati et al. 2008). In in vitro investigation, neem leaf extract led to apoptosis in oocytes of rat (Shail et al. 2006). These investigations specify that neem extracts influence the reproductive system by governing HPG hormones that control development of follicle of the ovary. Still, these extracts are very lethal at higher quantities.

11.3.3.4 Papaya (*Carica papaya* L.)

Papaya plants were grown in large scale in subtropical and tropical regions found in regions both subtropical and tropical. Extracts from various parts have been used as antioxidants, antibiotics, and contraceptives for both sexes. Seed extracts have harmful influences toward the reproduction system of females, inducing irregularity in estrous cycles, disturbing ovulation, and preventing implantation (Raji et al. 2005; Joshi and Chinoy 1996; Chinoy et al. 1997, 1995; Dosumu et al. 2008). Oral application of chloroform extract of papaya seeds, for a period of 14 days, led to noteworthy decrease in weight of the ovary, extended diestrous stage, and noticeable increase in atretic follicles number along with decrease in pregnancies (60%) number. Oral application of a benzene extract of plant seeds, in rats (female), for certain period led to irregularity of estrous cycles; important contrary variations in the uterus through changes in enzyme, glycogen, and protein levels; and 100% sterility subsequent one breeding valuation (Joshi and Chinoy 1996). Likewise, seed (an ethanol extract) when applied orally for a specific period led to anomalies in the estrous sequence and substantially decreases fertility rates (Chinoy et al. 1997), but papaya seed extracts have reversible effects after 30 days of treatment withdrawal (Joshi and Chinoy 1996).

11.3.3.5 *Melia azedarach* (Dharek)

It is concluded that the pulmonary fibrosis and NPF of *M. azedarach* and *A. indica* seed extracts meaningfully lessened normal follicles number in rats, with the maximum reduction occurring in *Azadirachta* extract. This is constant with its use in traditional medicine as an anti-conceptional mediator (Roop et al. 2005).

11.3.3.6 Peppermint Oil

Peppermint oil is supposed to prevent rats from entering treated zones. At high concentrations, oil of peppermint may exhibit some repellency properties. The major constraint in its application is the requirement that it should be applied at concentration that is unfriendly for the owner regarding smell. Once the aroma becomes tolerable to the individual, it is no longer exhibiting repellent characters.

11.3.3.7 Asafetida (*Ferula jaeschkeana* Vatke)

The extracts of parts (aerial) of asafetida including roots of the plant encouraged integration of ovarian system, variations in the reproductive system, and an anti-implantation impact in mice and feminine rats (Prakash and Jonathan 1996; Homady et al. 2002). Oral application of extract (hexane) of the plant parts to juvenile feminine rats for particular period led to noteworthy lessening in the number of emerging and maturating follicles (Prakash and Pathak 1994; Prakash and Sharma 1997). Likewise, female rats, when orally applied with the extract of hexane for some days, exhibited substantial variations in the uterine structure by increasing the luminal epithelium tissue height and gland number of uterine system and an absenteeism of follicles particularly the mature ones (Prakash and Jonathan 1996). However, a different species named *F. hormonis* seems to constrain the reproductive system in female mice. It is observed that when oral application of root extract of the plant *F. hormonis* (an ethanol) for a period of 42 days was induced, ovarian degeneration with reversion of corpora lutea is noted. Such treatment led to decreasing the quantity of pregnant creatures and female productiveness (Homady et al. 2002).

11.3.3.8 *Urginea maritima* (*Drimia maritima* Red Squill)

Urginea maritime is a species of flowering plant of family Asparagaceae. It has been used as a poison and also a medicinal remedy. It led to a digitalis-like activity and results in paralysis of the heart. This glucoside (present in plant) was first isolated from a marine plant called *Urginea maritima*. It was first industrialized and used in Switzerland but was later registered in many other countries like the United States. Nowadays this kind of product is not accessible as its use has stopped significantly in preference to the second-generation anticoagulant compounds. This product contains 0.05% active ingredient (Timm 1994; Buckle and Eason 2015).

11.3.3.9 Sodium Fluoroacetate (1080) Compound

Also identified as composite compound 1080, it was first isolated from an African plant named *Dichapetalum toxicarium* in Europe. It was reported that sodium fluoroacetate might be beneficial as a vertebrate poison. This chemical is quickly absorbed in the gastrointestinal system and blocks the tricarboxylic acid sequence which led to togetherness of citric acid and impediment of glucose metabolism process, which ultimately results in spasms and further more respiratory or circulatory system failure. It has toxic influence in most of vertebrates and in various nations, especially New Zealand and Australia. Presently, it is used at very limited scale in the United States as collars in livestock protection only. Such items usually contain 0.08% or 0.5% a.i. (Buckle and Eason 2015; Timm 1994).

11.3.3.10 *Strychnos* Spp.

Strychnine was firstly isolated from seeds of *Strychnos* spp., a tree. Being an alkaloid-natured antagonizing glycine compound and acetylcholine receptors, this led to muscle restlessness and twitching, trailed by sudden appropriations and fierce contractions and lastly death. Firstly, it was established as a rodenticide in Europe. It is used at large scale when it was produced commercially relatively than extraction from seeds of the plant. As it is highly toxic to many animal species, hence, its use is restricted in several nations. In the United States, it is only allowed as rodenticide in burrows and, in particular, in pocket gopher holes, not above the ground. Its use in Europe is banned. But its use is still prevalent in Australia for governing house mouse irruptions ("plagues"). Pelleted baits usually contain 2% a.i. but to coat grains, fruits, and vegetables at 0.5% active ingredient.

11.3.3.11 China Rose (*Hibiscus rosa-sinensis* L.)

This plant is indigenous to China and extensively grown and cultivated as a decorative plant throughout the subtropics and tropics. Extracts of various plant parts (China rose) adversely influence ovulation process and avoid process of implantation. In females, oral treatment of the extract (benzene) of *China rose* flowers for a specific period increased the quantity of atretic follicles and lessened uteri and the ovaries weights. Within application for 21 days, all females were in extended diestrous position; still, the influences were adjustable in a limit of 30 days of removal of application (Kholkute et al. 1976). In female mice, oral application of flower extract (benzene) creates unbalance diestrous cycles with extended metestrus and estrus and a considerable increased quantity of atretic follicles. These consequences proposed an antiovulatory influence of the flower extract (Murthy et al. 1997), but these influences were changeable after the termination of application.

11.3.3.12 White Cedar (*Melia azedarach* L.)

It is indigenous to Asia (Southeast), Australia, and Indochina. *Melia azedarach* has important antioxidant influences and protects from many skin diseases (Saleem et al. 2008; Samudram et al. 2009). Root extracts and seed of white cedar cause decreases in ovarian follicle and avoid gestation by disturbing implantation process (Keshri et al. 2003, 2004; Mandal and Dhaliwal 2007). Oral application of methanol and hexane fractions of white cedar seed extracts for a period of 18 days lessened the follicle number at different phases of advancement in ovaries of treated ones. An important decrease in quantity of follicle may have been due to disturbing of follicle enrollment owed to atresia (Roop et al. 2005). Moreover, extract mixture of chloroform + methanol (by volume, 9 + 1) of white cedar seeds when given orally to rats for a period of 18 days prohibited pregnancy in 50 and 100% of animals correspondingly. Consistent variations in metabolism of uterine led to reduction in the stature of the luminal epithelium system and a diminished uterine gland number (Mandal and Dhaliwal 2007).

11.3.3.13 Sensitive Plant (*Mimosa pudica* L.)

Mimosa pudica is native to South America and Central America. Root extracts affect the reproduction system of females both in rats and mice. These extracts alter and extend estrous cycles, increase degeneration of follicles, and lessened litter size in females (Valsala and Karpagaganapathy 2002; Ganguly et al. 2007). Root powder of the plant, when applied orally, extended the diestrous phase of estrous cycle. This application also resulted in a substantial lessening of ova number and increased the number of ova degeneration in the treated ones (Valsala and Karpagaganapathy 2002). Various outcomes submitted for Swiss albino mice (female) when oral treatment was given, that is, root extract for a period of 21 consecutive days (Ganguly et al. 2007), with lessening in litter number produced by treated rats, though the productiveness was reestablished 2–3 weeks after removal of application (Valsala and Karpagaganapathy 2002).

11.3.3.14 Bitter Gourd (*Momordica charantia* L.)

It is widely grown in Asia, Africa, and Caribbean as its fruit is edible and extensively cultivated in subtropical and tropical areas. It has been used as traditional remedy for controlling diabetes disorder and many other diseases, burns, birth control, and skin problems (Beloin et al. 2005; Sridhar et al. 2008). Seed extracts of bitter gourd lessen ovarian follicles number, modify normal estrous sequences, and prevent pregnancy (Sharanabasappa et al. 2002; Chan et al. 1984, 1986; Ng et al. 1988). Application of chloroform, benzene, and petroleum ether including ethanol seeds extract of bitter gourd orally for 30 days induced unbalance diestrous cycles, which led to substantial decrease in ovarian weight and in emerging follicles, Graafian follicles, and corpora lutea (Sharanabasappa et al. 2002).

11.3.3.15 Betel Vine

This plant is native to Southeast Asia. Leaves of this plant have been used as traditional remedies to cure various ailments like relieve toothache, headaches, and indigestion problem. Betel extracts show adverse effect against reproduction in rats (Adhikary et al. 1989; Choudhuri et al. 1991; Sharma et al. 2007; Pin et al. 2010; Al-Adhroey et al. 2011). Hypodermal application of ethanol extracts of the plant (stalk) for a period of 21 days increased the number of follicular atresia. The treated ones exhibited substantial reductions in weights of the ovaries and uterine systems and enlarged levels of cholesterol in the ovaries of treated ones. Treatment with the extract results estrous cycles but after application and repressed conception in treated rats (Adhikary et al. 1989). An ethanol extract of the leaves of vine brought analogous variations: oral application in rats for specified period extended estrous cycles and increased decreases in uterine weights. Conception rate was repressed in treated rats, and substantial reductions in litter size were also noted (Sharma et al. 2007).

11.3.3.16 Kudzu (*Pueraria tuberosa* DC.)

Kudzu is a woody tuberculated stem with climbing nature. Tubers of the plant are used as traditional drug to cure different ailments like malaria fever, cough sand rheumatism. Kudzu encourages antioxidant and antifertility actions in both rat sexes. Tubers extract lessened ovarian follicles number, increased the numbers of atretic follicles and evade implantation in females. Oral application of a butanol extract including tubers of kudzu in rats induced degeneration of the oocyte in follicles afterward for a period of 12 days (approximately) of the application. Subsequently in 18 and 24 days of application, a noteworthy increase in atretic number of follicles was noted. In addition these extracts led to a noteworthy increase in level of glycogen, protein absorption, and weight of the ovary of treated ones. *Pueraria tuberosa* plant extract seems to be inducing disturbance of hypothalamic–pituitary–gonadal axis responses rather than a straight influence at the follicular level.

11.3.3.17 Midnapore Creeper (*Rivea hypocrateriformis* Desr.)

Midnapore creeper is a climbing-type woody shrub commonly available in subtropical region. Plant extracts, from diverse parts, have good influence in the reproduction system of females by disturbing implantation and ovulation process. In female rats (albino type), oral application of an ethanol extract for 15 repeated days distressed estrous cycle, by decreasing the duration of estrous and metestrous stages and extension of the proestrous phase. Treated female rats had fewer Graafian follicles and expressively more atretic follicles than in control ones.

11.3.3.18 Soap Nuts (*Sapindus trifoliatus* L.)

Soap nuts are indigenous to warm tropical areas and cure asthma-like problem and also antifertility properties. Oral application of an acetone fraction of soap nuts extract for a period of 21 days disturbed cycle of estrous by encompassing the diestrous stage. This extract results in disintegration of emerging follicles and lesser pregnancy and rate of implantation. The treated rats exhibited decreased levels of luteinizing and follicle-stimulating hormone. Therefore, the extract led to reduction in gonadotrophin secretion and ultimately disturbance of follicular growth and the estrous cycle. These influences are reportedly rescindable once application is stopped.

11.3.3.19 Snake Gourd (*Trichosanthes cucumerina* L.)

Snake gourd is a subtropical as well as tropical vine. Various plant parts are used in traditional remedies like curing of inflammatory conditions, gastrointestinal problems, and liver disorders. The plant also disturbed ovulation process in rats. When oral treatment of an ethanol extract was applied for a period of 30 days, disturbed estrous process resulted in substantial increase in the metestrous and estrous stages. There was a noteworthy lessening of weight of the ovary, and the same was imitated in meaningfully fewer well follicles and an improved follicle number undergoing regression including atresia.

11.3.3.20 Thunder God Vine (*Tripterygium wilfordii* H.)

It is also known as an old Chinese drug, and its extracts have antitumor, immune suppressive, and anti-inflammatory effects. Thunder god vine has antifertility effects in females and males (Lue et al. 1998; Huynh et al. 2000). Oral application of triptolide (main ingredients of thunder god vine) to female rats for certain period results in a considerable increase in apoptotic secondary follicle number and led to continuation of the estrous cycle. In another experiment, oral higher doses of triptolide were given, for a period of 3 months (approximately); decreased serum levels of P and E2; and increased the levels of other hormones like FSH and LH. These applications resulted in a noteworthy decrease in the ovaries weights and uteri and ultimately decreased the number of emerging follicles, and atretic follicle number is increased. The results of this study suggested that triptolide (source: thunder god vine) had a straight influence especially on the advancement of ovarian follicles (secondary and tertiary).

11.3.3.21 Quickstick (*Gliricidia sepium*)

Commonly known as quick stick, it belongs to the family Fabaceae. It is a very important, multipurpose legume tree. It is very common in tropical and subtropical nations and used for various purposes such as fodder, live fencing, shade, coffee (Zhang et al. 2004) green manure, firewood, and rat poison. Live fences can be grown from *Gliricidia sepium* in just a month. It can be intercropped with maize crop.

11.4 Discussion

For governing fertility in rodents, targets comprise anticipation of the advancement and ovarian follicles development, hindering the pathway of ova process in oviduct, stoppage and fertilization of oocytes, prevention of implantation process, and intrusion with pregnancy process. Various studies regarding the effects of plant extracts on productiveness have revealed that these extracts may influence the reproductive system of females in both rats and mice. As a consequence, an applied extract may prevent fruitfulness by various mechanisms, such as follicle diminution, disturbance process of ovulation, fertilization, gestation or implantation, and abortion. Several plant extracts were evaluated following post-coitum application at various stages of pregnancy to estimate abortifacient/anti-implantation actions. Though for fertility control, plants encouraging inhibitory influences at the ovarian level should be theoretically better contenders since they may offer permanent and long-term influence on the reproductive system of the target. Indirect influences occur at the hypophysial or hypothalamic level, and they cause suppression of normal gonadotrophin secretion. These changes lead to prevention of consistent ovarian steroidogenesis (estrogen/progesterone), and thus, estrous cycles are disturbed, just like ovulation process. Otherwise, the influence may be straight and shortest if the extract of the plant constrains the regular advancement of the different phases of ovarian follicles. As females have limited number of primordial follicles since birth, the reproductive potential may be lessened if they are treated with various substances that may cause reduction or weakening of the primordial follicle system. But the effects are reversible when applications stopped, and influenced follicles will be substituted by other follicles. The initial follicular phases are targeted and washed out by another approach, the influences on productiveness will be overdue until the modest further (secondary and others) follicles are engaged and exhausted – prominently such result will be everlasting and cause sterility. Several experiments have described the effects on productiveness of female rodents by treatment with different plant extracts. Some of these plants are commonly available like *Azadirachta indica, Melia azedarach, Hibiscus rosa-sinensis, Trichosanthes cucumerina, Tripterygium wilfordii, and Momordica charantia.* Extracts of these plants demonstrate variable potential. The extracts of these plants at verified doses require constant and long periods of oral application to attain anticipated effects, but such influences are inverted soon after cessation of the application. While such plants are well intentioned for future search, inappropriately no one directly influences the non-regenerating primeval follicle numbers. A number of extracts, their duration and types, have been tested, and the result specifies that responses depend on the applied dose, but up to now, structured-dose-based experiments have not recognized the optimum doses obligatory. Overall, advanced doses will be required for testing purposes, but with this approach, palatability may be affected. Thus, recognition of the active component in such plant extract might be indispensable. This, though, would decrease the capability of farmers to formulate such extracts on his field. In spite of so much research related to such plant extracts, the precise active ingredient of compounds accountable for multiplicative influences has not been acknowledged, and synthesis of such chemicals is also posing a problem. While working, scholars should be conscious that the levels of such active compounds accessible in various plant materials may differ according to places, soil, and growth conditions. Further, extracts from the similar plant, but using dissimilar solvents for extraction of active compounds, may lead to dissimilar reactions. It is therefore recommended that further researches should lay emphasis on the recognition of contributing active ingredients in plant extracts and suitable solvents. Principally it is problematic to regulate distinct and effective responses; there is lack of consistency in terms of difficult valuation of various kinds of extracts and then in the range of practices used to evaluate generative influences. Research investigations have observed the influence on different stages of the estrous cycle, while others observed histological modifications in the uterus and ovary. The length of the treatment and course of treatment, as well as doses excerpts, are also contradictorily pragmatic. Consequently, recently a typical protocol is proposed for trying various extracts, which should be recognized to permit direct assessments. The protocol for female rats would include evaluating a range of verbal doses over applied periods of application/treatment along with assessment of different reproductive constraints. Finally, more comprehensive characterization of extracts, such as GLC or HPLC and mass spectrometry, should be accessible to permit other scholars for recurrence of published experiments/studies. Therefore, for practical application in productiveness control of such rodents, the anticipated testing procedure would commence to address the serious necessity for oral transfer of an agent that quickly and permanently diminishes productiveness. This would then result in the development of target-specific, edible, and persistent produce for use. In spite of recorded restrictions arising beforehand accessible to related research experiment, the use of plant extracts in productiveness could still deliver a number of benefits, and these substances are not penetrated in the subsequent trophic level. Furthermore, many plant species are commonly available and may be economical, particularly when such plant extracts are readied by local farmers. Still, there are some contests/challenges; for example, plant extracts need to be consumed on daily basis by the target species for achieving good results. Also, high concentrations of these extracts are mandatory to obtain their desired influences; this may influence their tastiness and decrease bait acceptance frequency. Lastly, a future prospect could present itself if a plant range of different ingredients are recognized and used in combination with various chemosterilants. These combinations may provide complementary influences and lead to quicker and everlasting alterations in ovarian system. Overall, these mediators may deliver an effective supplementary means for use in integrated administration of pest populaces.

11.5 Conclusion and Future Prospects

Some plants contain an effectively unexploited reservoir of chemicals with natural pesticidal properties that can be used straightly as prototypes for synthetic pesticides. Several factors have increased the interest of the pesticide business in such source of natural produce. These include lessening returns with traditional pesticide detection approaches, increased toxicological and environmental anxieties with synthetic insecticides, and increased level of dependence of modern agriculture on the pesticides. Despite the moderately small amount of earlier effort in advancement of plant-derived compounds as pesticides, they have made a huge influence in the related area. Immaterial rate of successes may be found in rodenticides, herbicides, and fungicides. Number of possibilities must be taken into account in advancement. Moreover, the molecular complexity, inadequate environmental loyalties, and low activity of several biocides from the plants, in comparison to synthetic pesticides, are depressing. Still, advances in biotechnology and chemicals are increasing the ease and speed with which mankind can formulate and advance secondary compounds of the plants as pesticides.

References

Adhikary P, Banerji J, Chowdhury D, Das AK, Deb CC, Mukherjee SR et al (1989) Antifertility effect of Piper betle Linn. extract on ovary and testis of albino-rats. Indian J Exp Biol 27:868–870

Afifi FA, Gomaa AA, Zidan ZH, Fam EZ, Ahmed SMSS (1992) Preliminary evaluation of chemical constituents in certain bioactive plant extracts. Arab Univ J Agric Sci Ain-Shams Univ Cairo 1(1):97–111

Al-Adhroey AH, Nor ZM, Al-Mekhlafi HM, Amran AA, Mahmud R (2011) Antimalarial activity of methanolic leaf extract of Piper betle L. Molecules 16:107–118

Angel A, Wanless RM, Cooper J (2009) Review of impacts of the introduced house mouse on islands in the Southern Ocean: are mice equivalent to rats. Biol Invasions 11:1743–1754

Beloin N, Gbeassor M, Akpagana K, Hudson J, de Soussa K, Koumaglo K et al (2005) Ethnomedicinal uses of *Momordica charantia* (Cucurbitaceae) in Togo and relation to its phytochemistry and biological activity. J Ethnopharmacol 96:49–55

Bordes F, Blasdell K, Morand S (2015) Transmission ecology of rodent borne diseases: new frontiers. Integr Zool 10:424–435

Buckle AP, Eason ET (2015) Control methods: chemical. In: Buckle AP, Smith RH (eds) Rodent pests and their control, 2nd edn. Cab International, Wallingford, pp 123–154

Buckle AP, Smith RH (2015) Rodent pests and their control, 2nd edn. CAB International, Wallingford, 422 pp

Chan WY, Tam PPL, Yeung HW (1984) The termination of early pregnancy in the mouse by beta-momorcharin. Contraception 29:91–100

Chan WY, Tam PPL, Choi HL, Ng TB, Yeung HW (1986) Effects of momorcharins on the mouse embryo at the early organogenesis stage. Contraception 34:537–544

Chinoy NJ, Dilip T, Harsha J (1995) Effect of *Carica papaya* seed extract on female rat ovaries and uteri. Phytother Res 9:169–175

Chinoy NJ, Joshi H, Ghosh S (1997) Antifertility investigations of alcoholic papaya seed extract in female rats. J Med Arom Plant Sci 19:422–426

Choudhuri D, Adhikary P, Banerji J, Jana S, Chatterjee A, Mukherjee SR (1991) Pregnancy interceptive effect of Piper betle. Fitoterapia 62:397–402

Conn PM, Crowley WFJ (1991) Gonadotropin-releasing hormone and its analogues. New Engl J Med 32:93–103

Coulson G, Nave CD, Shaw G, Renfree MB (2008) Long-term efficacy of levonorgestrel implants for fertility control of eastern grey kangaroos (*Macropus giganteus*). Wildl Res 35:520–524

Croxatto HB (2000) Progestin implants. Steroids 65:681–685

Delattre P, Giraudoux P, Baudry J, Musard P, Toussaint M, Truchetet D, Stahl P, Poule ML, Artois M, Damange JP, Quéré JP (1992) Land use patterns and types of common vole (*Microtus arvalis*) population kinetics. Agric Ecosyst Environ 39:153–168

Dhaliwal PK, Roop JK, Guraya SS (1998) Antifertility activity of neem seed oil in cyclic female rats. In: Dhaliwal P, Roop J, Guraya S (eds) Ecological Agriculture and Sustainable Development. Centre for Research in Rural and Industrial Development, Chandigarh, pp 340–346

Dhaliwal PK, Roop JK, Guraya SS (1999) Effect of neem-seed oil on the quantitative aspects of follicular development in cyclic female rats. Indian J Ecol 26:162–166

Dosumu OO, Akinola OB, Oremosu AA, Noronha CC, Okanlawon AO (2008) Antifertility effects of the aqueous extract of *Carica papaya* (Linn.) seeds on estrous cycle and ovulation of adult cyclic Sprague–Dawley rats. Niger J Hlth Biomed Sci 7:31–33

Doungboupha B, Aplin KP, Singleton GR (2003) Rodent outbreaks in the uplands of Laos: analysis of historical patterns and the identity of nuukhii. ACIAR Monogr Ser 96:103–111

Eason CT, Murphy EC, Wright GRG, Spurr EB (2002) Assessment of risks of brodifacoum to non-target birds and mammals in New Zealand. Ecotoxicology 11:35–48

Eisemann J, Swift C, Dunlevy P, Pitt W, Witmer G (2010) Regulatory and policy issues around non-target mortality and environmental fate of rodenticides. In: Timm RM, Fegerstone K

(eds) Proceedings of the 24th vertebrate Pest conference, 24th vertebrate pest conference, Sacramento, 22–25 Feb, pp 208–212
EL-Gengaihi SE, Dimetry NZ, Mohamed SM (1997) Chemical and biological investigation of harmal plant. 2-Alkaloidal investigation. J Appl Entomol 12(3):165–167
Fellows LE (1979) Plant secondary compounds as a chemical defense. In: Wright EN (ed) Bird problems in Agriculture, pp 188–195
Freedman BJ, Nowak J, Kwolek WF (1979) A bioassay for plant derived pest control agent using the European Comborer. J Econ Entomol 72:45–54
Gabr WM, Khidr FK, Hegab AM (2004) Evaluation of osher crude plant extract as a rodenticide under laboratory and field conditions. Zagazig J Agric Res 31(4A):1607–1616
Ganguly M, Devi N, Mahanta R, Borthakur MK (2007) Effect of Mimosa pudica root extract on vaginal estrous and serum hormones for screening of antifertility activity in albino mice. Contraception 76:482–485
Gbotolorun SC, Osinubi AA, Noronha CC, Okanlawon AO (2008) Antifertility potential of Neem flower extract on adult female Sprague–Dawley rats. Afr Hlth Sci 8:168–173
Geduhn A, Jacob J, Schenke D, Keller B, Kleinschmidt S, Esther A (2015) Relation between intensity of biocide practice and residues of anticoagulant rodenticides in red foxes (*Vulpes vulpes).* PLoS One 10:e0139191
Herbert CA, Trigg TE, Cooper DW (2006) Fertility control in female eastern grey kangaroos using the GnRH agonist deslorelin. 1. Effects on reproduction. Wild Res 33:41–46
Homady MH, Khleifat KM, Tarawneh KA, Al-Raheil IA (2002) Reproductive toxicity and infertility effect of *Ferula hormonis* extracts in mice. Theriogenology 57:2247–2256
Hussein HI (1991) Rodenticidal properties of cardenolides crude extracts isolated from sodom Apple calotropis procera. Ph. D. Thesis, Faculty of Agriculture Alex University
Huynh PN, Hikim APS, Wang C, Stefonovic K, Lue YH, Leung A et al (2000) Long term effects of Triptolide on spermatogenesis, epididymal sperm function, and fertility in male rats. *J Androl* 21:689–699
Jacob J, Buckle A (2018) Use of anticoagulant rodenticides in different applications around the world. In: van den Brink NW, Elliott JE, Shore RF, Rattner BA (eds) Anticoagulant rodenticides and wildlife. Springer, Cham, pp 11–43. https://doi.org/10.1007/978-3-319-64377-9_2
John A (2014) Rodent outbreaks and rice pre-harvest losses in Southeast Asia. Food Secur 6:249–260
Joshi H, Chinoy NJ (1996) Reversible antifertility effects of benzene extract of papaya seed on female rats. Phytother Res 10:327–328
Keshri G, Lakshmi V, Singh MM (2003) Pregnancy interceptive activity of Melia azedarach Linn. in adult female Sprague–Dawley rats. Contraception 68:303–306
Keshri G, Bajpai M, Lakshmi V, Setty BS, Gupta G (2004) Role of energy metabolism in the pregnancy interceptive action of *Ferula asafoetida* and *Melia azedarach* extracts in rat. Contraception 70:429–432
Kholkute SD, Chatterjee S, Udupa KN (1976) Effect of *Hibiscus rosa-sinensis* Linn on estrous-cycle and reproductive organs in rats. Indian J Exp Biol 14:703–704
Leirs H, Stenseth NC, Nichols JD, Hines JE, Verhagen R, Verheyen W (1997) Stochastic seasonality and nonlinear density-dependent factors regulate population size in an African rodent. Nature 389:176–180
Lue Y, Hikim APS, Wang C, Leung A, Baravarian S, Reutrakul V et al (1998) Triptolide: a potential male contraceptive. J Androl 19:479–486
Mandal R, Dhaliwal PK (2007) Antifertility effect of *Melia azedarach* Linn. (dharek) seed extract in female albino rats. Indian J Exp Biol 45:853–860
Marsh RE (1988) Rodent problems on the north American continent. In: Prakash I (ed) Rodent pest management. CRC Press, Inc., Boca Raton, pp 1–11
Meerburg BG, Singleton GR, Kijlstra A (2009) Rodent-borne diseases and their risks for public health. Crit Rev Microbiol 35:221–270
Miller LA, Johns BE, Elias DJ (1998) Immunocontraception as a wildlife management tool: some perspectives. Wild Soc Bull 26:237–243

Morovati M, Mahmoudi M, Ghazi-Khansari M, Aria AK, Jabbari L (2008) Sterility and abortive effects of the commercial neem (*Azadirachta indica* A. Juss.) extract Neem Azal-T/S (R) on female rat (Rattus norvegicus). Turk J Zool 32:155–162

Murthy DRK, Reddy CM, Patil SB (1997) Effect of benzene extract of *Hibiscus rosa-sinensis* on the estrous cycle and ovarian activity in albino mice. Biol Pharm Bull 20:756–758

Ng TB, Tam PPL, Hon WK, Choi HL, Yeung HW (1988) Effects of momorcharins on ovarian response to gonadotropin-induced superovulation in mice. Int J Fert 33:123–128

Nogeire T, Lawler J, Schumaker N, Cypher B, Phhillips S (2015) Land use as a driver of patterns of rodenticide exposure in modeled kit fox populations. PLoS One 10. https://doi.org/10.1371/journal.pone.0133351

Nowak R (1999) Walker's mammals of the world, vol II, 6th edn. Johns Hopkins University Press, Baltimore, 1936 pp

Oji O, Madubuik FN, Ojimelukwa PC, Ibed CM (1994) Rodenticide potential of Thervetia peruviana. J Herbs Species Med Plants 2(3):3–10

Parshad VR (1999) Rodent control in India. Integr Pest Manag Rev 4:97–126

Pelz H (2007) Spread of resistance to anticoagulant rodenticides in Germany. Int J Pest Manag 53(4):299–302

Pimentel D, Zuniga R, Morrison D (2005) Update on the environmental and economic costs associated with alien-invasive species in the United States. Ecol Econ 52:273–288

Pin KY, Chuah AL, Rashih AA, Mazura MP, Fadzureena J, Vimala S et al (2010) Antioxidant and anti-inflammatory activities of extracts of betel leaves (Piper beetle) from solvents with different polarities. J Trop For Sci 22:448–455

Pitt W, Berensten A, Shiels A, Volker S, Eisemann J, Wegmann A, Howald G (2015) Non-target species mortality and the measurement of brodifacoum rodenticide residues after a rat (*Rattus rattus*) eradication on Palmyra atoll, *Tropical Pacific*. Biol Conserv 185:36–46

Prakash I (1988) Rodent pest management. CRC Press, Boca Raton, 480 pp

Prakash AO, Jonathan S (1996) Effect of hexane extract of *Ferula jaeschkeana* on the reproductive organs in cyclic female rats. Philipp J Sci 125:129–146

Prakash AO, Pathak S (1994) Inhibition of beta-glucuronidase activity in rat uterus during implantation by feeding Ferula jaeschkeana extract. Int J Pharmacognosy 32:362–365

Prakash AO, Sharma M (1997) Morphological and functional consequences in the reproductive organs of immature rats exposed to multiple administration of *Ferula jaeschkeana* extract. Phytother Res 11:1–5

Proulx G (2014) On the misuse of pesticides to control northern pocket gophers and Richardson's ground squirrels in agriculture and the pressing need for sustainable solutions. In: Holroyd GL, Trefry AJ, Crockett B (eds) Proceedings of the 10th prairie conservation and endangered species conference, engaging people in conservation, 10th prairie conservation and endangered species conference, Red Deer, 19–22 Feb, pp 134–157

Qureshi MA, Qureshi NM, Arshad R, Begum R (1991) A study on the antisperm activity in extracts from different parts of Calotropis procera. Pak J Zool 23(2):161–166

Raji Y, Morakinyo AO, Oloyo AK, Akinsomisoye OS, Olufadekemi T, Kunle-Alabi T et al (2005) Impact of the chloroform extract of Carica papaya seed on oestrous cycle and fertility in female albino rats. J Med Sci (Pakistan) 5:337–343

Rattner B, Horak K, Lazarus R, Eisenreich K, Meteyer C, Volker S, Campton C, Eisemann J, Johnston J (2012) Assessment of toxicity and potential risk of the anticoagulant rodenticide diphacinone using eastern screech-owls (Megascopsasio). Ecotoxicology 21:832–846

Rattner BA, Horak KE, Lazarus RS, Goldade DA, Johnston J (2014a) Toxicokinetics and coagulopathy threshold of the rodenticide diphacinone in eastern screech-owls (Megascopsasio). Environ Toxicol Chem 33(1):74–81

Rattner BA, Lazarus RS, Elliott JE, Shore RF, van den Brink N (2014b) Adverse outcome pathway and risks of anticoagulant rodenticides to predatory wildlife. Environ Sci Technol 48:8422–8445

Roop JK, Dhaliwal PK, Guraya SS (2005) Extracts of *Azadirachta indica* and *Melia azedarach* seeds inhibit folliculogenesis in albino rats. Braz J Med Biol Res 38:943–947

Saleem R, Rani R, Ahmed M, Sadaf F, Ahmad SI, ul Zafar N et al (2008) Effect of cream containing Melia azedarach flowers on skin diseases in children. Phytomedicine 15:231–236

Saleh MA, Ibrahim NA, EL-bolok MM, Abdel-salam EA (1986) Insecticidal activity of selected Egyptian wild plants. Bull Fac Agric Cairo Univ 37(1):517–525

Samudram P, Vasuki R, Rajeshwari H, Geetha A, Moorthi PS (2009) Antioxidant and antihepatotoxic activities of ethanolic crude extract of Melia azedarach and Piper longum. J Med Plants Res 3:1078–1083

Schmutterer H (1981) Some properties of components of the neem tree (Azadirachta indica) and their use in pest control in developing countries. *Mededelingen Van de facultetlandbouweten, Schappen*. Rijksuniversiteit Gent 46(1):39–47

Shail KC, Pramod VP, Beena K, Tulsidas GS (2006) Extract of *Azadirachta indica* (neem) leaf induces apoptosis in rat oocytes cultured in vitro. Fert Sterility 85:1223–1231

Sharanabasappa A, Vijayakumar B, Patil SB (2002) Effect of *Momordica charantia* seed extracts on ovarian and uterine activities in albino rats. Pharm Biol 40:501–507

Sharma JD, Sharma L, Yadav P (2007) Antifertility efficacy of Piper beetle Linn (petiole) on female albino rats. Asian J Exp Sci 21:145–150

Singleton GR, Leirs H, Hinds LA, Zhang Z (1999) Ecologically-based management of rodent pests – re-evaluating our approach to an old problem. In: Singleton GR, Hinds LA, Leirs H, Zhang Z (eds) Ecologically-based management of rodent pests. Australian Centre for International Agricultural Research, Canberra, pp 17–29

Singleton GR, Hinds LA, Krebs CJ, Spratt DM (2003) Rats, mice and people: rodent biology and management. Clarus Design, Canberra, 564 pp

Singleton GR, Brown PR, Jacob J, Aplin KP (2007) Unwanted and unintended effects of culling: a case for ecologically-based rodent management. Integr Zool 2:247–259

Singleton GR, Belmain S, Brown P, Hardy B (eds) (2010) Rodent outbreaks: ecology and impacts. International Rice Institute, Los Banos, 289 pp

Sridhar MG, Vinayagamoorthi R, Suyambunathan VA, Bobby Z, Selvaraj N (2008) Bitter gourd (*Momordica charantia)* improves insulin sensitivity by increasing skeletal muscle insulin-stimulated IRS-1 tyrosine phosphorylation in high-fat-fed rats. Br J Nutr 99:806–812

Tigner JR (1966) Chemically treated multiwall tarps and bags tested for rat repellency. J Environ Manag 30(1):180–184

Timm R (1994). Active ingredients. In: Hygnstrom SE, Timm RM, Larson GE (eds) Prevention and control of wildlife damage, Cooperative Extension Division, University of Nebraska, Lincoln, pp G-23–G-61

Tran TT, Hinds LA (2012) Fertility control of rodent pests: a review of the inhibitory effects of plant extracts on ovarian function. Pest Manag Sci 69:342–354

Valsala S, Karpagaganapathy PR (2002) Effect of Mimosa pudica root powder on oestrous cycle and ovulation in cycling female albino rat, Rattus norvegicus. Phytother Res 16:190–192

Weeks JR (1959) Laboratory testing of rat repellents, quantitative evaluation of repellency of chemical coatings on paperboard. J Agric Food Chem 7(3):193–196

Witmer G, Eisemann JD (2007) Rodenticide use in rodent management in the United States: an overview. In: Nolte DL, Arjo WM, Stalman D (eds) Proceedings of the 12th wildlife damage management conference, 12th wildlife damage management conference, Corpus Christi, 9–12 Apr 2007, pp 114–118

Witmer G, Pitt W (2012) Invasive rodents in the United States: ecology, impacts, and management. In: Blanco J, Fernandes A (eds) Invasive species: threats, ecological impact and control methods. Nova Science Publishers, New York, pp 47–75

Witmer G, Singleton G (2012) Sustained agriculture: the need to manage rodent damage. In: Triunveri A, Scalise D (eds) Rodents: habitat, pathology, and environmental impact. Nova Science Publishers, Inc., New York

Witmer G, Eisemann JD, Howald G (2007) The use of rodenticides for conservation efforts. In: Nolte DL, Arjo WM, Stalman D (eds) Proceedings of the 12th wildlife damage management conference, 12th wildlife damage management conference, Corpus Christi, 9–12 Apr 2007, pp 160–167

World Health Organization (WHO) Geneva (1988) Chapter 15 Environmental control on board ship. In: International Medical Guide for Ships Including the Ship's medicine chest. 2nd ed. pp 283–295. Switzerland

Zhang Z-B, Liao L-F, Wang S-Q, Cao X-P, Wang F-S, Wang C et al (2004) Effect of a contraceptive compound (EP-1) on fertility of female Brandt's voles, gray hamsters and mid-day gerbils. Acta ZoolSinica 50:341–347

Natural Product as Avermectins and Milbemycins for Agriculture Perspectives

12

Kanchan Bhardwaj, Jayanthi Abraham, and Simran Kaur

Abstract

The very first development of avermectins was done from isolation in Kitasato Institute laboratories, using a novel soil-dwelling bacterium. It was transmittal to the Merck & Co. research laboratory incorporation. These belong to the 16-membered, closely related family, which are macrocyclic lactones constituting of four major and four minor homologous compounds. One of the macrocyclic lactone compounds is milbemycin, which is a group of chemical related to the avermectins and was first isolated from *Streptomyces hygroscopicus* in 1972. They are a group of macrolides chemically associated to the ivermectins. Milbemycin, a commercially available insecticide, consists of milbemycin A3 and milbemycin A4 (30% and 70%, respectively). Milbemycin and avermectin anthelmintic groups share a common action mechanism, but the moxidectin molecular structure differs from avermectin anthelmintics, which afford much potency and high lipid solubility and therefore perseverance. Abamectin is the only compound that belongs to the family avermectin and has some application in crop protection from parasites. Apart from it, abamectin also causes oral and dermal toxicity. Other members of the macrolide group are used in antiparasitic medicines, in order to inhibit the animal products contaminating parasites. Remains of macrocyclic lactones compound including avermectin and milbemycin are used in veterinary medicines to inhibit parasites found in meat and milk.

K. Bhardwaj (✉)
Department of Botany, Shoolini University of Biotechnology and Management Sciences, Solan, Himachal Pradesh, India

J. Abraham
School of Bio Sciences and Technology, Vellore Institute of Technology VIT, Vellore, Tamil Nadu, India

S. Kaur
Department of Biotechnology, Lovely Professional University, Phagwara, Punjab, India

J. Singh, A. N. Yadav (eds.), *Natural Bioactive Products in Sustainable Agriculture*, https://doi.org/10.1007/978-981-15-3024-1_12

Keywords
Streptomyces hygroscopicus · Avermectin · Crop protection · Milbemycin veterinary medicines · Emamectin benzoate

12.1 Introduction

Satoshi Ōmura, a microbiologist, had isolated and cultivated a Gram-positive bacterium (sample NRRL 8165), later identified as an unknown species of *Streptomyces*; the sample was taken from soil and collected from the woods nearby a golf course in Kawana, on the southeast coast of Honshu, Japan. The bacterial isolates were further referred to William Campbell at Merck to check for the antiparasitic activity of the strains (which differed in morphological and culture characteristics). The bioactive compound isolated from the NRRL 8165 cultures shows potent activity against *Nematospiroides dubius* (now known as *Heligmosomoides polygyrus*), which causes mice infection. The purified compound was reported to belong to a macrocyclic lactones (MLs) family. The bioactive compound were so-called avermectins from the bacterium *Streptomyces avermitilis* because of the helminth-free conditions they produced (Burg et al. 1979; Campbell 1981). There are four compound mixtures present in the naturally occurring bioactive compound. There are four compounds consists in the naturally occurring bioactive avermectins. These compounds are A1, A2, B1, and B2, and all of these compounds consist of two variants each, that is, a and b, as shown in Fig. 12.1a (Campbell 1981; Campbell et al. 1983). Due to presence of isopropyl group C25 and chemical structure difference at C22 and C23 position, the bioactivity of avermectins against sheep gastrointestinal nematodes has been proved to have the highest anthelmintic property.

Following the discovery of avermectin, milbemycins were discovered in the year 1973 for protecting crops (Takiguchi et al. 1980). Till 1980s, the ivermectin (IVM) were not used even after these were discovered before. It has been observed that ivermectin and milbemycin helps in preventing infections in dogs (mainly caused by *Dirofilaria immitis*), as it contains bioactivities such as anti-helmintic activity (Rawlings et al. 2001) Sakamoto et al. 1984).

Milbemycin D is commercially being used in Japan only, with the oral dosage 1000 pg/kg for dogs. Milbemycin A has been removed from the Japanese market. From milbemycin, moxidectin (MOX) was formed, which was at molecule C25 at an unsaturated side-chain molecule illustrated in Fig. 12.1b, and was sold in market in year 1990 in Argentina. Moxidectin is being commercialized all over the world as it has a number of applications against cattle parasites in injectable form, wherein the oral therapeutic dosage is maintained at 200 pg/kg for sheep parasites. The milbemycin spectral activity in nematodes is more effective than in arthropods. Apart from it, moxidectin has shown efficacy in both host species, against both ecto- and endoparasites (Ranjan et al. 1992; Webb et al. 1991; Williams et al. 1992). In nematodes, the milbemycin's spectral activity is more effective in comparison to arthropods, and also, license approval of moxidectin proved efficacy against

a

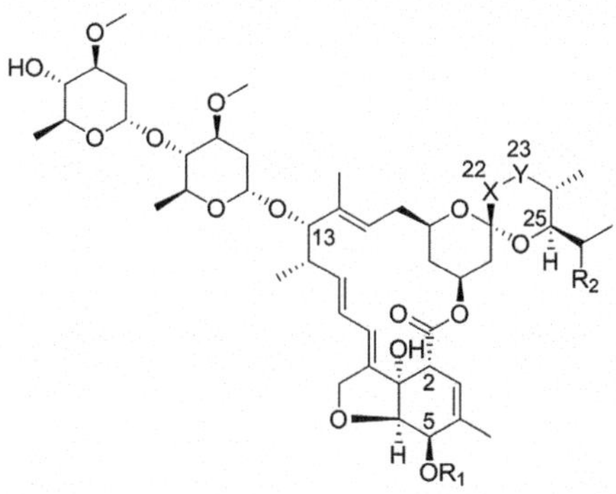

Avermectins	R_1	R_2	X-Y
A1a	CH_3	CH_2CH_3	$CH{=}CH$
A1b	CH_3	CH_3	$CH{=}CH$
A2a	CH_3	CH_2CH_3	CH_2-$CH(OH)$
A2b	CH_3	CH_3	CH_2-$CH(OH)$
B1a	H	CH_2CH_3	$CH{=}CH$
B1b	H	CH_3	$CH{=}CH$
B2a	H	CH_2CH_3	CH_2-$CH(OH)$
B2b	H	CH_3	$CH_2.CH(OH)$
Ivermectins			
B1a	H	CH_2CH_3	CH_2-CH_2
B1b	H	CH_3	CH_2-CH_2

b

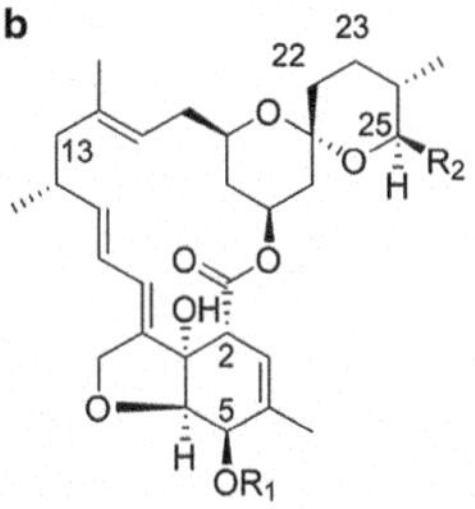

Milbemycins	R_1	R_2
A3	H	CH_3
A4	H	CH_2CH_3
D	H	$CH(CH_3)_2$
B2	CH_3	CH_3
B3	CH_3	CH_2CH_3
G	CH_3	$CH(CH_3)_2$

Fig. 12.1 The structures of (**a**) avermectins and ivermectins and (**b**) milbemycins. (Adopted from Kim et al. 2017)

endo- and ectoparasites in both host species (Ranjan et al. 1992; Webb et al. 1991; Williams et al. 1992). Since 1990, in USA, milbemycin oxime has also been licensed and used for the prevention of *D. immitis* in dogs and as a healing agent for adult *Ancylostoma caninum*. The amount needed for optimal efficiency is 500 pg/kg (Bowman et al. 1990; Grieve et al. 1989; Stansfield and Hepler 1991). Large doses for dogs have shown safer effects in collies that are sensitive to ivermectin. It's been done by substituting ketoxime at C5 position, which changes position and helps in the reduction of distribution in the central nervous system (CNS) (Shoop and Mrozik 1994; Tranqulli et al. 1991).

12.2 Biosynthesis of Avermectin

Biosynthesis pathway elucidation includes three steps:

1. Initial aglycon (6, 8a-seco-6, 8a-deoxy-5-oxoavermectin aglycons) derived from Polyketide formation.
2. Formation of aglycon to avermectin aglycons.
3. Generation of avermectin by avermectin aglycon's glycosylation.

Table 12.1 Macrocyclic lactones commercially available formulations

Macrocyclic lactones	Formulations
Doramectin	Injectable
Ivermectin	Tablets, oral liquid and paste, injectable, topical
Milbemycin	Tablets
Moxidectin	Tablets, injectable, oral drench, topical
Selamectin	Topical

Adopted from Merola and Eubig (2012)

It has been observed that aglycon moiety of the bioactive avermectins is the result of some fatty acids. In order to generate initial aglycon to form avermectin, alteration in the aliphatic polyketide-derived precursor such as lactonization happened. O-glycosylation happens at positions C13 and C4 using deoxythymidine diphosphate-oleandros, which is converted into avermectins as an end product in the last step of avermectin biosynthesis.

12.3 Formulations

Macrocyclic lactones (MLs) including avermectin and milbemectin are used to kill parasites. These macrocyclic lactones are present in a large number of formulations with different varieties as shown in Table 12.1 (Merola and Eubig 2012). There are some common tablet formulated veterinary products including milbemycin, avermectin, and moxidectin. One of the best examples of macrolide is selamectin (SLM), which is basically used to treat infections caused by heartworm, and also, it is used against endo- and ectoparasites. Some other compounds of avermectin including milbemycin, moxidectin, and doramectin (DRM) help in the treatment of sarcoptic and demodectic mange (Plumb 2005) The symptoms of milbemycin and avermectin intoxication are related to the CNS, for example, blindness, ataxia, depression, mydriasis, and hypertension. If not treated, the symptoms worsen over time.

12.4 Uses in Veterinary and Livestock

Avermectins were firstly developed against onchocerciasis in humans. After some time, it was observed that a number of parasites are being inhibited in humans including enterobiasis, ancylostomiasis, trichuriasis, scabies, head lice, lymphatic filariasis, sea lice, and strongyloidiasis in the presence of avermectins (Canga et al. 2008; Davies and Rodger 2000; Geary 2005; Ottesen and Campbell 1994; Patra 2010). Abamectin (ABM), ivermectin (IVM), doramectin (DRM), and eprinomectin (EPM) all are used against lung nematodes, gastrointestinal nematodes, and cattle ectoparasites. The best proficient route of injecting in relations of availability of drug for livestock as well as some other species is ML subcutaneous injection, in comparison to topical and oral administration (Gayrard et al. 1999; Laffont et al. 2001; Lespine et al. 2003). ML administrating dosage for animals to pour-on formulations and injectables should be maintained at 0.5 mg/kg and 0.2 mg/kg, respectively. As the ingredient is less absorbable in the gastrointestinal track and skin, and because of that, the dosage required for injections is less than that of the dosage required for pour-on formulations (Bousquet-Melou et al. 2011). Route of administration shows some adverse effects on efficiency of MLs against ectoparasites due to 14–28 days activity of avermectins. For example, injectable or pour-on administration is much effective as compared to oral administration in case of mange mites, while injectable administration is less effective against several red lice, such as *Bovicola bovis* (Benz et al. 1989; Chick et al. 1993). DRM has achieved high activity (for 27 days) (Muniz et al. 1995). Generally, oral administration is best suitable for ruminants that are small. Even though there are rare controls, while EPM is registered for small dairy ruminant's usage, the formulations of cattle are frequently used due to low residues present in milk (Prichard et al. 2012).

IVM is registered for horse use. It is commonly administered as a paste orally or as a liquid oral formulation. IVM has very much efficacy against many nematode parasites that are found on horses as well as bots (Prichard et al. 2012). The IVM and selamectin (SLM) has efficacy against heartworm disease present in cats and dogs. SLM is also effective against lice, fleas, and mange mites. Usually, in heartworm, preventatives are given after 30 days during transmission season of the heartworm. IVM (at 50 μg/kg) act against and kill microfilariae of *Dirofilaria immitis*. IVM take almost 17–37 months for eliminating all different stages of *D. immitis*, including the adult worms (McCall 2005).

Milbemycin derivatives are valuable as agricultural and horticultural acaricidal, anthelmintics, and insecticidal agents (Zhao et al. 2011). Moxidectin (MOX) is basically used to control lung and gastrointestinal nematodes and ectoparasites found in cattle. MOX when given as a single dose has activity for long time based on species type. This is due to MOX's higher persistent efficiency present in the host. Usually, IVM and MOX must not be given to milk-producing dairy cattle, because the milk is excreted very minute. However, in the case of dairy cattle that are lactating, pour-on MOX is registered with no withdrawal of milk, which is due to less toxicity level of MOX as compared to IVM. Injection for long-term lactation of MOX (1 mg/kg) provides 50 days of protection against *Boophilus microplus* (Davey et al. 2011). Generally, MOX commonly shows good effect, when it is taken at recommended dose and dosage to control parasitic resistance development having avermectin as sheep and goats. MOX is also registered and orally given as paste to horses. MOX is highly effective against encysted small strongyle larvae. MOX is also used to eradicate adult stage, as well as the stage that is not mature in the case of hookworms as well as roundworms in cats and dogs, lungworm mites (*Otodectes cynotis*), demodex mites, and sarcoptic mange mites. The MOX injectable preparations provide long-acting defense against *D. immitis* that causes heartworm diseases.

12.5 Uses in Crop Protection

Avermectins were the first reported for their insecticidal activity (Ostlind et al. 1979). Later studies revealed that the avermectins' broad spectrum activity includes its use in agricultural science (Putter et al. 1981). Avermectin B_1 (abamectin) is useful in protecting crop because it is very toxic for arthropods (Dybas et al. 1989). Avermectin B_1 possesses unique effectiveness property in contrary to phytophagous insects, with a value of LC_{90} ranges from 0.02 to 0.24 ppm Eriophyidae members, Tarsonemidae and Tetranychidae (Lasota and Dybas 1991). Avermectin B_1 is deadly poisonous for eriophyid mites (Dybas 1989). It is considerably not as much poisonous to other phytophagous mites, for example, citrus red mite (Dybas 1989). Avermectin B_1's poisonousness to pests and insects are more inconstant. Even though abamectin is more poisonous to tobacco hornworm *Manduca sexta*, tomato pinworm *Keiferia lycopersicella*, Colorado potato beetle *Leptinotarsa decemlineata*, moth diamond back *Plutella xylostella*, budworm *Heliothis virescens*, and *Liriomyza trifolii*, as well as the serpentine leaf miner (LC_{90} values range between 0.02 and 0.19 ppm), it is not much effective against many of Homoptera and Lepidoptera (LC_{90} values range between 1 and > 25 ppm) (Dybas 1989; Lasota and Dybas 1991). Due to its lesser potency against many coleoptera, homoptera and lepidoptera restricted its chances for later development. Even though avermectin B_1 is poisonous to certain aphids, for example, LC_{90} values against *Aphis fabae* are in between 0.2 and 0.5 ppm and also against cotton aphid, *Aphis gossypii*, are in range between 0.4 and 1.5 ppm, avermectin B_1 are not showing the efficacy in regulating aphids in translaminar assays (Dybas and Green 1984; Putter et al. 1981). The efficiency that is reduced at inhibiting aphids is perhaps caused by toxic (or lower) concentrations of avermectin B_1 in phloem tissue where aphids feed actively. Generally, avermectin B_1 is not much toxic to valuable arthropods, particularly when they are introduced after 1 day of application. LC values contrary to many of valuable arthropods are maximum than those for the key target pests (Dybas 1989; Zhang and Sanderson 1990). The spectrum and simultaneous proficiency of emamectin benzoate have not been broadly studied as those for avermectin B_1. Generally, emamectin benzoate is more efficient to a wide-ranging lepidoptera spectrum. It is a highly efficient insect repellent compound registered and developed for use in agricultural practices. The range of LC_{90} emamectin benzoate value, against a lepidoptera family variety, lies between 0.002 and 0.89 ppm (Dybas 1989). Emamectin hydrochloride has high efficacy against armyworm species, for example, beet armyworm *Spodoptera exigua*, than abamectin and is also more potent against southern armyworm *Spodoptera eridania*, than fenvalerate, methomyl, and thiodicarb, respectively (Dybas et al. 1989; Mrozik et al. 1989; Trumble et al. 1987).

Emamectin benzoate is less lethal for different valuable organisms in comparison of abamectin. Emamectin benzoate foliar filtrates were marginally toxic (<20% mortality) to many useful insects, comprising honeybee, and numerous predators and parasitoids, within a day or few hours after application. The reason for its low toxicity is related to the short half-life of emamectin benzoate on foliage. The half-life of foliar dislodgeable residues was likely to be approximately 0.66 days on celery. After 24 h of application, still 1.3 ng/cm^2 residues were found in celery and alfalfa crops. Emamectin benzoate wet residues are usually toxic in nature to many arthropods; convergent lady beetle *Hippodamia convergens* and common green lacewing *Chrysopa carnea* show tolerance against emamectin benzoate, while they are exposed to wet residues. Emamectin hydrochloride (MK-243) showed slightly adversative effects counter to parasitoids (*Cotesia orobenae* and *Pteromalus puparum*) (Kok et al. 1996).

Milbemectin shows substantial miticidal activity, which affects so many kinds of important mite pest such as the citrus red mite (*Panonychus citri* McGregor), the carmine spider mite (*Tetranychus cinnabarinus* Boisd.), and two-spotted spider mite (*Tetranychus urticae* Koch). Milbemycin D shows insecticidal effects on the gypsy moth *Lymantria dispar* (L). It has also been found that milbemectin shows potential activity in controlling *Bemisia tabaci* population, and its effectiveness is improved by using mineral oils against both whitefly larvae and adults (Pluschkell et al. 1999).

12.6 Mode of Action

The avermectins and milbemycins (A/M) targeted ligand-gated chloride channels. This receptor family is found in both CNS of invertebrates and vertebrates. These receptors pass a number of transmitters. The main transmitter is the one that is gated by glutamate, known as GluCls, and it is known as avermectin (Wolstenholme and Rogers 2005). In invertebrates, the GluCl channels are the only channels found in targeted phyla of avermectin as well as in mollusks. The pharynx of nematode mainly performs a role to intake as well as to process the food partially before it is transferred to the gut. Arrangement of organ differs broadly among the species of nematodes. However, it is considered as the most specific characteristic in their morphology. It includes muscle, gland cells, separate nerve, and other self-regulating system. Pharyngeal pumping is quite sensitive to A/M in nematodes. Apart from it, the values of EC_{50} are also sensitive to the effect caused by drugs, which ranges from 0.2 to 10 nM (Wolstenholme and Rogers 2005). The A/M affects mainly nematodes, and it results in paralysis and ultimately leads to the death of the worm. In the case of nematodes, when GluCl is present on pharyngeal muscle cell, the pumping is inhibited (Fig. 12.2) (Martin 1996). These receptors' irreversible activation in ivermectin causes muscle depolarization, likely due to high internal [Clx] and an ending of pumping (Pemberton et al. 2001). Another major effect caused by A/M on worm is a specious body wall muscle paralysis, which results into immobilization. Though, from evidence point of view it is an indirect effect instead of a direct inhibition of neuromuscular transmission also GluCl indicate no sign in cells of muscle. Nematodes' motion is controlled by both neurons' excitatory and inhibitory motors, which are organized into dorsal and ventral nerve cords. Reciprocal excitation waves and waves of inhibition go down in the body, which lead to relaxation of dorsal muscles and contraction of ventral muscles and vice versa. This causes characteristic sinusoidal swimming motion (Wolstenholme and Rogers 2005). In

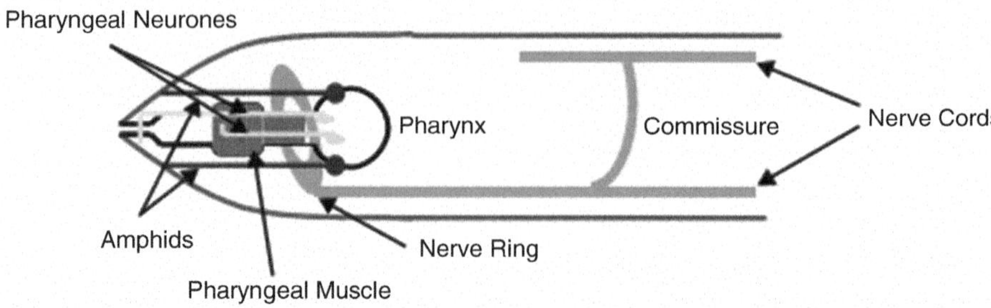

Fig. 12.2 Schematic representation of the distribution of GluCl in nematodes. The cuticle is outlined in gray and the pharynx in black. Structures reported to express GluCl are indicated by arrows. (Adopted from Wolstenholme and Rogers 2005)

Caenorhabditis elegans, studies revealed that interneuron in the head of the worm commands and controlled the motor neurons that regulate the worm's reverse and backward locomotion (Zheng et al. 1999). Initial study reveals that, in *Ascaris suum*, ventral cord avermectin obstructs transmission between interneurons and excitatory motor neurons and also inhibits ventral transmission (Kass et al. 1984). It is expected that the GluCl subunits are expressed on motor neurons, and using reporter gene constructs, it is set in *C. elegans*, while in case of *Haemonchus contortus*, it is set using antibiotics (Fig. 12.2). In *C. elegans*, motor neurons avr-14 and avr-15 are expressed (Dent et al. 1997, 2000). However, the HcGluCla, a3A, a3B, and b subunits all have been identified in *H. contortus* motor neurons (Wolstenholme and Rogers 2005). The anti-GABA antibodies are proposed as motor neuron inhibitors (Portillo et al. 2003). The use of A/M on these inhibitory motor neuron channels would thus apparently result in irreversible hyperpolarization of cell and their consequential incapability to produce action potentials. This avoids inhibition of transmission at the neuromuscular junction and thus waves elimination of muscular relaxation vital for movement.

Even though macrocyclic lactones mainly target glutamate-gated chloride channels, evidence suggest that these drugs, such as moxidectin and ivermectin, can also target cys-loop GABA receptors in *Ascaris*, *C. elegans*, *H. contortus*, and *Trichinella spiralis* causing either a potentiation or receptor activity inhibition (Boisvenue et al. 1983; Brown et al. 2012; Feng et al. 2002; Holden-Dye et al. 1988; Holden- Dye and Walker 1990; Kass et al. 1980; Ros-Moreno et al. 1999).

12.7 Food Contamination

Avermectin family member residues are used against veterinary parasites contained in products that are obtained from animals, for example, meat and milk. The maximum residue limit (MRL) of ivermectin and abamectin in livestock presents as 0.01 mg per kg and 0.005 mg per kg, respectively (Bai and Ogbourne 2016). Ivermectin and abamectin half-life in milk ranges between 4 days and 2 days, respectively (Cerkvenik-Flajs et al. 2007; Imperiale et al. 2004). However, the presence of abamectin and ivermectin has been observed in milk for 23 days and 1 day, respectively, (Cerkvenik-Flajs et al. 2007; Imperiale et al. 2004). Hence, it is suggested not to use milk and products made from it after 30 days of cattle treatment (Cerkvenik-Flajs et al. 2007). However, withholding period for food products with exposure of abamectin has not been established. As described earlier, it is important to gain approval for abamectin, so that they can be easily used, according to suitable labeling procedures, containing holding period (Moreno et al. 2015).

Avermectin residues in food can be reduced by food processing, though the degree differs under some conditions, e.g., heating milk under low warm conditions at 75 °C and 65 °C for 15 s and 30 min, respectively. Levels of ivermectin did not reduce as they belong to avermectin family, which are lipophilic drugs (Imperiale et al. 2009). Though, in later studies, cheese were obtained from processed milk kept for ripening for 58–61 days, the residues of ivermectin were detected in lower levels, that is, 5–25 days (Cerkvenik et al. 2004). In Europe, a study has been done on beef samples (approx. 1061 beef samples), and around 2.45% of the sample showed detectable veterinary drug residues (0.2–171 μg/kg). However it has been studied that the overall risk of exposure of the European consumer to anthelmintic-drug residues in beef is less than 0.02% which is negligible. These were within the acceptable European maximum residue limits (Cooper et al. 2012). Residues present in meat are capable of lowering up to 50% by boiling or by frying (Slanina et al. 1989).

The maximum residue limit (MRL) acceptable for abamectin in vegetables and fruit is up to 0.01–0.02 mg per kg (Bai and Ogbourne 2016). On the other hand, very less assessment has been done for the assessment of abamectin in food items; nevertheless, abamectin is used to be used as a veterinary treatment or as an acaricide (Kamel et al. 2007; Palmer et al. 1997). Abamectin residue toxicity is found in a number of crops including apricot, celery, Chinese cabbage, cucurbits, and peach. An increase in MRL by 0.05 mg per kg has been reported in certain cases that were investigated by the European Food Safety Authority. However, it was found that the residue presence will not lead to consumer exposure to toxicological reference limits and improbably it was a public health concern (EFSA 2010, 2015). MRL for MOX in milk 40 μg kg^{-1} are established. The maximum residue limit of various citrus fruits for milbemycin A3 is 20 ppb, for milbemycin A4 in pome fruits is 20 ppb, for (Z) 8,9 milbemycin A3 in stone fruits is 20 ppb, and for (Z) 8,9 milbemycin A4 in strawberries is 20 ppb (Food Notice 2018).

12.8 Conclusion

Avermectins and milbemycins are natural products that are further synthesized for marketable supply to use as veterinary therapeutics, pest repellent (insecticides), and pharmacological drugs. These are capable and are utilized for other uses to protect animal, as well as crop, and also in health sector. Other than veterinary and human medicine, there is a lot more about IVM. Although the mode of action of helminth parasite is not yet known, the drug efficacy and host immunity relationship is worth further study. In the case of food contamination, facts recommend that even the residue of avermectin at high concentrations is enough to cause substantial risk in human health. The main cause of this is short half-life as well as decrement in residue throughout food processing. More important, guidelines that need to be followed on how to carry out residue analysis that requires to develop action mechanism in both nontargeted species and targeted species necessity should be systematically understood.

References

Bai SH, Ogbourne S (2016) Eco-toxicological effects of the avermectin family with a focus on abamectin and ivermectin. Chemosphere 154:204–214

Benz GW, Roncalli RA, Gross SJ (1989) Use of ivermectin in cattle, sheep, goats, and swine. In: Campbell WC (ed) Ivermectin and abamectin. Springer-Verlag, New York, pp 215–229

Boisvenue RJ, Brandt MC, Galloway RB et al (1983) In vitro activity of various anthelmintic compounds against *Haemonchus contortus* larvae. Vet Parasitol 13:341–347

Bousquet-Melou A, Jacquiet P, Hoste H et al (2011) Licking behaviour induces partial anthelmintic efficacy of ivermectin pour-on formulation in untreated cattle. Int J Parasitol 41:563–569

Bowman DD, Johnson RC, Hepler DI (1990) Effects of milbemycin oxime on adult hookworms in dogs with naturally acquired infections. Am J Vet Res 51:487–490

Brown DD, Siddiqui SZ, Kaji MD et al (2012) Pharmacological characterization of the *Haemonchus contortus* GABAgated chloride channel, Hco-UNC-49: modulation by macrocyclic lactone anthelmintics and a receptor for piperazine. Vet Parasitol 185:201–209

Burg RW, Miller BM, Baker EE et al (1979) Avermectins, new family of potent anthelmintic agents: producing organism and fermentation. Antimicrob Agents Chemother 15:361–367

Campbell WC (1981) An introduction to the avermectins. N Z Vet J 29:174–178

Campbell WC, Fisher MH, Stapley EO et al (1983) Ivermectin: a potent new antiparasitic agent. Science 221:823–828

Canga AG, Prieto AMS, MJD L et al (2008) The pharmacokinetics and interactions of ivermectin in humans-a mini-review. AAPS J 10:42–46

Cerkvenik V, Perko B, Rogelj I et al (2004) Fate of ivermectin residues in ewes' milk and derived products. J Dairy Res 71:39–45

Cerkvenik-Flajs V, Grabnar I, Kozuh Erzen N et al (2007) Kinetics of abamectin disposition in blood plasma and milk of lactating dairy sheep and suckling lambs. J Agric Food Chem 55:9733–9738

Chick BF, McDonald D, Cobb R et al (1993) The efficacy of injectable and pour-on formulations of moxidectin against lice on cattle. Aust Vet J 70:212–213

Cooper K, Whelan M, Kennedy D et al (2012) Anthelmintic drug residues in beef: UPLCMS/MS method validation, European retail beef survey, and associated exposure and risk assessments. Food Addit Contam Part A 29:746–760

Davey RB, Pound JM, Klavons JA et al (2011) Efficacy and blood sera analysis of a long-acting formulation of moxidectin against *Rhipicephalus* (*Boophilus*) microplus (Acari:Ixodidae) on treated cattle. J Med Entomol 48:314–321

Davies I, Rodger G (2000) A review of the use of ivermectin as a treatment for sea lice [*Lepeophtheirus salmonis* (Krøyer) and *Caligus elongatus* Nordmann] infestation in farmed Atlantic salmon (*Salmo salar* L.). Aquacul Res 31:869–883

Dent JA, Davis MW, Avery L (1997) Avr-15 encodes a chloride channel subunit that mediates inhibitory glutamatergic neurotransmission and ivermectin sensitivity in *Caenorhabditis elegans*. EMBO J 16:5867–5879

Dent JA, Smith MM, Vassilatis DK et al (2000) The genetics of avermectin resistance in *Caenorhabditis elegans*. Proc Natl Acad Sci U S A 97:2674–2679

Dybas RA (1989) Abamectin use in crop protection. In: Campbell WC (ed) Ivermectin and abamectin. Springer, Berlin/Heidelberg/New York, pp 287–310

Dybas RA, Green ASJ (1984) Avermectins: their chemistry and pesticidal activity. In: British crop protection conference pests and diseases, vol. 9B-3. British Crop Protection Council, Croydon, pp 947–954

Dybas RA, Hilton NJ, Babu JR et al (1989) Novel second-generation avermectin insecticides and miticides for crop protection. In: Demain AL, Somkuti GA, HunterCevera IC, Rossmoore HW (eds) Novel microbial products for medicine and agriculture. Elsevier, New York, pp 203–212

EFSA (2010) Modification of the existing MRLs for abamectin in apricots and peaches (including nectarines). EFSA J 8:1683

EFSA (2015) Reasoned opinion on the modification of the existing MRLs for abamectin in various crops. EFSA J 13:4189

Feng XP, Hayashi J, Beech RN et al (2002) Study of the nematode putative GABA type-A receptor subunits: evidence for modulation by ivermectin. J Neurochem 83:870–878

Food Notice (2018) Maximum Residue Levels for Agricultural Compounds https://www.mpi.govt.nz/dmsdocument/19550/direct. Accessed 10 June 2019

Gayrard V, Alvinerie M, Toutain PL (1999) Comparison of pharmacokinetic profiles of doramectin and ivermectin pour-on formulations in cattle. Vet Parasitol 81:47–55

Geary TG (2005) Ivermectin 20 years on: maturation of a wonder drug. Trends Parasitol 21:530–532

Grieve RB, Frank GR, Stewart VA et al (1989) Effect of dosage and dose timing on heartworm (*Diroflaria immitis*) chemoprophylaxis with milbemycin. In: Otto GF (ed) Proceedings of heartworm symposium '89. American Heartworn Society, Washington DC, pp 121–124

Holden-Dye L, Walker RJ (1990) Avermectin and avermectin derivatives are antagonists at the 4-aminobutyric acid (GABA) receptor on the somatic muscle cells of *Ascaris* – is this the site of anthelmintic action? Parasitology 101:265–271

Holden-Dye L, Hewitt GM, Wann KT et al (1988) Studies involving avermectin and the 4-aminobutyric acid (GABA) receptor of *Ascaris suum* muscle. Pest Sci 24:231–245

Imperiale F, Lifschitz A, Sallovitz J et al (2004) Comparative depletion of ivermectin and moxidectin milk residues in dairy sheep after oral and subcutaneous administration. J Dairy Res 71:427–433

Imperiale F, Farias C, Pis A et al (2009) Thermal stability of antiparasitic macrocyclic lactones milk residues during industrial processing. Food Addit Contam 26:57–62

Kamel A, Al-Dosary S, Ibrahim S et al (2007) Degradation of the acaricides abamectin, flufenoxuron and amitraz on Saudi Arabian dates. Food Chem 100:1590–1593

Kass IS, Wang CC, Walrow JP et al (1980) Avermectin b1A, a paralysing anthelmintic that affects interneurons and inhibitory motorneurons in *Ascaris*. Proc Natl Acad Sci U S A 77:6211–6215

Kass IS, Stretton AOW, Wang CC (1984) The effects of avermectin and drugs related to acetylcholine and 4-aminobutyric acid on neurotransmitters in *Ascaris suum*. Mol Biochem Parasitol 13:213–225

Kim MS, Cho WJ, Song MC et al (2017) Engineered biosynthesis of milbemycins in the avermectin high-producing strain *Streptomyces avermitilis*. Microb Cell Factories 16:9

Kok LT, Lasota JA, McAvoy TJ et al (1996) Residual foliar toxicity of 4″ epimethylamino-4″-deoxyavermectin BI hydrochlaride (MK-243) and selected commercial insecticides to adult hymenopterous parasites, *Pteromalus puparum* (Hymenoptera: Pteromalidae) and *Cotesia orobenae* (Hymenoptera: Braconidae). J Econ Entomol 89:63–67

Laffont CM, Alvinerie M, Bousquet-Mélou A et al (2001) Licking behaviour and environmental contamination arising from pour-on ivermectin for cattle. Int J Parasitol 31:1687–1692

Lasota JA, Dybas RA (1991) Avermectins, a novel class of compounds: implications for use in arthropod pest control. Annu Rev Entomol 36:91–117

Lespine A, Sutra JF, Dupuy J et al (2003) Eprinomectin in goat: assessment of subcutaneous administration. Parasitol Res 89:120–122

Martin RJ (1996) An electrophysiological preparation of *Ascaris suum* pharyngeal muscle reveals a glutamategated chloride channel sensitive to the avermectin analogue, milbemycin D. Parasitology 112:247–252

McCall JW (2005) The safety-net story about macrocyclic lactone heartworm preventives: a review, an update, and recommendations. Vet Parasitol 133:197–206

Merola VM, Eubig PA (2012) Toxicology of avermectins and milbemycins (macrocylic lactones) and the role of P-glycoprotein in dogs and cats. Vet Clin North Am Small Anim Pract 42:313–333

Moreno L, Dominguez P, Farias C et al (2015) Ivermectin pharmacokinetics, metabolism, and tissue/egg residue profiles in laying hens. J Agric Food Chem 63:10327–10332

Mrozik H, Eskola P, Linn BO et al (1989) Discovery of novel avermectins with unprecedented insecticidal activity. Experientia 45:315–316

Muniz RA, Hernandez F, Lombardero O et al (1995) Efficacy of injectable doramectin against natural infestations *Boophilus microplus* in cattle. Am J Vet Res 56:460–463

Ostlind DA, Cifelli S, Lang R (1979) Insecticidal activity of the anti-parasitic avermectins. Vet Res 105:168

Ottesen EA, Campbell W (1994) Ivermectin in human medicine. J Antimic Chemoth 34:195–203

Palmer R, Coyne R, Davey S et al (1997) Case notes on adverse reactions associated with ivermectin therapy of Atlantic salmon. Bull EU Assoc Fish Pathol 17:62–67

Patra KC (2010) Ivermectin the antiparasitic antibiotic. Pediatr Infect Dis 2:14–20

Pemberton DJ, Franks CJ, Walker RJ et al (2001) Characterization of glutamate-gated chloride channels in the pharynx of wild-type and mutant *Caenorhabditis elegans* delineates the role of the subunit GluCl-alpha 2 in the function of the native receptor. Mol Pharmacol 59:1037–1043

Plumb DC (2005) Plumb's veterinary drug handbook. Blackwell, USA

Pluschkell U, Horowitz AR, Ishaaya I (1999) Effect of milbemectin on sweetpotato whitefly, *Bemisia tabaci*. Phytoparasitica 27(3):183–191

Portillo V, Jagannathan S, Wolstenholme AJ (2003) Distribution of glutamate-gated chloride channel subunits in the parasitic nematode *Haemonchus contortus*. J Comp Neurol 462:213–222

Prichard R, Ménez C, Lespine A (2012) Moxidectin and the avermectins: consanguinity but not identity. Int J Parasitol Drugs Drug Resist 2:134–153

Putter I, MacConnell JG, Preiser FA (1981) Avermectins: novel insecticides, acaricides, and nematicides from a soil microorganism. Exp Dermatol 37:963–964

Ranjan S, Trudeau C, Prichard RK et al (1992) Efficacy of moxidectin against naturally acquired nematode infections in cattle. Vet Parasitol 41:227–231

Rawlings CA, Bowman DD, Howerth EW, Stansfield DG, Legg W, Luempert LG (2001) Response of dogs treated with ivermectin or milbemycin starting at various intervals after *Dirofilaria immitis* infection. Vet Ther 2(3):193–207

Ros-Moreno RM, Moreno-Guzman MJ, Jimenez-Gonzalez A et al (1999) Interaction of ivermectin with gammaaminobutyric acid receptors in *Trichinella spiralis* muscle larvae. Parasitol Res 85:320–323

Sakamoto T, Seki I, Dikuchi D et al (1984) Anthelmintic effect of milbemycin D on parasites of dogs. J Fac Agric Iwate U17:69–81

Shoop WL, Mrozik H (1994) veterinary pharmacology in animal health discovery: structure and activity of avermectins and milbemycins in animal health. In: Veterinary pharmacology in the pharmaceutical industry, proceedings 9th biennial symposium, The American Academy of Veterinary Pharmacology and Therapeutics. Jernigan AD, Sadler LM, Brown SA Kalamazoo MI (eds), pp 35–51

Slanina P, Kuivinen J, Ohlsen C et al (1989) Ivermectin residues in the edible tissues of swine and cattle: effect Kalamazoo MI of cooking and toxicological evaluation. Food Addit Contam 6:475–481

Stansfield DG, Hepler DI (1991) Safety and efficacy of milbemycin oxime for parasite control. Canine Practice 16:11–16

Takiguchi Y, Mishna H, Okuda M et al (1980) Milbemycins, a new family of macrolide antibiotics: fermentation, isolation and physico-chemical properties. J Antibiot 33:1120–1127

Tranqullli W, Paul AJ, Todd KS (1991) Assessment of toxicosis induced by high-dose administration of milbemycin oxime. Am J Vet Res 52:1170–1172

Trumble JT, Moar WJ, Babu JR et al (1987) Laboratory bioassays of the acute and antifeedant effects of avermectin BI and a related analogue on *Spodoptera exigua* (Hübner). J Agric Entomol 4:21–28

Webb JD, Burg JG, Knapp FW (1991) Moxidectin evaluation against *Solenoptes cupillatus* (Anoplura: Linognathidae), *Bovicola bovis* (Mallophaga: Trichodectidae) and *Muscll autumnalls* (Diptera: Muscidae) on cattle. J Econ Entomol 84:1266–1269

Williams JC, Barras SA, Wang GT (1992) Efficacy of moxidectin against gastrointestinal nematodes of cattle. The Vet Rec 131:345–347

Wolstenholme AJ, Rogers AT (2005) Glutamate-gated chloride channels and the mode of action of the avermectin/milbemycin anthelmintics. Parasitology 131:S85–S95

Zhang Z, Sanderson JP (1990) Relative toxicity of abamectin to the predatory mite *Phytoseiulus persimilis* (Acari: Phytoseiidae) and the twospotted spider mite (Acari: Tetranychidae). J Econ Entomol 83:1783–1790

Zhao JH, Xu XJ, Ji MH et al (2011) Design, synthesis, and biological activities of milbemycin analogues. J Agric Food Chem 59:4836–4850

Zheng Y, Brockie PJ, Mellem JE et al (1999) Neuronal control of locomotion in *C. elegans* is modified by a dominant mutation in the GLR-1 ionotropic glutamate receptor. Neuron 24:347–361

Phytogenic Feed Additives in Animal Nutrition

13

Jatinder Singh and Dhananjay Suresh Gaikwad

Abstract

India is very rich in fauna and diversified flora. It is established that synthetic drugs could pose serious problems; besides this, they are toxic and costly. In contrast to this, herbal medicines are relatively nontoxic, economical, and eco-friendly. Moreover, people have been using them for generations. Herbal feed additives, enzymes, probiotics and prebiotics vitamins, and trace minerals with herbal extracts are commonly used as an herbal animal health product. The range of herbal supplements offered is well known for its nutritious value, effectiveness, and ability to resist different diseases, etc. They have also been used in managing problems of health-related issues of animals. Animal health products protect the sustainable food supply chain and are dynamic enough to prevent disease, etc., in animals. They are a wide variety of herbs, and products are derived from them. Although several reports have confirmed antimicrobial and antioxidative and stimulation of immune efficacy in vitro, a particular experiment under vivo conditions is still not sufficient. Less number of experimental comparisons of herbal plants has suggested similar effects on the animal gut microflora. It is mandatory to use herbal products to keep fit our farm animals and to get healthy and quality products. However, the future of using herb products in animal feed will in a great degree depend on the information of chemical structure, value, and features of practical herbs.

Keywords

Herbal · Natural · Animal · Product · Health and disease

J. Singh · D. S. Gaikwad (✉)
School of Agriculture Lovely Professional University (LPU), Phagwara, Punjab, India

J. Singh, A. N. Yadav (eds.), *Natural Bioactive Products in Sustainable Agriculture*, https://doi.org/10.1007/978-981-15-3024-1_13

13.1 Introduction

Today, phytogenic is a term not used homogeneously. The phytogenic compound contains phytochemicals, photobiotic, and secondary plant compounds having a great deal for interpretation. Some people used the term essential oil as a phytogenics. However, they show the only plant-derived compounds category. The fundamental idea was used to a natural plant extract and its products. The natural plant extracts and their products are used as potential scours in sustainable livestock farming (Phytogenius.com; Mayrhuber 2018). In the animal feed industry, phytogenic feed additives acquire significant awareness and become a greater expanse producer which incorporates them into their feeding projects. Today, 70–80% of animal feed manufacturing industries include phytogenic feed additives in broilers and pig nutrition (World Poultry 2008). Phytogenic feed additives are alternative to antibiotics, and they increase the number of key roles in the animal body. Phytogenic feed additives may be included in poultry ration, that is, focus to emphatically affect body weight gain, feed consumption, blood profile, immune/heath, and carcass quality in poultry birds. Phytogenic feed additives are classified into four groups, which are as follows: Sensory feed additives mean those feed additives which affect scenery/organoleptic properties of animal products and increase the palatability of commercial feed. Technological feed additives are the antioxidants which reduce the mycotoxin in animal feed. Zootechnical feed additives are the substances which affect immunomodulation; improve digestion, promoting the growth from the nonmicrobial origin; improve the quality or performance of animal products, etc. The feed additives contain different plant enzymes from plant sources, minerals, and vitamins. The first three classes are important as a source of animal feed additives (Karaskova et al. 2015). In human nutrition and medicine, commonly, compounds from plants have been used as spices and flavors. Growth promoters have reported producing meat for the last few decades to increase parameters like gain in feed conversion weight and ratio. In poultry feed, huge numbers of herbs and spices are used for natural growth promoters, of which importance is illustrated in Table 13.1.

Phytogenics are derived from herbs, spices, or other plant parts. An Austrian multinational feed additive company named Delacon coined the term phytogenic feed additives and introduced in the market in the 1980s. It is noticeable that the class of feed additives includes a huge type of substances containing greater number of active ingredients, including vitamin c, saponins, gingerols, allicin, cinnamic aldehyde, and thymol; these are some important examples. The phytogenic feed additives are commercialized in the form of extracts, essential oils, and dried powder. However, the phytogenic feed additives are chemical-based ingredients, based on their configuration and effect of environmental conditions, places, and harvesting time. Therefore, dissimilarity in capability between phytogenic products is commercialized at fields assigned to differences in their chemical ingredients (Steiner 2006).

Table 13.1 Phytogenic feed additives in animals and poultry

Herb/ species	Latin name	Plant family	Plant part used	Active principle	Structure	Key benefits
Amla	*Emblica officinalis*	Phyllanthaceae	Fruits	Vitamin C	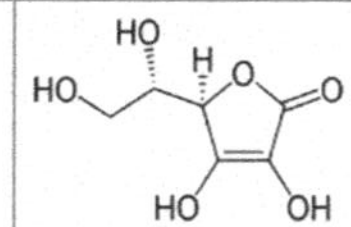	Rejuvenator and immunomodulator
Shatavari	*Asparagus racemosus*	Asparagaceae	Roots	Saponins	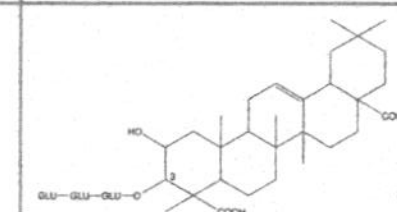	Antioxidant properties, boost immune system
Ginger	*Zingiber officinale*	Zingiberaceae	Rhizome	Gingerols and shogaols	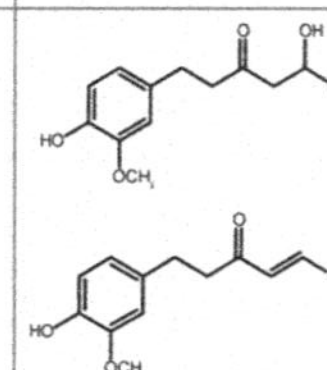	A gain in weight and antioxidants
Garlic	*Allium sativum*	Amaryllidaceae	Bulb	Allicin (diallyl-thiosulfinate)	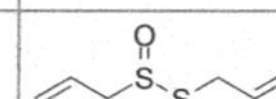	Improve nutrient digestibility, antimicrobial, anti-inflammatory, anti-oxidant and immunostimulant
Cinnamon	*Cinnamomum verum*	Laurels	Bark	Cinnamaldehyde	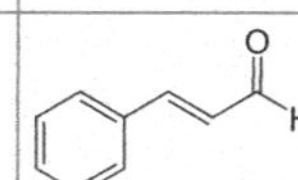	Enhanced activity of gut microflora, improvement of immune response and feed efficiency
Thyme	*Thymus vulgaris*	Lamiaceae	Leaves	Thymol	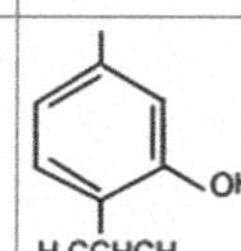	Antimicrobial, anti-inflammatory, Antimelanoma, antioxidant, antiseptic, carminatives, and flavors

13.2 Amla (*Emblica officinalis*)

Out of several Indian traditional medicines, Indian gooseberry or amla fruit has a distinct concern. Its composition is as follows (Table 13.2).

This fruit is also reported as the richest source of vitamin C content and encompasses 700 mg. Besides this, some active tannoid compounds, namely, emblicanin B, emblicanin A, pedunculagin, and punigluconin, have been reported in amla fruit. It is used as an eyewash, restorative tonic, appetite stimulant, as a laxative, and to cure various kinds of diseases like anemia, jaundice, diarrhea, indigestion, anorexia, etc. Amla tree is deciduous and belongs to family Euphorbiaceae. Its significance is also highlighted in old Ayurvedic books like *Sushruta Samhita* and *Charaka Samhita*. It has various kinds of medicinal as well as pharmaceutical properties including antihepatotoxic, anti-inflammatory, hypolipidemic, antibacterial, antifungal, immunomodulating, etc. (Kiritikar and Basu 1935). Besides the abovementioned diseases, it has a very important role in curing cancer, liver, heart problems, and many other diseases.

Furthermore, amla also helps in decreasing cholesterol levels of the human body. It has an important role in the dyeing industry. Extract of amla is used also during the preparation of ink. The wood of amla plant is used in a firework (Kumar Sampath 2012). The fruit of amla is advantageous in case of various skin diseases, conjunctivitis, emaciation, dyspepsia, gastrohelicosis, intermittent fever, hepatopathy dysentery, menorrhagia, hemorrhage, asthma, cephalgia, bronchitis, colic jaundice, and ophthalmopathy (Anjaria et al. 2002). Medical studies recommend that *Emblica officinalis* has antiviral features and is able to act as an antifungal and antibacterial agent. Better ability for controlling bacteria be due to stimulation of phagocytic cell is recorded in case of animals which are fed with amla products, especially powder. Indian gooseberry (amla) has also shown some good antiviral and antimicrobial properties but in in vitro conditions. Some introductory indications also recorded that amla extracts encourage apoptosis process and alter gene expression in osteoclasts involved in osteoporosis and rheumatoid arthritis.

Table 13.2 Different components present in Amla (*Emblica officinalis*) and their approximate range

Biochemical Contents	Range (approximately)
Moisture content	5.05–6.78%
Fat content	0.23–0.59%
Calcium content	79.6 mg
Phosphorous content	12.38 mg
Iron content	88.03 mg per 100 g
Mineral content	Not specified

13.2.1 Chemistry of Amla Fruit

Indian scientists have identified an important pharmacological action of amla and nominated it as "phyllemblin." Amla fruit has a significant amount of pectin, flavonoids, tannins, quercetin, gallic acid, phyllaemblic compounds, and vitamin C. Various phytochemical components and polyphenolic compounds are also recorded in its composition. A wide range of tannins, including flavonoids, terpenoids, and alkaloids, has been revealed to possess advantageous biological accomplishment (Kim et al. 2005). Various parts of the amla treelike leaves and bark including fruits are also rich in tannins. Even roots of amla tree contain lupeol and ellagic acid. Moreover, the bark of the plant contains leucodelphinidin. Above all seeds yield a fixed amount of oil (16%) and yellow. Oil of amla contains linoleic (44.0%), linolenic (8.8%), stearic (2.15%), oleic (28.4%), myristic (1.0%), and palmitic acid (3.0%) (Thakur et al. 1989). Amla plant has a significant amount of hydrolysable tannins, alkaloids, along with some flavonoids, that is, phyllantine and phyllantidine (Ghosal et al. 1996). From amla fruit various kinds of chemicals like chebulinic acid, gallic acid, quercetin, ellagic acid, chebulagic acid, corilagin, and isostrictinnin are isolated in good amount (Zhang et al. 2003). Its fruit also is known to contain two tannins, having a molecular weight less than 1000, known as emblicanin B (33%), emblicanin A (37%), pedunculagin (14%), and punigluconin (12%) (Ghosal et al. 1996). Newly invented tannins, namely, emblicanins A and B, are known for their strong antioxidant action (Santos et al. 1999; Bhattacharya et al. 1999), and these are also recognized to defend erythrocytes from oxidative stress persuaded by generation of superoxide (ROS) radical or asbestos by a distinct procedure extract of amla powder which is prepared from fresh amla. By doing so, all the properties of amla fruit are sustained. This kind of preparation has good solubility in water, and it is free-flowing powder, and color is pale green to deep brown and involves enough polyphenols. This powder helps in rebuilding new cells and improves the count of blood cells particularly RBC. Throughout stress conditions, there is an acute shortage of vitamin C (North 1984). Vitamin C acts as an intoxicating agent against the leukocyte phagocytic activity, thus leading to the reticuloendothelial arrangement, and as a result, antibodies are developed. The following are the major benefits of feeding as feed supplement of amla powder to broilers:

(i) It is very rich in antioxidant and anti-stress properties.
(ii) Amla extract encourages the growth process.
(iii) It helps stimulate the immune system.
(iv) Advances efficiency of feed conversion (Fig. 13.1).

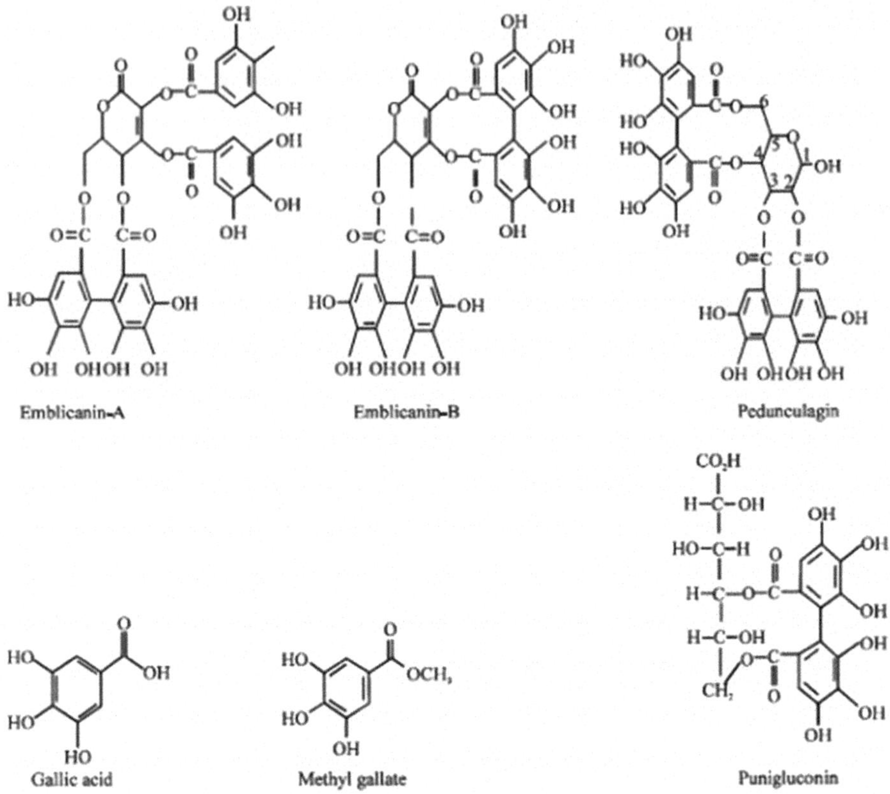

Fig. 13.1 Important phytochemicals of amla plant

13.3 Shatavari (*Asparagus racemosus*)

Shatavari (*Asparagus racemosus*) belongs to family Asparagaceae in order Asparagales. In Ayurveda books, this remarkable plant is recognized as the "queen of herbs." Roots (Rhizome) along with leaves are beneficial to plant parts, and they are used frequently for various purposes.

Plants of Shatavari contain the main active ingredient, Shatavarins, which is having properties like an anti-stress and nutritive tonic (Kamat et al. 2000; Rege et al. 1989). The root of Shatavari plant is reported to add herbal ingredient as an addition in feeds of poultry. It improves appetite and thus stimulates the function of the liver. Shatavari is highly effective in several complications associated with the female reproductive system. In Charak and Ashtanga Hridayam, *Asparagus racemosus* (*A. racemosus*) is listed as an important part of the formulations that are used to cure women's health disorders. Shatavari has a vital role in various diseases, being anticancerous, astringent, antiseptic (Moharana 2008), immunomodulative (Seena and Kuttan 1993). Cowan (1999) described that plant extract of Shatavari has an extensive range of pharmacological actions and encompass an essential oil that exhibits antimicrobial property. Shatavari has tuberous roots which have a bitter and sweet taste and performs as a potential antioxidant.

The antioxidant property of Shatavari may be used in purifying to capture reactive oxygen species produced during stress conditions, leading to extreme damage to cells in the body. Botanical preparations are also used as growth-promoting substances. Above all, time-bound weakening or deteriorating property of antibiotic growth-promoting substances offers a good possibility for other growth-promoting substances plus herbal preparations for using them. Several herbal preparations support the body system to fight against stress ascending due to various reasons related to birds. Many herbs adaptogenic in nature like Amla, Ashwagandha, Ginseng, Tulsi, etc. have been used against anti-stress factors since earlier times in medicines for human as well as for animal, with well-known and good results (Ranade and Desai 2005). In this concern, several herbs can be known as they have a good influence on gut health. But their mode of action on organs of the body is specific due to active ingredients they contain. It is supposed that they have improved influence on several factors which encourage health and growth processes including physiological and immunological functions. Accordingly, such herbal preparations may be supposed as long-term solutions and are more sustained. When these herbal growth-promoting substances are compared with other antibiotic growth promoters, they are found to encourage feed consumption, feed conversion ratio, and consequently body weight (Prasad 2004).

Furthermore, several herbs are supposed to improve various processes of the body including digestion and assimilation of nutrients and general performance (Dobretsberger et al. 1997). These herbal growth-promoting substances are multidirectional in their actions. These are helpful in governing harmful bacteria without the probability of advancement of resistance, and moreover herbal preparations are considered as safe for humans and animals. In this way, they also act as antimicrobial compounds with supplementary benefits such as cost-effective, free from prospect of resistance development, sustaining normal activities, enhancing production levels particularly in broilers, improvement in size of egg, quantity of lymphocyte, and weight of egg in layers (Shon et al. 2002 and Reddy et al. 2002). Moreover, such feed supplemented with herbs is deprived of any side effect, while synthetic-natured additives may cause some contrary effects like toxicity, anxiety-reduced growth rate, and lastly suppression of the immune system. In the end, we can conclude that such herbal preparation may be declared safe and recommend in poultry rearing. The demand for meat is ever increasing in the world especially in India. Under such circumstances, it is expected that poultry industry growth will be accelerated at the same speed. These conditions lead to continuous stress. Therefore, the herbal growth-promoter substances, especially for poultry, may create the optimal condition for normal and dynamic growth in several ways. If these herbal growth promoters are used successfully, it will generate more revenue to poultry owners by effective conversion of feed digested to the body system. The use of herb products in feed acting as growth-promoting substances may encourage the activity of broiler by improving their FCR and weight gain (Prasad and Sen 1993; Samarth et al. 2002). Tanwar et al. (2008) recorded that Shatavari has lactogenic characters to encourage the milk production and improvement in economy for feeding of dairy animals. The same studies were documented by Samkuwar et al. (2005) who described that Shatavari results in a noteworthy increase in daily milk production in animals (Fig. 13.2).

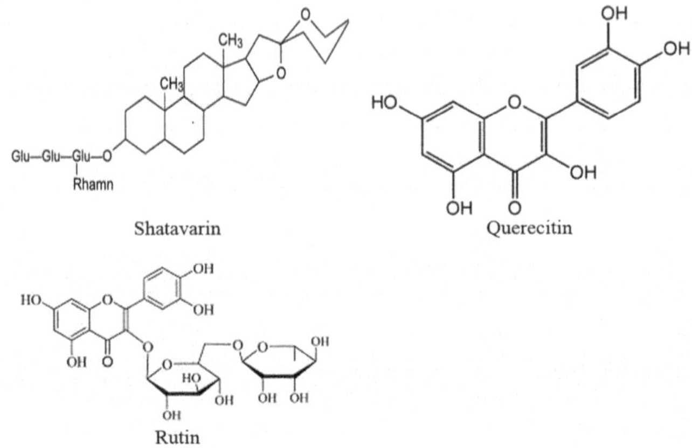

Fig. 13.2 Structures of some reported phytoconstituents of *Asparagus racemosus*

13.4 Ginger (*Zingiber officinale*)

Ginger having distinct medicinal properties comes under the family of *Zingiberaceae* and is a perennial herb. Ginger plant flowers having greenish-yellow petals streaked with purple color look like an orchid flower. Ginger is grown where the annual rainfall is high. Ginger is originated from Asia and is grown in tropical areas including Jamaica, Haiti, Nigeria, and in India; it's mainly grown for a spice crop (Bajaj 1989). Roots of ginger are commonly utilized for spice flavoring (Larsen et al. 1999) and in the treatment for relief from different ailments. Ginger dried powder and extracts act on stimulating tonic and stomachic and raise gastric juice secretion. It is highly capable in constant flatulence, hysteria, atonic dyspepsia, and relaxed and enfeebled habits, especially of gouty and old individuals and act as relieve pain, nausea, cramps of the bowels and stomach, obviate tenesmus, and from colds. Ginger is used from time to time in case of fevers, especially if there in problem in salivary secretions and it is role in curing gases in the intestine cannot be ignored. Ginger extract takes over the antioxidative property since it can scavenge hydroxyl radicals and superoxide anion (Krishnakantha and Lokesh 1993). The ethanolic extract of ginger remarkably decreases the triglyceride's total cholesterol (serum) and increases the level of HDL cholesterol. The extract of ginger prevents cells from lipid peroxidation and low lipid activity in diabetic rats (Bhandari et al. 2005). Ginger also decreased triglycerides levels, VLDL cholesterol, and LDL cholesterol, in mice deficient of apolipoprotein-E was also reported (Fuhrman et al. 2000). Mohammed and Yusuf (2011) reported that ginger plays an important role as a pro-nutrient because it contains active ingredients (Mohammed and Yusuf 2011).

13.5 Cinnamon (*Cinnamomum verum*)

Cinnamon (*Cinnamomum cassia*) is a member of Lauraceae family originated from South India and Sri Lanka. It is also called as "dalchini" and mostly used worldwide (Jakhetia et al. 2010). It contains digestion and appetite stimulant properties (Tung et al. 2008). Cinnamic acids, cinnamate, cinnamaldehyde, caryophyllene oxide, and L-borneol eugenol are the oils extracted from cinnamon (Tabak et al. 1999). Cinnamon contains cinnamaldehyde which contains antibacterial (Singh et al. 2007) and antioxidant properties (Chang et al. 2001), and it also contains anti-ulcer, antidiabetic, and anti-inflammatory properties (Jakhetia et al. 2010). Cinnamaldehyde, 2-hydroxy cinnamaldehyde, cinnamyl acetate, coumarin, eugenol, and caryophyllin are present in cinnamon oil. In cinnamon oil, four to ten phenolic compounds (mainly eugenol), hydrocarbons (alpha-pinene and caryophyllin), and small amounts of ketones, alcohols, and esters are present. Cinnamon is rich in antioxidants, and it enhances the meat quality of broilers (Ciftci et al. 2010). Many antibacterial activities found in cinnamaldehyde and eugenol are against various pathogenic microbes (Gurdip et al. 2007). The ginger extract shows appetite and digestion stimulating properties besides antimicrobial effects (Yu et al. 2007; Kamel 2001). Medicinal plants like essential oils and cinnamon are used as in replacement of antibiotics because of their antimicrobial effects which stimulate positive response on the digestive system (Lee et al. 2004). Utilization of cinnamon and its oils in broiler diet improves the growth of broilers (Osman et al. 2005). In cinnamon, cinnamaldehyde and eugenol are the active ingredients which increase the gain in weight of the body. The bark of cinnamon is also used as hypoglycemic and cholesterol-reducing agent (Khan et al. 2003); it improves wound healing (Kamath et al. 2003), is an antimicrobial agent (Simic et al. 2004), and is also an anti-inflammatory compound (Chao et al. 2005). Many studies carried out on cinnamon on meat quality and broiler performance, but the results are not significant (Chen et al. 2008; Park 2008) (Table 13.3).

Table 13.3 Chemical components of various parts of cinnamon (Vangalapati et al. 2012)

Morphological Plant parts	Chemical Component
Leaves	Eugenol (70.0–95.0%), cinnamaldehyde (1.0–5.0%),
Bark	Eugenol (5.0–10.0%), cinnamaldehyde (65.0–80.0%),
Root bark	Camphor (60.0%)
Fruit	Trans-cinnamyl acetate (42.0–54.0%), caryophyllene (9.0–14.0%)
C. zeylanicum buds	Terpene hydrocarbons (78.0%), oxygenated terpenoids (9.0%), alpha-bergamotene (27.38%) alpha-copaene (23.05%)
C. zeylanicum flowers	(E)-Cinnamyl acetate (41.98%), caryophyllene oxide (7.20%), trans-alpha-bergamotene (7.97%)

13.6 Thyme (*Thymus vulgaris*)

In the world, thyme is used for medical as well as spice purpose. It is covered through the Mediterranean area and contains in hot and dry summer weather. In thyme, the strong scent is present that people used these plants and have used it in industrial utilization sector (Stahl-Biskup and Saez 2002). Thyme belongs to genus *Thymus* and in perennial evergreen aromatic herbs of family Lamiaceae. They have ornamental uses, culinary, and in medicines also. Thyme is used as a medicinal plant or as a spice; any herb having a property of used multipurpose functions to meet present and future challenges.

Thymol contains monoterpene [5-methyl-2-(1-methyleth-yl) phenol; C10H14O]. It acts on different gram-positive and gram-negative bacteria, and it has well-analyzed active components of essential oils. The European Union (Reg. no. 1831/2003/EC) imposed a ban on the usage of growth promoter-enriched antibiotics in animal feed, which increase the health problem and losses (Attia et al. 2016). Probiotics, prebiotics, symbiotic, medicinal plants, and essential oil are used in replacement of antibiotic growth promoter (Bozkurt et al. 2012; Masek et al. 2014; Hady et al. 2016).

13.7 Mode of Action of Phytogenic Feed Additives

Phytogenic feed additives have a multiplex action mechanism, with a history of it being used in animal feed utilized as a feed substance in animal nutrition. Hence the uses of these in animal nutrition improve the performance and health of the animal (Hippenstiel et al. 2011; Mathe 2009; Windisch et al. 2008).

The actions of essential oils on microbial cells differ in the presence of functional groups. For example, carvacrol and common thymol terpenoids have the same antimicrobial effects but have different acting mechanisms on gram-negative and gram-positive bacteria. Phenolic terpenoids contain hydroxyl group which acts as antimicrobial (Ultee et al. 2002; Lambert et al. 2001). The presence of functional groups determines their level of activity which works on different bacteria. Hellander et al. (1998) reported that carvacrol and thymol disintegrate the membrane and release lipopolysaccharides (Hellander et al. (1998) (Table 13.4).

13.8 Antimicrobial Effects of Phytogenic Compounds

One way by which antibiotics can kill bacteria consists in the inhibition of peptidoglycan synthesis of the bacterial cell wall. Another important class of antibiotics, namely, tetracyclines and macrolides, inhibit bacterial growth via the inhibition of bacterial protein biosynthesis at the 30S-ribosomal or the 50S-ribosomal subunit, respectively. Two further important modes of action are inhibition of DNA topoisomerase or RNA polymerase, inhibiting DNA or RNA synthesis. And finally, inhibition of folic acid synthesis will reduce bacterial enumeration. Bacteria can develop

Table 13.4 Mode of action of different feed additives in animal and poultry nutrition

S. no	Phytogenic feed additives	Mode of action	References
1.	Shatavari (*Asparagus racemosus*)	Acidity disorders, improves indigestion, enhancing blood prolactin and quality of milk and cellular division	Choudhary et al. (2010) and Upadhaya and Ho Kim (2017)
2.	Amla (*Phyllanthus emblica*)	Antimicrobial effects, anticancer, antiulcer, and gastroprotective effects and effects	Kumar et al. (2018), Kumar et al. (2014a) and Dalal et al. (2018)
3.	Ginger (*Zingiber officinale*)	Protein-digesting enzyme, improves digestion, antimicrobial, and kills parasites and their eggs	George et al. (2015)
4.	Cinnamon (Cinnamomum *cassia*)	Antimicrobial and appetizing	Kumar et al. (2014b) and Ali Ahmad Alaw Qotbi (2016)
5.	Thyme (*Thymus vulgare*)	Antioxidant and antimicrobial	Haselmeyer et al. (2014) and ALsafa and AL-Faragi (2017)

resistance mechanisms against all antibiotic target sites, in which risk is increased in said use as antimicrobial growth promotion (Blair et al. 2015). Regarding bactericidal effects of phytogenic substances, it has been frequently postulated that essential oils can penetrate or damage the bacterial cell wall and cell membrane. Once inside the bacterium, essential oils are assumed to trigger the blockage of cytosolic proteins and the efflux of essential intracellular compounds and with it the destruction of bacteria.

13.9 Feed Palatability Improvement, Feed Digestibility, and Feed Intake

The efficiency of phytogenic feed additives on the characteristics of broilers are attributed to different types, although they are used further as an antimicrobial agent and reported by various studies (Cross et al. 2007).

These compounds are significant as a mediator between the environment and plants. Out of various phytogenic compounds, essential oils are regarded as the most studied compounds because of strong odors (Burt 2004; Navarrete et al. 2010; Faleiro 2011; Yap et al. 2014).

The essential oils and their derived compounds act on cytoplasm, and cell wall changes the shape of cells as they disintegrate the membrane (Kim et al. 1995; Lambert et al. 2001; Delaquis et al. 2002). EOs kill the pathogenic bacteria by leaking cell contents, damaging cytoplasmic membrane, destructing membrane proteins (Conner and Beuchat 1984; Cox et al. 1998; Hellander et al. 1998; Ultee et al. 2002), and reducing proton force (Nazzaro et al. 2013). Plant active compounds have beneficial effects as they are therapeutically active (Martins et al. 2000).

Plant-derived compounds have low molecular weight and contain alkaloids, polyphenolic, glucosides, saponins, essential oils, flavonoids, etc. (Huyghebaert 2003; Martins et al. 2000).

These compounds are also reported to act against various stresses and as protection from pathogens, predators, and environmental factors (Huyghebaert 2003). Essential oils are extracted by steam extraction procedure (Oetting 2005). The act mechanism of essential oils is not fully understood, but some hypothesis revealed that it has antimicrobial properties (Dorman and Deans 2000), antioxidant properties (Hui 1996), and enzyme stimulation properties (Brugalli 2003; Kalemba and Kunicka 2003; Kamel 2001).

Phytopathogens have bacteriostatic effects on protozoa, virus, fungi, and bacteria, and their activity is dependent on dosage (Smith-Palmer et al. 1998). Gram-negative bacteria are less sensitive than gram-positive in relation to essential oils (Dorman and Deans 2000; Kohlert et al. 2000).

Further studies on the action mechanism of the active principles and their effects in vivo are required. In addition, significant improvement of animal performance must be shown before plant extracts are effectively adopted in animal nutrition. This study aimed at evaluating the efficacy of cinnamon, clove, oregano, and red pepper plant extracts as alternatives to antimicrobial growth promoters in broiler diets.

Further examinations on the activity of the dynamic standards and their belongings in vivo are required. Furthermore, critical improvement of creature execution must have appeared before plants removed are successfully embraced in nutritional of animals. This examination planned for assessing the adequacy of red pepper cinnamon, oregano, and clove plant extricates against antimicrobial agents in poultry systems.

13.10 Challenges of Phytogenic Compounds

The minimum inhibitory amount of these compounds mandatory for governing enteric germs may not provide assurance concerning the best feed consumption, constant immunity, and cost-wise efficiency. Mostly photogenic materials which are lipophilic may pose a severe challenge in imposing delivery. The effects on anti-inflammation and probable interruption of sensing minimum bacterial number could be explained with various animals, for the enhanced production performance, that have fed with phytogenic additives. Moreover, the erratic consequences and doubtful way of action are prime concerns. Such compounds may also contain regulatory obstacles, side effects, and above all cost-efficiency. The reliability of such compounds during processing is often doubtful. In animal feed, a combination of various types of antibiotics including their alternatives is the better solution substitute to them. In this concern, three prime reasons can be considered:

(i) The action of an individual substitute is doubtful for various performance-enhancing activities of antibiotics.
(ii) The positive influence between various substitutes to diminish essential effective quantities may occur.

(iii) Substituting antibiotics should be a unified technique including nutrition management, and biosecurity, rather than supplementation of additives lonely.

Moreover, the results of recent studies revealed that the combination of various antibiotic substitutes had improved effects on health and performance. A clear understanding of action, including the influence of these different alternatives, will help design better additives (phytogenic) to advance feed competence along with animal growth. It is important to advance practically reasonable delivery approach for the use of such essential oils. Thus, it is clear that proper and novel technologies can decrease the quantity of dose of a phytogenic compound to be used and reduce costs.

13.11 Conclusion

The phytogenic feed additives like cinnamon, ginger, garlic, amla, Shatavari, and thymol influence the performance of broilers like the feed conversion ratio, body weight gain, feed efficiency, meat quality, health, and blood profile. Improvements were observed in different studies for the impact of phytogenic feed additives on inflammatory process, digestibility, and morphology in the digestive tract. Future aspects should be to conduct standardized experiments indicating the composition of the phytogenic feed additives employed in the investigation so that the observation could be easily compared. This is of increasing importance in view of the global trend to restrict the overall use of antibiotic growth promoters and antimicrobial in animal products.

References

Ali Ahmad Alaw Qotbi (2016) The effect of cinnamon powder and cinnamon extract on performance, blood parameters and microbial population of broiler chicks. J Babylon Univ/Pure Appl Sci 24(9):2654

ALsafah AH, AL-Faragi JK (2017) Influence of thyme (*Thymus vulgaris*) as feed additives on growth performance and antifungal activity on *Saprolegnia* spp. in Cyprinus carpio L. J Entomol Zool Stud 5(6):1598–1602

Anjaria J, Parabiam M, Dwivedi S (2002) Ethnovet heritage, 1st edn. Prathik Enterprises, Ahmedabad, p 45

Attia YA, Bovera F, Abd El-Hamid AE, Tag El-Din AE, Al-Harthi MA, El-Shafy AS (2016) Effect of zinc bacitracin and phytase on growth performance, nutrient digestibility, carcass and meat traits of broilers. J Anim Physiol Anim Nutr (Berl) 100:485–491

Bajaj YPS (1989) Biotechnology in agriculture and forestry: medicinal and aromatic plants, vol VI. Springer, Berlin

Bhandari U, Kanoji R, Pillai KK (2005) Effect of ethanolic extract of *Zingiber officinale* on dyslipidaemia in diabetic rats. J Ethnopharmacol 97:227–230

Bhattacharya A, Chatterjee A, Ghosal S, Bhattacharya SK (1999) Antioxidant activity of tannoid of *Emblica officinalis* (amla) in chronic stress induced changes in rat brain. Indian J Exp Biol 37:676–680

Blair JMA, Webber MA, Baylay AJ, Ogbulu DO, Piddock LJV (2015) Molecular mechanism of antibiotic resistance. Nat Rev Microbiol 13:42–51

Bozkurt M, Kücükyilmaz K, Pamukcu M, Cabuk M, Alcicek A, Catli AU (2012) Long-term effects of dietary supplementation with an essential oil mixture on the growth and laying performance of two layer strains. Ital J Anim Sci 11:23–28

Brugalli I (2003) Alimentação alternativa: a utilização de fitoterápicos ou nutracêuticos como moduladores da imunidade e desempenho animal. Anais do Simposio sobre Manejo e Nutrição de Aves e Suínos, Campinas, São Paulo. CBNA, Brasil Campinas, pp 167–182

Burt S (2004) Essential oils: their antibacterial properties and potential applications in foods – a review. Int J Food Microbiol 94:223–253

Chang ST, Chen PF, Chang SC (2001) Antibacterial activity of leaf essential oils and their constituents from *Cinnamomun osmophloeum*. J Ethnopharmacol 77:123–127

Chao LK, Hua KF, Hsu HY (2005) Study on the antiinflammatory activity of essential oil from leaves of *Cinnamomum osmophloeum*. J Agric Food Chem 53:7274–7278

Chen A, Xu J, Yang C, Hong Q (2008) Effects of cinnamon extracts on growth performance and excreta urease activity and nitrogen loss in broilers. In: Proceedings of the 8th livestock environment conference, Iguassu Falls, Brazil, pp 331–338

Choudhary PL, Khare A, Saxena RR, Choudhary KK, Shukla N, Dibya KK (2010) Shatavari (*Asparagus racemosus* willd.) used as a feed supplement for enhancing production and quality of indigenous cross breed cow's milk. Indira Gandhi Krishi Vishwavidyalaya, Raipur

Ciftci M, Simsek UG, Yuce A, Yilmaz O, Dalkilic B (2010) Effects of dietary antibiotic and cinnamon oil supplementation on antioxidant enzyme activities, cholesterol levels and fatty acid compositions of serum and meat in broiler chickens. Acta Vet Brno 79:33–40

Conner DE, Beuchat IR (1984) Effects of essential oils from plants on growth of food spoilage yeasts. Food Sci 49:429–434

Cowan MM (1999) Plant products as antimicrobial agent. Clin Microbiol Rev 12:564–582

Cox SD, Gustafson JE, Mann CM, Markham JL, Liew YC, Hartland RP, Bell HC, Warmington JR, Wyllie SG (1998) Teat tree oil causes K+ leakage and inhibits respiration in *Escherichia coli*. Lett Appl Microbiol 26:355–358

Cross DE, McDevitt RM, Hillman K, Acamovic T (2007) The effect of herbs and their associated essential oils on performance, dietary digestibility and gut microflora in chickens from 7 to 28 days of age. Br Poult Sci 48:496–506

Dalal R, Panwar VS, Ahlawat PK, Tewatia BS, Sheoran N (2018) Effect of supplementation of amla (*Emblica officinalis*) fruit powder on growth performance during different growth phases of broiler chicken. J Anim Res 8(4):621–628

Delaquis RJ, Stanich K, Girard B, Massa G (2002) Anti-microbial activity of individual and mixed fractions of dill, cilantro, coriander and eucalyptus essential oils. Int J Food Microbiol 74:101–109

Dobretsberger M, Glatzl M, Krafek WG (1997) Natural products for enhanced poultry productivity. Indian J Indg Med 18:101–108

Dorman HJD, Deans SG (2000) Antimicrobial agents from plants: antibacterial activity of plant volatile oils. J Appl Microbiol 88:308–316

Faleiro ML (2011) The mode of antibacterial action of essential oils. In: Mendez-Villas A (ed) Science against microbial pathogens: communicating current research and technological advances. Formatex, Badajoz, pp 1143–1156

Fuhrman B, Roseblate M, Hayek T, Coleman R, Aviram M (2000) Ginger extract consumption reduces plasma cholesterol, inhibits LDL oxidation and attenuates development of atherosclerosis in atherosclerotic, apolipoprotein E-Deficient mice. J Nutr 130:1124–1131

George OS, Kaegon SG, Igbokwe AA (2015) Feed additive effects of graded levels of ginger (*Zingiber officinale*) on serum metabolites of broilers. J Agric Vet Sci 8(3):59–62. ISSN: 2319-2372

Ghosal S, Rao G, Sarvana V, Nira Mishra M, Dipak R (1996) A possible chemical mechanism of the bioactivities of mangiferin. Indian J Chem 35(6):561

Gurdip S, Sumitra M, DeLampasona MP, Cesar ANC (2007) A comparison of chemical, antioxidant and antimicrobial studies of cinnamon leaf and bark volatile oils, oleoresins and their constituents. Food Chem Toxicol 45:1650–1661

Hady MM, Zaki MM, Abd EL-Ghany W, Korany Reda MS (2016) Assessment of the broilers performance, gut healthiness and carcass characteristics in response to dietary inclusion of dried coriander, turmeric and thyme. Int J Environ Agric Res 2:153–159

Haselmeyer A, Zentek J, Chizzola R (2014) Effects of thyme as a feed additive in broiler chickens on thymol in gut contents, blood plasma, liver and muscle. J Sci Food Agric 95(3):504–508. https://doi.org/10.1002/jsfa.6758

Hellander IM, Alakomi HL, Latva K, Mattila-Sandholm T, Pol I, Smid EJ, Gorris LGM, von Wright A (1998) Characterization of the action of selected essential oil components on Gram-negative bacteria. Agric Food Chem 46:3590–3595

Hippenstiel F, Abdel-Wareth AAA, Kehraus S, Sudekum KH (2011) Effects of selected herbs and essential pils, and their active components on feed intake and performance of broilers – a review. Arch Geflugelkd 75(4):226–234

Hui YH (1996) Oleoresins and essential oils. In: Hui YH (ed) Bailey's industrial oil and fat products, cap. 6. Wiley-Interscience Publication, New York, pp 145–153

Huyghebaert G (2003) Replacement of antibiotics in poultry. In: Proceedings of the Eastern nutrition conference, Quebec, Canada. UON, Quebec

Jakhetia V, Patel R, Khatri P, Pahuja N, Garg S, Pandey A, Sharma S (2010) Cinnamon: a pharmacological review. J Adv Sci Res 1:19–23

Im H, Alakomi HL, Latva-kala K, Mattila-Sandholam TPI, Smid EJ, Gorris LGM, Wright A (1998) Haracterization of the action of selected essential oil components on Gram-negative bacteria. J Agric Food Chem 46:3590–3595

Kalemba D, Kunicka A (2003) Antibacterial and antifungal properties of essential oils. Curr Med Chem 10:813–829

Kamat JP, Boloor KK, Devasagayam TP, Venkatachalam SR (2000) Antioxidant properties of *Asparagus racemosus* against damage induced by gamma-radiation in rat liver mitochondria. J Ethno Pharmacol 71(3):425–435

Kamath JV, Rana AC, Chowdhury AR (2003) Prohealing effect of *Cinnamomum zeylanicum* bark. Phytother Res 17:970–972

Kamel C (2001) Tracing methods of action and roles of plant extracts in non-ruminants. In: Garnsworthy PC, Wiseman J (eds) Recent advances in animal nutrition. Nottingham University Press, Nottingham, pp 135–150

Karaskova K, Suchy P, trakova E (2015) Current use of phytogenic feed additives in animal nutrition: a review. Czeh J Anim Sci 60(12):521–530

Khan A, Khattak KN, Safdar M (2003) Cinnamon improves glucose and lipids of people with type two diabetes. Diabetes Care 26:3215–3218

Kim JM, Marshall MR, Wei C (1995) Antibacterial activity of some essential of some essential oil components against 5 foodborne pathogens. Agric Food Chem 43:2839–2845

Kim HJ, Yokozawa T, Kim YH, Tohda C, Rao TP, Juneja LR (2005) Influence of amla (*Emblica officinalis* Gaertn.) on hypocholesterolemia and lipid peroxidation in cholesterol-fed rats. J Nutr Sci Vitaminol 51(6):413–418

Kiritikar KR, Basu BD (1935) M/s. periodical experts, D-42, Vivek Vihar. Indian Med Plants 3(2):2220–2222

Kohlert C, Van Rensen I, Marz R, Schindler G, Graefe EU, Veit M (2000) Bioavailability and pharmokinetics of natural volatile terpenes in animal and humans. Planta Med 66:495–505

Krishnakantha TP, Lokesh BR (1993) Scavenging of superoxide anions by spice principles. Indian J Biochem Biophys 30:133–134

Kumar Sampath KP (2012) Recent trends in potential traditional Indian herbs Emblica officinalis and its medicinal importance. J Pharmacogn Phytochem 1(1):18–28

Kumar M, Kumar V, Roy D, Kushwaha R, Vaiswani S (2014a) Application of herbal feed additives in animal nutrition – a review. Int J Livestock Res 4(9). ISSN 2277

Kumar M, Kumar P, Sethi APS, Sikka SS (2014b) Effect of cinnamon (*Cinnamomum cassia*) powder as a phytobiotic growth promoter in commercial broiler chickens. Anim Nutr Feed Technol 14(3):471–479

Kumar S, Kumar D, Yadav PK, Bal LM, Singh BP (2018) Amla as phytogenic feed additive for efficient livestock production. J Pharmacogn Phytochem 7(4):1030–1036

Lambert RJW, Skandamis PN, Coote PJ, Nychas GJE (2001) A study of the minimum inhibitory concentration and mode of action of oregano essential oil, thymol and carvacrol. Appl Microbiol 91:453–462

Larsen K, Ibrahim H, Khaw SH, Saw LG (1999) Gingers of Peninsular Malaysia and Singapore. Natural History Publications (Borneo), Kota Kinabalu

Lee KW, Everts H, Beynen AC (2004) Essential oils in broiler nutrition. Int J Poult Sci 3(12):738–752

Martins ER, Castro DM, Castellani DC, Dias JE (2000) Plantas medicinais. Editora UFV: Universidade Federal de Viçosa, Viçosa

Mašek T, Starčević K, Mikulec Ž (2014) The influence of the addition of thymol, tannic acid or gallic acid to broiler diet on growth performance, serum malondialdehyde value and cecal fermentation. Eur Poult Sci 78:2014. https://doi.org/10.1399/eps.2014.64

Mathe A (2009) Essential oils: biochemistry, production and utilization. In: Steiner T (ed) Phytogenic in animal nutrition: natural concepts to optimize gut health and performance. Nottingham University press, Nottingham, pp 1–18

Mayrhuber K (2018) Phytogenic feed additives are (NOT) just essential oils. https://www.phytogenius.com/en/. Accessed 11 Dec 2018

Mohammed A, Yusuf M (2011) Evaluation of ginger (*Zingiber officinale*) as a feed additive in broiler diets. Livest Res Rural Dev 23(9):12

Moharana D (2008) Shatavari, Jastimadhu and Ashwagandha the ayurvedic therapy. Orissa Ayurvedic Sebasadana, Cuttack

Navarrete P, Toledo I, Mardones P, Opazo R, Espejo R, Romero J (2010) Effect of *Thymus vulgaris* essential oil on intestinal bacterial microbiota of rainbow trout, *Oncorhynchus mykiss* (Walbaum) and bacterial isolates. Aquac Res 41:e667–e678

Nazzaro F, Fratianni F, de Martino L, Coppola R, de Feo V (2013) Effect of essential oils on pathogenic bacteria. Pharmaceuticals 6:1451–1474

North MO (1984) Commercial chicken production measure, 3rd edn. The Avi Publishing Company, Inc, Westport

Oetting LL (2005) Extratos vegetais como promotores do crescimento de leitões recém-desmamados [tese]. Piracicaba (SP): Escola Superior de Agricultura "Luiz de Queiroz"

Osman N, Talat G, Mehmet C, Bestami D, Simsek G (2005) The effect of an essential oil mix derived from oregano, clove and aniseed on broiler performance. Int J Poult Sci 4:879–884

Park BS (2008) Effect of dietary cinnamon powder on savor and quality of chicken meat in broiler chickens. J Korean Soc Food Sci Nutr 37(5):618–624

Prasad A (2004) Unpublished data, VHL Commercial Farm, Hyderabad

Prasad CM, Sen AK (1993) Study of the effect of four different growth promoters on performance of broiler. Poult Advis 26(7):49

Ranade AS, Desai DN (2005) Natural products for enhanced poultry productivity. IPSACON – 2005. Available from poultvet.com

Reddy R, Reddy V, Reddy KS (2002) Unpublished data, Nagarjuna Poultry Farm, Shanshabad, Hyderabad

Rege NN, Nazareth HM, Isacc A, Karandikar SM, Dahanukar SA (1989) Immunotherapeutic modulation of intraperitoneal adhesion by *Asparagus racemosus*. J Postgrad Med 35:199–203

Samarth VR, Jagtap DG, Dakshinkar NP, Nimbalkar MV, Kothekar MD (2002) Effect of Ashwagandha root powder (*Wtihania somnifera)* on performance of broiler. Indian Vet J 79:733–734

Santos AR, De-Campos RO, Miguel OG, Cechinel FV, Yunes RA, Calixto JB (1999) The involvement of k+ channels and gi/o protein in the antinociceptive action of the gallic acid ethyl ester. Eur J Pharmacol 379:7–17

Seena KG, Kuttan R (1993) Amla. Res Bull 13(8):41–45

Shon KS, Knon OS, Min BJ, Cho JH, Chen YJ, Kim IH, Kim HS (2002) Unpublished data, Department of animal resources and sciences, Dankook University, Cheonan, Korea

Simic A, Sokovic MD, Ristic M (2004) The chemical composition of some Lauraceae essential oils and their antifungal activities. Phytother Res 18:713–717
Singh G, Maurya S, Cesar MP, Catalan AM (2007) A comparison of chemical, antioxidant and anti-microbial studies of cinnamon leaf and bark volatile oils, oleoresins and their constituents. Food Chem Toxicol 45:1650–1661
Smith-Palmer A, Stewart J, Fyfe L (1998) Antimicrobial properties of plant essential oils and essences against five import food-borne pathogens. Lett Appl Microbiol 26:118–122
Somkuwar AP, Khadtare CM, Pawar SD, Gatne MM (2005) Influence of shatavari feeding on milk production in buffaloes. Pashudhan 31(2):3
Stahl-Biskup E, Saez F (2002) Thyme, the genus *Thymus*, 1st edn. Taylor and Francis, London
Steiner T (2006) Managing gut health – natural growth promoters as a key to animal performance. Nottingham University Press, Nottingham
Tabak M, Armon R, Neeman I (1999) Cinnamon extracts inhibitory effect on Helicobacter pylori. J Ethnopharmacol 67:269–277
Tanwar PS, Rathore SS, Kumar Y (2008) Effect of shatavari (*Asparagus recemosus*) on milk production in dairy animals. Indian J Anim Res 42(3):232–233
Thakur RS, Puri HS, Akhtar H (1989) Major medicinal plants of India. Central Institute of Medicinal and Aromatic Plants, Lucknow
Tung YT, Chua MT, Wang SY, Chang ST (2008) Anti-inflammatory activities of essential oil and its constituents from indigenous cinnamon (*Cinnamon osmophloeum*) twigs. Bioresour Technol 99:3908–3913
Ultee A, Bennik MH, Moezelaar R (2002) The phenolic hydroxyl group of carvacrol is essential for action against the food-borne pathogen *Bacillus cereus*. Appl Environ Microbiol 68:1561–1568
Upadhaya SD, Ho Kim I (2017) Efficacy of phytogenic feed additive on performance, production and health status of monogastric animals – a review. Ann Anim Sci 17(4):929–948
Vangalapati M, Sree Satya M, Surya PD, Avanigadda S (2012) A review on pharmacological activities and clinical effects of cinnamon species. Res J Pharm, Biol Chem Sci 3(1):653–663
World Poultry (2008) Plant extracts popular in poultry. World Poult 24(3):9
Windisch W, Schedle K, Plitzner C, Kroismayr A (2008) Use of phytogenic products as feed additives for swine and poultry. J Anim Sci 86(E. Supp):E140–E148
Yap PSX, Yiap BC, Ping HC, Lim SHE (2014) Essential oils, a new horizon in combating bacterial anti-biotic resistance. Open Microbiol J 8:6–14
Yu HS, Lee SY, Jang CG (2007) Involvement of 5-HT1A and GABAA receptors in the anxiolytic-like effects of *Cinnamomum cassia* in mice. Pharmacol Biochem Behav 87:164–170
Zhang LZ, Zhao WH, Guo YJ, Tu GZ, Lin S, Xin LG (2003) Studies on chemical constituents in fruits of Tibetan medicine *Phyllanthus emblica*. Zhongguo Zhong Yao Za Zhi 28:940–943

Global Scenario of Natural Products for Sustainable Agriculture

14

Saket Jha, Sharad Kumar Tripathi, Ravikant Singh, Anupam Dikshit, and Anand Pandey

Abstract

Agronomic practices and agricultural goods can be only compete with growing populations. Better crop productivity as well as transparency in market are essential for any farmer's business model. In modern agriculture system, the use of chemicals to fertilize soils and plants is widespread. Such compounds are also used to kill phytopathogens and several plant pests that are limiting factors to optimum productivity. There is an estimation of the increasing world population expected to be approximately 10 billion in the coming three to four decades for fulfilling the demands of growing populations; it needs to improve the agronomic products as well as agriculture practices to encounter the demand of this growing population. Apart from this, another focusing area is to increase agricultural products with high quality. Generally, significant damage to crop production is carried out by many diseases caused by several groups of phytopathogens, namely, bacteria, fungi, viruses, etc., which mutually epitomize a substantial encumbrance of the production of crops. The problem gets more aggravated with the evolution of resistant phytopathogenic microbes which causes even more severe threat to the crops as well as stored plant products and leads to severe damage to the production and storage of crops. In recent times, the use of natural entities such as plants metabolites, microbes, nanomaterials, and viruses against these pathogens reduces the loss of crop productivity and storage damage. The commonly used approach is to introduce the biological compounds to food directly or indirectly, which further shows antimicrobial activity. To avoid undesirable inactivation, application of these natural compounds in the form of

S. Jha · S. K. Tripathi · R. Singh · A. Dikshit
Biological Product Laboratory, Department of Botany, University of Allahabad, Prayagraj, India

A. Pandey (✉)
Department of Plant Sciences, Avvaiyar Government College for Women, University of Puducherry, Karaikal, Puducherry, India

J. Singh, A. N. Yadav (eds.), *Natural Bioactive Products in Sustainable Agriculture*, https://doi.org/10.1007/978-981-15-3024-1_14

fabricated nanoparticles seems to be more productive. Biologically synthesized nanoparticles can also be acceptable nowadays to control plant pathogens as antibacterial agents. These too can enhance seed germination and increase growth parameters. Bacteriophages can be used potently to control bacterial diseases. Phages are newly used to manage the pathogenic bacterial population and provide a promising tool to cope with bacterial diseases instead of antibiotics. Bacteriophages have a unique property where they feed on specific bacteria only. This specification makes them future antibacterial agents. Still, there is a need to work on several experiments regarding the use of bacteriophages as antibacterial agents. The present chapter is promising and focusing one to reveal the new options and trends in agriculture.

Keywords
Bactericides · Sustainable farming · Natural compounds · Nanoparticles · Antimicrobial

14.1 Introduction

During recent centuries, a significant loss in productivity leads to constructing calamitous and causes drastic hunger globally. Due to limited disease control methods in some of the less developed countries, an annual loss of 30–50% of major crops was recorded. This type of forfeiture of diseases in agro-plants affects global hunger and famishment. These pests and diseases move rapidly to threaten neighboring countries and continents. A total crop failure is the severe effect of such conditions. There is annually around US$220 billion losses in agro-economics which estimates 20–40% of agro-products due to global diseases and pest infections (FAO 2017).

14.1.1 Historical Use of Botanical Products in Antimicrobials

As mentioned earlier, the population of plant species were reported with approx. 2.5 to 5.0 lakh species present globally (Cowan 1999), but the humans and other land dwellers consumed only a minute fraction. There are so many possibilities of development of phytomedicines using plants (Moerman 1996). Most of the European countries now using botanicals act as another option from mainstream medicinal practices because the present allopathic system has side effects and other reactions. Plant products are used at a greater extent now in mainstream medicine too, because of the ineffectiveness of traditional antibiotics. The rate of extinction of plant species was hastened during the past 20 years. Several microbiologists have a belief that the assemblies of active phytoconstituents having the configurations, manufactured chemically, have a risk to extinct permanently (Borris 1996).

14.1.2 Bacterial Diseases

Around 200 species of bacteria are known for their phytopathogenic nature. These bacterial pathogens may be associated with either parasitic or superficial plants. Occasionally they are present in dead plant debris or found in soils mixed with decaying organic matters. Spreading of bacteria can be proficient by numerous means. It is very important to know the endurance physiognomies of phytopathogenic bacteria as fewer can subsist on inanimate substances, in water or inside creepy-crawlies. Sometimes infected seeds and transplants create a cradle of a banquet. For the entry into host tissue, bacteria require an opening (either artificial or natural). For the establishment of a bacterial colony, they need warm and moist conditions. A high velocity of wind-blown soil and sand causes wounds in a plant which provide a platform for bacterial infection. Bacteria get nutrition from leaked nutrients of intercellular space and colonize themselves inside the host. They may also grow within the vascular tissue of the plant. Bacterial species usually produce and release enzymes that cause alteration in the normal physiology of plant (Abdallah 2011).

14.1.3 Diagnosis of Plant Bacterial Diseases

Indications of bacterial infection and fungal infection in plants are almost similar. Some bacteria-associated plant diseases are as follows (Cooper and Gardener 2006).

14.1.3.1 Bacterial Spots

Leaf spots are the best and typical example of bacterial disease. These spots may give the impression on all parts of plant, namely, leaves and surface of fruits and speedily extend throughout the area then the disease is well-thought-out blight. In the leaves of dicotyledonous plants, spots show a foul or fishy order. These are water-soaked and are initially found in the veins and veinlets with angular appearance. The presence of bacterial exude is the diagnostic feature for bacterial infection. Many times it is observed that infected leaf possesses a chlorotic halo near the bacterial lesion. Leaf spots may amalgamate causing large areas of necrotic tissue. But on a monocot plant, bacterial spots will give the impression as streaks or stripes. Bacterial canker generally appears on stems and branches. The brownish color cortical tissue appeared, and gum was produced in the cankered branches and twigs. Those spotted regions mainly diseased with canker are sour and have odor with soft and sunken appearance if they do not produce any gums. These stems or twigs were concluding dieback of the portion of the tree distal to the canker (Cooper and Gardener 2006).

14.1.3.2 Bacterial Galls

Gall or tumors are caused by bacteria, namely, *Rhodococcus*, *Agrobacterium*, *Arthrobacter*, and *Pseudomonas*. Galls in more than 390 genera of plants worldwide have been caused by several species of genera *Agrobacterium* and generally called as crown gall or root gall. In most of the agricultural soil, members of

Agrobacterium species with other microbes present in the soil. Damage caused by gall may be benign to fatal. The orange-brown color was observed due to the infection of *A. tumefaciens*. Gummosis can be caused due to infection of *Pseudomonas* sp. (Cooper and Gardener 2006).

14.1.3.3 Bacterial Vascular Wilts

Herbaceous plants are attacked by pathogen bacteria to cause vascular wilt. Xylem and phloem were accountable for the ascent of sap and translocation of solutes, respectively. Hence, the causal pathogen moves through the xylem vessels of the host plant and multiply themselves. They can interfere in the transportation of water and minerals by developing gum in vessels of xylem (Cooper and Gardener 2006).

14.1.3.4 Bacterial Soft Rots

Several species of *Bacillus*, *Pseudomonas*, *Clostridium*, and *Erwinia* causes rot tissue of plants. Some of the saprophytes which are non-phytopathogenic bacteria may cause soft rots. Soft rots are known to be corpulent and have spongy structure in plant tissue (Cooper and Gardener 2006).

14.1.3.5 Bacterial Scabs

Underground parts of the plants show infection in this type of disease. The most common example of this category is potato. Scab disease is mostly caused by *Streptomyces* sp. in potato. In this case, lesions occurring in scabs appear in the superficial surface. Below and around the lesion, a typical corky tissue will appear. Some of the pathogens like rot pathogens can enter through these lesions into the host plant tissue and degrade the host (Cooper and Gardener 2006).

14.2 Vectors of Pathogens

Vectors of pathogens are categorized as fastidious vascular-colonizing bacteria shown in Table 14.1 and non-fastidious bacteria shown in Table 14.2 (Mitchell 2004).

14.2.1 Fastidious Vascular-Colonizing Bacteria

Fastidious colonizing bacteria (FVC) are constrained to elements of xylem or phloem. These bacteria are rod-shaped with cell-walled structure. Spittlebugs and sharpshooters are xylem feeders and known as the best vector for fastidious xylem-limited bacteria (Fletcher et al. 1998). Bacteria get accumulated in the

Table 14.1 List of vectors of fastidious vascular-colonizing bacteria

Vector species	Plants	Plant disease	Descriptions
Anasa tristis	Cucurbits	Yellow wine	Noncirculative
Piesma quadratum	Beet	Latent rosette	Persistent, propagative

Source: Mitchell (2004)

Table 14.2 List of vectors of non-fastidious vascular-colonizing bacteria

Vector species	Plants	Plant disease	Phytopathogens	Descriptions
Hypselonotus fulvus	Cotton	Boll rot	*X. campestris malvacearum*	Transmission
Lygus lineolaris	Potato	Ring rot	*Clavibacter michiganensis sepedonicus*	Associated
Lygus lineolaris, L. elisus, Adelphocoris rapidus, Campylomma verbasci, Heterocordylus malinus, Lygidea mendax, Lygocoris communis	Potato and family Solanaceae	Fire blight	*Erwinia amylovora*	Associated
Lygus lineolaris	Celery	Soft rot	*Erwinia carotovora*	Vector associated with punctures
Lygus rugulipennis, Orthotylus flavosparsus	Beet	Beet bacterial disease	*Pseudomonas syringae aptata*	Associated
Pseudatomoscelis seriata, Helopeltis sp., *Taylorilygus vosseleri*	Cotton	Bacterial blight of cotton, angular leaf spot	*X. campestris malvacearum*	Transmission, associated
Nezara viridula	Cowpea	Stem canker	*X. campestris phaseoli*	Transmission
Nezara viridula	Soybean	Leaf spot and vein necrosis	*Pseudomonas* spp., *Curtobacterium* spp.	Isolated
Nezara viridula, Edessa meditabunda, Dysdercus fasciatus, D. honestus, D. intermedius, D. mendesi, D. nigrofasciatus, D. ruficollis	Cotton	Cotton ball rot	*Xanthomonas campestris malvacearum*	Transmission, associated

Source: Mitchell (2004)

foregut of the insect and are ingested into the host. Several times in *Xylella fastidiosa*, the transmission is noncirculative. *Serratia marcescens* is reported to cause yellow wine disease in several cucurbits (Bruton et al. 2003). Transmission of *S. marcescens* has been carried out by vector *Anasa tristis*, known as the squash bug (Mitchell 2004).

14.2.2 Non-fastidious Bacteria (NFVC)

Insects as a vector is not an essential part of the transmission of the majority of pathogenic bacteria. Bacteria cannot protrude inside the plant tissue directly.

They penetrate via stomata or by small wounds (Goto 1992). Thus, a bacterial infection is mainly facilitated when the insect feeds on plant parts and causes feeding punctures (Mitchell 2004).

14.3 Natural Products as Bactericides

Majority of the aromatic compounds have phenolic components (Geissman 1963). Most of the aromatic compounds are secondary metabolites, of which approx. 12,000 have been isolated. Still, the total counts are less than 10% and need to enlighten on plant-based natural products (Schultes 1978). Plant extract compositions were produced in the response of stress and work in plant defense mechanisms. These metabolites provide defense against pathogens and pests. In comparison to all metabolites, terpenoids produce odors, whereas quinones and tannins were responsible for the synthesis of phyto-pigments. Various biological active components are used as a flavoring agent, for example, capsaicin (terpenoid) produced from *Capsicum* sps. Useful medicinal compounds can also be extracted from the various herbs and many spices.

14.3.1 Phenolics and Polyphenols

These are simple biologically active phytoconstituents with a single phenolic ring. For example, Cowan (1999) reported that cinnamic acid and caffeic acids partake phenyl-propane-derived compounds and have maximum oxidation property. Brantner and his coworkers (1996) extracted the caffeic acid from tarragon and thyme plants and showed prominent antibacterial activity. Both have 2- and 3-hydroxyl group, respectively, in their structure and found that these groups increased in number and cause toxicity to many microbes, namely, bacteria, fungi, etc. (Geissman 1963). Phenolics were reported as highly oxidized and showed maximum toxicological effects (Scalbert 1991). These oxidized phenols and hydroxylated compounds believed to interact either with groups of hydrosulfide or nonspecific with the proteins and lead to cause an inhibitory effect on the enzymatic mechanism (Mason and Wasserman 1987). Eugenol found in clove showed the best example for enzymatic inhibition and proved to be good bacteriostatic against bacteria (Fig. 14.1).

14.3.2 Quinones

These compounds are a major cause to form browning color after reaction on the cut or injured fruits and vegetables (Schmidt 1988). These were also present in the henna plant by which gives its dying property (Fessenden and Fessenden 1982).

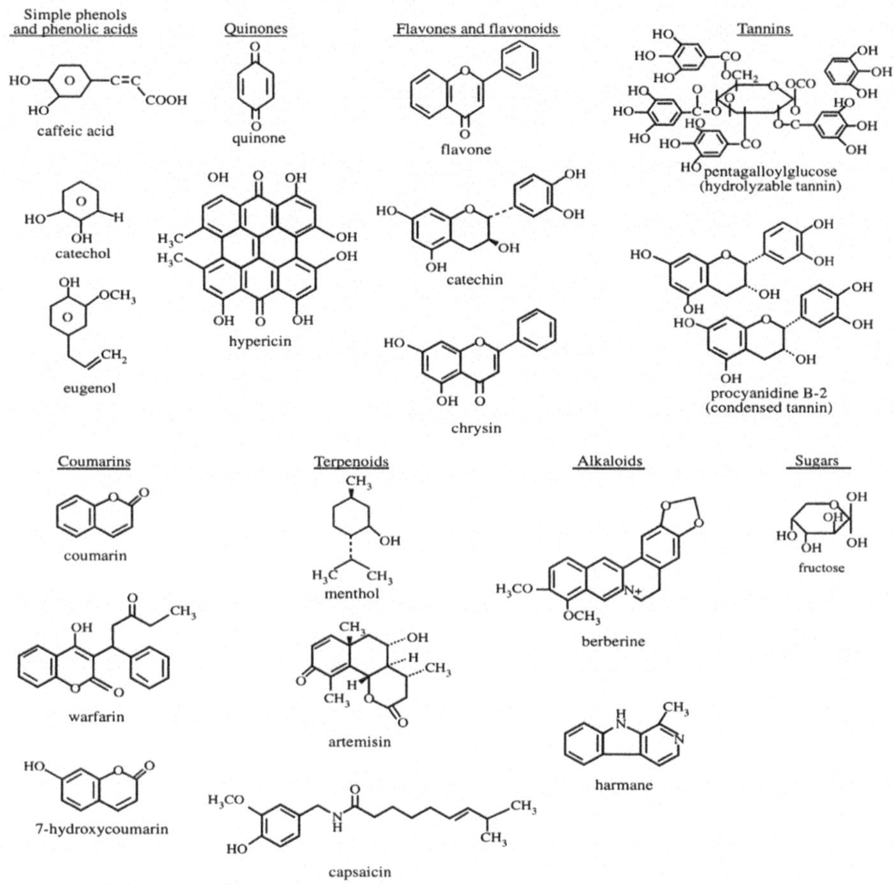

Fig. 14.1 Examples of some biologically active compounds. (Source: Cowan 1999)

14.3.3 Flavones and Its Derivatives

Flavones contain one carbonyl group, and flavonols have an additional 3-hydroxyl group (Fessenden and Fessenden 1982). Flavonoids have aromatic rings linked with C3–C6 unit. Dixon and his coworkers (1983) reported that flavonoids synthesized in the plants have antimicrobial activity. Tsuchiya and his colleagues (1996) found that these flavonoids can penetrate into the cell wall and disrupt microbial membranes by changing the internal chemical environment. Toda et al.'s (1989) reports suggested that reduced form of flavonoid was catechins found in tea plants which have high antimicrobial property.

14.3.4 Tannins

Tannins are polymeric phenolic compounds used for tanning of leather or hastening gelatin from a solution called as astringency. These were found almost in all plants with various ranges of molecular weights (Scalbert 1991). They are mainly found in two forms, namely, hydrolysable and condensed tannins. The monomeric type of flavonoid produces the condensed tannins (Geissman 1963).

14.3.5 Essential Oils

These were hydro-distillate constituents of plants and have great fragrances. They are also called terpenoids when contains additional elements usually oxygen atom. These have antibacterial properties (Amaral et al. 1998; Chaurasia and Vyas 1977).

14.3.6 Alkaloids

These were firstly reported in opium plant named morphine (Fessenden and Fessenden 1982). There are many more now known alkaloids, and they have high antimicrobial activity and used widely in medical practices to prevent disease-causing microbes.

14.3.7 Simple Polypeptides and Lectins

Balls and his colleagues (1942) reported that peptides had antimicrobial activity. They contain disulfide bonds with a positively charged skeleton. The acting mechanism of peptides is still to be discovered, but some reported showed that they interact with microbial cell membrane cytosolic constituents and form ion channels (Zhang and Lewis 1997). Contemporary research findings showed biological activity of several extracted metabolites, and their specific constituents have antifungal and antibacterial properties (Kumar and Malik 2011; Bajpai et al. 2008). These are used as food or plant sample preservations against various fungal and bacterial pathogens. Phytochemical analysis revealed that plants synthesized these bioactive compounds during stress conditions, evincing to have many inhibitory activities against microbes, pests, and herbs (Matos and Ricardo 2006).

14.3.8 Plants with Antibacterial Properties

Since forever, plants were used for curing the bacterial infections, and till today a majority of the plants are known to have antibacterial properties. Several plants were investigated and found that all plants aqueous and ethanolic extracts were effective against various phytopathogenic bacteria (Ghosh et al. 2000; Krupinski and Sobiczewski 2001).

Table 14.3 List of peptide-based antimicrobial components

Targeted plant pathogens	Antimicrobial peptides
Erwinia amylovora	Pseudopeptides
Erwinia carotovora	Cyclopeptides
Clavibacter michiganensis	Peptaibols
Pseudomonas syringe	Cyclopeptides
Rhodococcus fascians	Cyclopeptides

Source: Dubey (2011)

14.3.8.1 Phyto-protein

Several reports suggest that some proteins synthesized by plants showed immune response to various diseases and took part in defense mechanisms of plants. Thus, scientists were focused on inducing or enhancing the production of these protein molecules to increase the phyto-resistivity against phytopathogens (Dubey 2011).

14.3.8.2 Antimicrobial Peptides

Some plants were found to produces defensive peptides to cope with several diseases (Table 14.3). These peptides were small in size by which they can easily penetrate inside the microbial cell membrane and disrupt their cytoplasmic constituents and lead to killing these pathogens (Park et al. 2009). Biologically synthesized peptides were having pesticidal as well as antimicrobial activity as these inhibited the nucleic acid and protein synthesis. These were also found to inhibit enzymatic pathways (Huang 2000).

14.4 New Approach of Agricultural Bactericides

14.4.1 Phytoalexins and Phytoanticipins

Phytoalexins are the antimicrobial compounds which are expressed by the enzymes when elicitation occurs (Grayer and Kokubun 2001). Therefore once pathogen was detected, plants initiate the transcriptional and translational process for phytoalexins synthesis. Many pieces of evidence prove that these antimicrobial compounds which were formed due to induction and also autonomous process comprise resistivity to several diseases (Lamothe et al. 2009).

14.4.2 Response by Phytoanticipins

Saponins glycosylated phytoanticipins which were widely present in many plants. These compounds have impressive antimicrobial activity. The saponins were studied very well concerning plant defense molecule, and majority are avenacin and α-tomatine. The major avenacin, that is, avenacin A-1, was confined in layers of an epidermal cell of the oat root tip and makes an appearance in the form of lateral root initials and form a chemical blockade (Osbourn et al. 1994). Furthermore, the ability of *Gaeumannomyces graminis* var. *aveanae* to suppress the toxic effects of avenacin A-1 was compulsory for interaction with oat. Excitingly avenacin biosynthetic pathway mediates callose accumulation in oat root tip which is known for defense mechanism and indicates the role of phytoanticipin in several defense responses (Mylona et al. 2008). All healthy plants contain phytoanticipin in biologically active form. Excitingly the degraded product of α-tomatine was capable in restriction of defense response (Bouarab et al. 2002).

14.4.3 Phytoalexins: Some Biological Examples

Phytoalexin plays a role in defense mechanism in rice and crucifers. Also, it showed the biological processes present in the root emission. Valle and his coworkers (1997) reported that the important phytoalexins in tobacco are the hydroxyl-coumarin scopoletin (6-methoxy-7-hydroxycoumarin) having antimicrobial properties. Tissues near necrotic lesions appeared bright blue in color under UV radiation due to phytoalexins (Costet et al. 2002; Chong et al. 1999). A forager of ROS produced in abundance after stimulating the hypersensitive response (Lamothe et al. 2009).

Camalexin was not metabolically synthesized by deforming of mutant pad3 in its last step of the synthesis, which leads to accumulating in mutant pad3 (Rogers et al. 1996). Camalexin was only known phytochemical to disturb the permeability of bacterial plasma membrane at lower concentration. Due to this property of camalexin, it is involved in the defense metabolism of plants against bacterial infections (Lamothe et al. 2009).

14.5 Mechanism of Bioactive Components

Sikkema and his colleagues in 1995 found that the biological metabolism of terpenes and its mode of action are still unknown, but some hypothesis was given as it disrupts the lipophilic components of the microbial cell. The oils of the tea plant are capable of interfering the penetrability barrier of plasma membrane structure and conduct damage of chemiosmotic mechanism leading to cause the death of *Candida albicans* and *E. coli* (Cox et al. 2000). The susceptibility to lysis of bacterial cells showed that oil extracted from tea plant interferes in the membrane cytosolic constituents (Carson et al. 2002) (Fig. 14.2).

Chemical analysis showed that antimicrobial phytochemicals contain phenolic structures connected with –OH. Diallyl thiosulfonate (allicin) found in garlic having the antimicrobial properties. Low doses of allicin (0.2–0.5 mM) showed high efficacy of bacteriostatic on the growth of several bacterial pathogens that was due to the action of suppression and inhibition of DNA and protein (Rasmussen et al. 2005). Allicin also inhibits quorum sensing in bacteria (Bazaka et al. 2015).

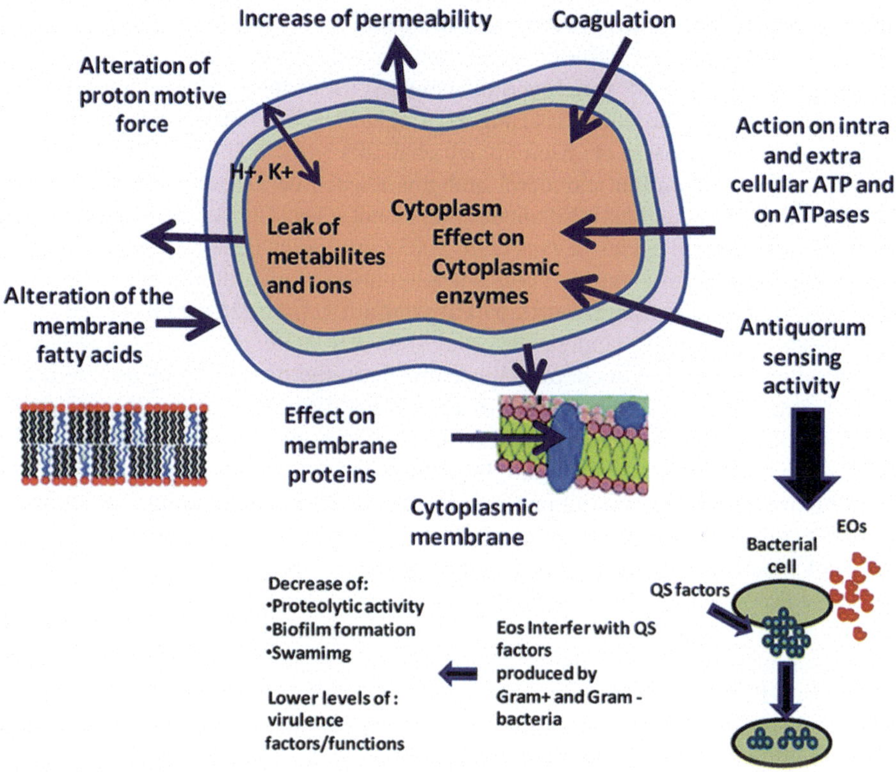

Fig. 14.2 Regulation of bio-constituents and its antimicrobial activity (Bazaka et al. 2015)

14.6 Nanoparticles as Bactericides

Nanotechnology application in agriculture sectors allows more well-organized and sustainable agriculture by plummeting the probabilities of enhancement resistant to various diseases and pests (Nair and Kumar 2013). Some of the important uses of nanomaterials are in sewage treatment, purification of water, remediation of toxic heavy metals in soil and water, packaging and processing of food materials, coating of capsules and other pharmacological activity and household purposes (Huang et al. 2015).

All the approaches of recent researches were mainly focusing on the application of agricultural practices and improving crop quality and productivity. Nanomaterials were further approaching toward the agriculture sector and in the protection and storage of crops (Sharon et al. 2010). The toxicological efficacy of several nanoparticles was tested against various pathogens (bacteria and fungi) and showed high antimicrobial activity (Huang et al. 2015).

14.6.1 Metal, Metalloid, and Nonmetal NPs

NP of Ag was the first NP which was investigated for different aspects of plant disease management. The efficacy of Ag NPs against bacteria has been studied by many researchers (Park et al. 2016; Ocsoy et al. 2013).

But more investigation about NP of Ag is needed as its exact function about defense mechanism and activity of ionic Ag is not clear. Copper nanoparticles are used as antimicrobial agents (Evans et al. 2007). Biologically as well as chemically synthesized Cu NP in controlled concentration showed great inhibitory impact on microbes without affecting plants. Several scientific reports showed the antimicrobial properties of Zn NPs (Indhumathy and Mala 2013). Recently, many zinc-based nanoproducts were used in controlling various bacterial diseases such as Zinkicide used for bacteria *Xanthomonas citri* causing canker disease in lemons (Young et al. 2017). Zinkicide was showed high effectiveness during field trials as compared with chemical bactericides in scab disease in lemons.

Many investigations showed that a wide range of bacterial and fungal pathogens can be controlled by nanoparticles of Zn, Au, Ti, Fe, Si, etc. (Czajkowski et al. 2015). A piece of proper information about working mechanism and efficacy must be needed as many of the reports have arisen from a single study. NP of Si has received more attention as it shows some consequences on plant health when supplied continuously. To maintain the plant defense system, a continuous and constant supply of NP of Si is needed. The use of NP Al_2O_3, MgO, NP TiO_2, NP S, and NP CeO_2 showed effectiveness against several phytopathogens, namely, fungi and bacteria (Mallmann and Hemstreet 1924; Czajkowski et al. 2015) (Fig. 14.3).

14.7 Bacteriophages in Crop Production

Bacteriophages were used as an alternative for pesticides as well as potent bactericide. The antibacterial activity was first reported by Twort (1915) and further confirmed in 1917 by Felix D'Herelle. Both experimentally found antibacterial properties done by these phages and found more specific to host bacteria, that is, *Xanthomonas campestris*, which caused rot disease in cabbage. Phages showed no negative effects on cabbage plants (Svircev et al. 2018).

14.7.1 Blight and Bacterial Wilt

Phages belong to Myoviridae used as biocontrol agents in controlling the soft rot in potato. These phages inhibit the growth of *Dickeya solani* and *Pectobacterium* spp. causing rot disease in the storage of potato (Svircev et al. 2018).

Phages were singly or consortia treated on *Xanthomonas perforans* and *Pseudomonas syringae* on potato and tomato plants, respectively. These phages were found to have potent inhibitory effects on these two pathogens which are

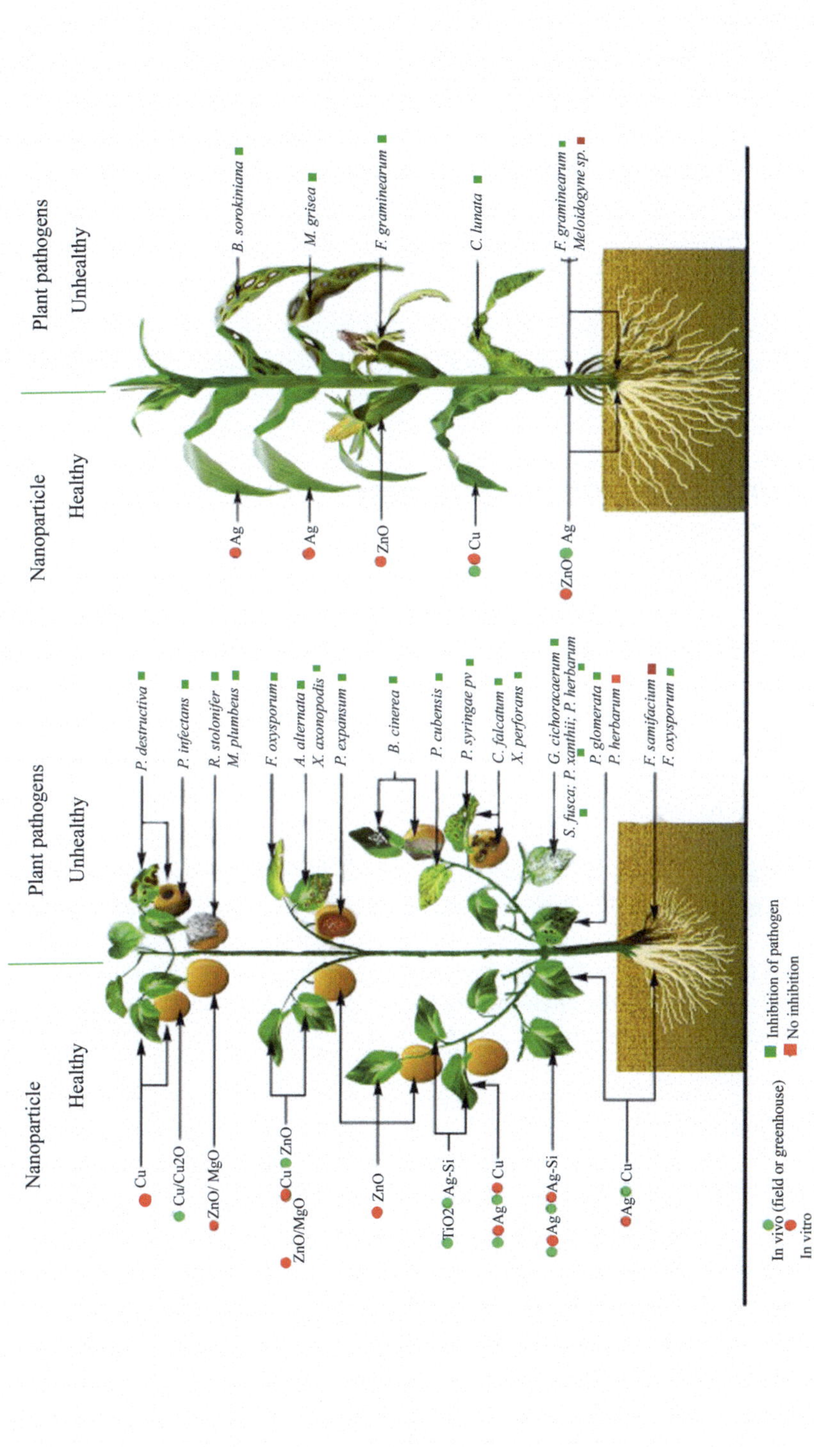

Fig. 14.3 Interaction and antibacterial activity against some bacteria on the plant (Elmer et al. 2018)

responsible for serious threat to plant disease and cause significant damage to potato and tomato production (Hirano and Upper 1990).

Recently, one experimental study revealed that phages also control the growth of *P. syringae* pv. *actinidae*. *P. syringae* causes blight disease on plants and causes severe loss in kiwi and leek production every year (Lehman 2007). These phages proved to be potentially used to control phytopathogens (Di Lallo et al. 2014).

14.7.2 Citrus Bacterial Canker and Spot

Recent investigation found that these phages also showed inhibitory activity against several bacterial species which causes bacterial canker and spot diseases in grape plants.

14.8 Conclusion and Future Prospects

There is ruthless demand for crops as population increases. It is very unfortunate that the demand still does not achieve the satisfaction point. Major causes are disease in crops and ill storage strategy. In agriculture, the bacterial disease causes severe loss after fungal diseases. Recently the world focused on natural bactericides and also developing hybrids that are genetically disease resistant. The secondary metabolites from plants, namely, oil, terpenoids, alkaloids, etc., are potent bactericidal.

The world introduces a new concept of biologically synthesized nanoparticles used to control bacterial diseases. They have very high potential to alter the mechanism in growth and also inhibit the wide range of phytopathogens. Some researchers found that use of viruses like bacteriophage can be one of the excellent tools to tackle bacterial and fungal pathogens. Instead of using chemicals or other expensive tools and techniques, agronomic practices must be more natural and biological. Natural and phyto-products must be used as bactericides. There is the only one way to develop a safe and healthy environment for agriculture and society and strengthen the base of sustainable agriculture.

References

Abdallah EM (2011) Plants: an alternative source for antimicrobials. J Appl Pharm Sci 01(06):16–20

Amaral JA, Ekins A, Richards SR, Knowles R (1998) Effect of selected monoterpenes on methane oxidation, denitrification, and aerobic metabolism by bacteria in pure culture. Appl Environ Microbiol 64:520–525

Bajpai V, Shukla S, Kang S (2008) Chemical composition and antifungal activity of essential oil and various extract of *Silene armeria* L. Bioresour Technol 99:8903–8908

Balls AK, Hale WS, Harris TH (1942) A crystalline protein obtained from a lipoprotein of wheat flour. Cereal Chem 19:279–288

Bazaka K, Jacob MV, Chrzanowskic W, Ostrikova K (2015) Anti-bacterial surfaces: natural agents, mechanisms of action, and plasma surface modification. RSC Adv 5:48739–48759. https://doi.org/10.1039/c4ra17244b

Borris RP (1996) Natural products research: perspectives from a major pharmaceutical company. J Ethnopharmacol 51:29–38

Bouarab K, Melton R, Peart J, Baulcombe D, Osbourn A (2002) A saponin-detoxifying enzyme mediates suppression of plant defences. Nature 418(6900):889–892

Brantner A, Males Z, Pepeljnjak S, Antolic A (1996) Antimicrobial activity of *Paliurus spina-christi* mill. J Ethnopharmacol 52:119–122

Bruton BD, Mitchell F, Fletcher J, Pair SD, Wayadande A, Melcher U, Brady J, Bextine B, Popham TW (2003) *Serratia marcescens*, a phloem-colonizing, squash-bug transmitted bacterium: causal agent of cucurbit yellow vine disease. Plant Dis 87:937–944

Carson CF, Mee BJ, Riley TV (2002) Mechanism of action of *Melaleuca alternifolia* (tea tree) oil on *Staphylococcus aureus* determined by time-kill, lysis, leakage, and salt tolerance assays and electron microscopy. Antimicrob Agents Chemother 46(6):1914–1920. https://doi.org/10.1128/AAC.46.6.1914-1920.2002

Chaurasia SC, Vyas KK (1977) In vitro effect of some volatile oil against *Phytophthora parasitica var. piperina*. J Res Indian Med Yoga Homeopath 1977:24–26

Chong J, Baltz R, Fritig B, Saindrenan P (1999) An early salicylic acid-, pathogen- and elicitorinducible tobacco glucosyltransferase: role in compartmentalization of phenolics and H2O2 metabolism. FEBS Lett 458:204–208

Cooper J, Gardener M (2006) WSU county extension, SJI Edited by Dr. Tom Schultz

Costet L, Fritig B, Kauffmann S (2002) Scopoletin expression in elicitor-treated and tobacco mosaic virus-infected tobacco plants. Physiol Plant 115:228–235

Cowan MM (1999) Plant products as antimicrobial agents. Clin Microbiol Rev 12(4):564–582

Cox SD, Mann CM, Markham JL, Bell HC, Gustafson JE, Warmington JR, Wyllie SG (2000) The mode of antimicrobial action of the essential oil of *Melaleuca alternifolia* (tea tree oil). J Appl Microbiol 88(1):170–175. https://doi.org/10.1046/j.1365-2672.2000.00943.x

Czajkowski R, Ozymko Z, de Jager V, Siwinska J, Smolarska A, Ossowicki A, Narajczyk M, Lojkowska E (2015) Genomic, proteomic and morphological characterization of two novel broad host lytic bacteriophages PhiPD10.3 and PhiPD23.1 infecting pectinolytic *Pectobacterium spp.* and *Dickeya spp*. PLoS One 10(3):e0119812

D'Herelle F (1917) Sur un microbe invisible antagoniste des *Bacillus dysentérique*. Acad Sci Paris 165:373–375

Di Lallo G, Evangelisti M, Mancuso F, Ferrante P, Marcelletti S, Tinari A, Superti F, Migliore L, D'Addabbo P, Frezza D (2014) Isolation and partial characterization of bacteriophages infecting *Pseudomonas syringae pv. actinidiae*, causal agent of kiwifruit bacterial canker. J Basic Microbiol 54:1210–1221

Dixon RA, Dey PM, Lamb CJ (1983) Phytoalexins: enzymology and molecular biology. Adv Enzymol 55:1–69

Dubey NK (2011) Natural products in plant pest management. CABI International Publications, Wallingford. ISBN-13: 978 1 84593 671 6

Elmer WH, De La Torre-Roche R, Pagano L, Majumdar S, Zuverza-Mena N, Dimpka C, Gardea-Torresdey J, White W (2018) Effect of metalloid and metallic oxide nanoparticles on *Fusarium* wilt of watermelon. Plant Dis 102(7):1394–1401

Evans I, Solberg E, Huber DM (2007) Copper and plant disease. In: Datnoff LE, Elmer WH, Huber DN (eds) Mineral nutrition and plant disease. APS Press, St. Paul, pp 177–188

FAO (2017) The future of food and agriculture – trends and challenges. FAO, Rome

Fessenden RJ, Fessenden JS (1982) Organic chemistry, 2nd edn. Willard Grant Press, Boston

Fletcher J, Wayadande A, Melcher U, Ye F (1998) The phytopathogenic mollicute-insect vector interface: a closer look. Phytopathology 88:1351–1358

Geissman TA (1963) Flavonoid compounds, tannins, lignins and related compounds. In: Florkin M, Stotz EH (eds) Pyrrole pigments, isoprenoid compounds and phenolic plant constituents, vol 9. Elsevier, New York, p 265

Ghosh SK, Sanyal B, Ghosh S, Gupta S, Sanyal B, Ghosh S, Gupta S (2000) Screening of some angiospermic plants for antimicrobial activity. J Mycopathol Res 38:19–22
Goto M (1992) Fundamentals of bacterial plant pathology. Academic, San Diego, p 342
Grayer RJ, Kokubun T (2001) Plant-fungal interactions: the search for phytoalexins and other antifungal compounds from higher plants. Phytochemistry 56:253–263
Hirano S, Upper C (1990) Population biology and epidemiology of *Pseudomonas syringae*. Annu Rev Phytopathol 28:155–177
Huang HW (2000) Action of antimicrobial peptides: two-state model. Biochemistry 39:8347–8352
Huang S, Ling Wang L, Liu L, Hou Y, Li L (2015) Nanotechnology in agriculture, livestock, and aquaculture in China. A Rev Agron Sustain Dev 35:369–400
Indhumathy M, Mala R (2013) Photocatalytic activity of zinc sulphate nano material on phytopathogens. Int J Agric Environ Biotechnol 6(4S):737–743
Krupinski G, Sobiczewski P (2001) The influence of plant extracts on growth of *Erwinia amylovora* – the causal agent of fire blight. Acta Agrobot 54:81–91
Kumar A, Malik A (2011) Antimicrobial potential and chemical composition of *Mentha piperita* oil in liquid and vapour phase against food spoiling microorganisms. Food Control 22:1707–1714
Lamothe RG, Mitchell G, Gattuso M, Diarra MS, Malouin F, Bouarab K (2009) Plant antimicrobial agents and their effects on plant and human pathogens. Int J Mol Sci 10:3400–3419. https://doi.org/10.3390/ijms10083400
Lehman SM (2007) Development of a bacteriophage-based biopesticide for fire blight. PhD thesis, Brock University, St. Catharines, ON, Canada
Mallmann W, Hemstreet C (1924) Isolation of an inhibitory substance from plants. Agric Res 28:599–602
Mason TL, Wasserman BP (1987) Inactivation of red beet betaglucan synthase by native and oxidized phenolic compounds. Phytochemistry 26:2197–2202
Matos O, Ricardo C (2006) Screening of plants against fungi affecting crops and stored foods. Adv Phytomed 3:139–170
Mitchell PL (2004) *Heteroptera* as vectors of plant pathogens. Neotrop Entomol 33(5):519–545
Moerman DE (1996) An analysis of the food plants and drug plants of native North America. J Ethnopharmacol 52:1–22
Mylona P, Owatworakit A, Papadopoulou K, Jenner H, Qin B, Findlay K, Hill L, Qi X, Bakht S, Melton R, Osbourn A (2008) Sad3 and sad4 are required for saponin biosynthesis and root development in oat. Plant Cell 20:201–212
Nair R, Kumar DS (2013) Chapter 10: plant diseases-control and remedy through nanotechnology. In: Tuteja N, Gill SS (eds) Crop improvement under adverse conditions. Springer, New York, pp 231–243
Ocsoy I, Paret ML, Ocsoy MA, Kunwar S, Chen T, You M, Tan W (2013) Nanotechnology in plant disease management: DNA-directed silver nanoparticles on graphene oxide as an antibacterial against *Xanthomonas perforans*. ACS Nano 7(10):8972–8980
Osbourn AE, Clarke BR, Lunness P, Scott PR, Daniels MJ (1994) An oat species lacking avenacin is susceptible to infection by *Gaeumannomyces graminis var. tritici*. Physiol Mol Plant Pathol 45:457–467
Park HY, Park HC, Yoon MY (2009) Screening for peptides binding on *Phytophthora capsici* extracts by phage display. J Microbiol Methods 78:54–58
Park HJ, Kim SH, Kim HJ, Choi SH (2016) A new composition of nanosized silica-silver for control of various plant diseases. Plant Pathol J 22:295–302
Rasmussen TB, Bjarnsholt T, Skindersoe ME, Hentzer M, Kristoffersen P, Köte M, Nielsen J, Eberl L, Givskov M (2005) Screening for quorum-sensing inhibitors (QSI) by use of a novel genetic system, the QSI selector. J Bacteriol 187:1799–1814
Rogers EE, Glazebrook J, Ausubel FM (1996) Mode of action of the *Arabidopsis thaliana* phytoalexins camalexin and its role in Arabidopsis-pathogen interactions. Mol Plant-Microbe Interact 9:748–757
Scalbert A (1991) Antimicrobial properties of tannins. Phytochemistry 30:3875–3883

Schmidt H (1988) Phenol oxidase (E.I.14.18.1), a marker enzyme for defence cells. Prog Histochem Cytochem 17(3):1–194. PMID: 3127860

Schultes RE (1978) The kingdom of plants. In: Thomson WAR (ed) Medicines from the earth. McGraw-Hill Book Co, New York, p 208

Sharon M, Choudhary A, Kumar R (2010) Nanotechnology in agricultural diseases and food safety. J Phytology 4:83–92

Sikkema J, de Bont JA, Poolman B (1995) Mechanisms of membrane toxicity of hydrocarbons. Microbiol Rev 59:201–222

Svircev A, Roach D, Castle A (2018) Framing the future with bacteriophages in agriculture. Viruses 10:218

Toda M, Okubo S, Ohnishi R, Shimamura T (1989) Antibacterial and bactericidal activities of Japanese green tea. Jpn J Bacteriol 45:561–566

Tsuchiya H, Sato M, Miyazaki T, Fujiwara S, Tanigaki S, Ohyama M, Tanaka T, Iinuma M (1996) Comparative study on the antibacterial activity of phytochemical flavanones against methicillin-resistant *Staphylococcus aureus*. J Ethnopharmacol 50:27–34

Twort F (1915) An investigation on the nature of ultra-microscopic viruses. Lancet 186:4814

Valle T, Lopez JL, Hernandez JM, Corchete P (1997) Antifungal activity of scopoletin and its differential accumulation in *Ulmus pumila* and *Ulmus campestris* cell suspension cultures infected with *Ophiostoma ulmi* spores. Plant Sci 125:97–101

Young M, Ozcan A, Myers ME, Johnson EG, Graham JH, Santra S (2017) Multimodal generally recognized as safe ZnO/nanocopper composite: a novel antimicrobial material for the management of citrus phytopathogens. J Agric Food Chem 66(26):6604–6608

Zhang Y, Lewis K (1997) Fabatins: new antimicrobial plant peptides. FEMS Microbiol Lett 149:59–64

The manufacturer's authorised representative in the EU is Springer Nature Customer Service Centre GmbH, Europaplatz 3, 69115 Heidelberg, Germany. If you have any concerns regarding our products, please contact ProductSafety@springernature.com

Printed and bound by CPI Group (UK) Ltd, Croydon, CR0 4YY
30/09/2026
02225248-0005